NANOTECHNOLOGY

Present Status and Future Prospects in Developing Countries

EDITORS

Harish Padh, *Ph.D.*

Director, B. V. Pharmaceutical Education and Research Development (PERD) Centre Ahmedabad, INDIA

Email: perd@perdcentre.com

Bernard Odhiambo Aduda, *Ph.D.*

Department of Physics, University of Nairobi, KENYA

Email: boaduda@uonbi.ac.ke, adudab@yahoo.com

Ajith P. de Alwis, *Ph.D.*

Professor of Chemical and Process Engineering, University of Moratuwa, Science Team Leader, Sri Lanka Institute of Nanotechnology, SRI LANKA
Email: ajithd@susnanotec.lk

EDITORIAL ASSISTANCE

Sweta A. Patel, *M.Phil.*

B.V. Pharmaceutical Education and Research Development (PERD) Centre Ahmedabad, INDIA

NANOTECHNOLOGY

Present Status and Future Prospects in Developing Countries

– Editors –

Harish Padh

Bernard O. Aduda

Ajith P. de Alwis

CENTRE FOR SCIENCE & TECHNOLOGY OF THE NON-ALIGNED AND OTHER DEVELOPING COUNTRIES (NAM S&T CENTRE)

2011

DAYA PUBLISHING HOUSE®

Delhi - 110 002

ISBN 9789351241225

Centre for Science and Technology of the Non-Aligned and Other Developing Countries (NAM S&T Centre)
Core-6A, 2nd Floor, India Habitat Centre, Lodhi Road,
New Delhi-110 003 (India)
Phone: +91-11-24644974, 24645134, Fax: +91-11-24644973
E-mail: namstct@gmail.com
Website: www.namstct.org

Published by : **Daya Publishing House®**
A Division of
Astral International Pvt. Ltd.
– ISO 9001:2008 Certified Company –
4760-61/23, Ansari Road, Darya Ganj
New Delhi-110 002
Ph. 011-43549197, 23278134
E-mail: info@astralint.com
Website: www.astralint.com

Laser Typesetting : **Classic Computer Services**
Delhi - 110 035

Printed at : **Chawla Offset Printers**
Delhi - 110 052

PRINTED IN INDIA

Foreword

In order to review state-of-the-art research and development in nano-scale phenomenon, devices, systems and manufacturing, the NAM S&T Center along with the Iranian Research Organization for Science and Technology (IROST), Iranian Nanotechnology Initiative Council (INIC) and IOR-ARC Regional Centre for Science and Technology Transfer (RCSTT) organised an international workshop on 'Nanotechnology–Present Status And Future Prospects In Developing Countries' from 18-20 May 2009 at Kashan, Iran. Scientists from developing and developed countries actively participated in its deliberations. The latest developments in the areas related to nanoscience and nanotechnology were reviewed by well known experts besides presentation of several country reports. This publication includes 21 articles and review papers on various aspects of nanotechnology and would be useful for the nanotechnology researchers, experts and practitioners in this fast emerging field.

India has always had an open attitude to scientific collaboration particularly with developing countries and I am very happy at this endeavour. Nanotechnology has tremendous applications to widespread industrial development and development of new products and cost reduction. It is a scale neutral global technology and is compatible with industrial organizations of different types including relatively small hi tech firms with mobility between 'worker' and 'employer' and encompasses both competition and cooperation. It can lead to Lean Production with emphasis on costs, competitiveness and profitability. Application to medicaland infrastructure areas like drinking water shows it is possible to integrate the strengths with social organisation issues? The papers show the need for balancing of objectives with a major focus on growth and profitability but with synergy with social concerns.

I am sure this effort will lead to others in a globally cooperative form.

I congratulate Dr. H.Padh and his colleagues.

Yoginder K. Alagh

Chairman
MSME Foundation, Chancellor, Nagaland University; and
Former Minister of Science and Technology of India

From the Editors' Desk

It has been said that nowadays any significant problem in science requires inter-disciplinary or rather more accurately a multi-disciplinary approach. More than any other areas of science, this holds true for the science of 'NANOTECHNOLOGY'. Science has once again amazed us that the same chemical compound in nano-size particles and micron size particles can have surprisingly very different properties which can be very useful in as diverse areas as electronics, forensic sciences, drug delivery, agriculture and pesticide control.

Realising the significance of this emerging area of science and its implication to developing countries, NAM S&T Centre in New Delhi organised an international workshop on 'Nanotechnology: Present Status and Future Prospects in Developing Countries' at Kashan, Iran, from May18-20, which was attended by 37 experts and professionals from 16 countries. This workshop had two major objectives: (1) To review the science of nanotechnology and its applications, and (2) To review the status of nanotechnology in member countries of NAM S&T. This workshop also had the objective of fostering collaboration among member state and to come out with some common directions for development of this aspect of science in the member countries.

The three-day workshop took place in serene atmosphere of Kashan, approximately 240 km south of Tehran in the foothills of Zagros Mountains. Whether by design or by chance, the location had a lot to do with the open and friendly atmosphere in which intense deliberations resulted. Several participants presented technical aspects of this new branch of science and its applications in varied fields. These presentations are summarized in total of 11 chapters presented in the first two sections of this book. Section I includes six chapters on the theme of "Characterization and Synthesis of Nanomaterials", while Section II having five chapters attempts to sample a few of the applications of nanomaterials. Needless to emphasise, nanoscience has wide spread applicability–these five chapters represent only a small sample. There were many more presentations reviewing the status of this area in several developing countries. These have been effectively captured in Section III having ten chapters–giving us flavour of country status and various government initiatives to bring benefits to people. Wide gap exists among the developing countries in terms of progress and planning for this new aspect of technology. The deliberations resulted in KASHAN DECLARATION, appended here, and outlining common platform for the member countries to develop the science and applications of the NANOTECHNOLOGY for the benefits of developing nations and their people. It also outlines scope of collaboration among the member countries.

This book represents efforts of the NAM S&T Centre, its member countries and other developing countries. Particularly reference is made to the host country, Iran, also a member of the NAM S&T Centre, which took painstaking efforts in making this workshop a meaningful one. We acknowledge the NAM S&T Centre and the team from Iran for effective organisation of this workshop. Thanks are due to all the authors for timely submission and revision of their respective chapters. Our special thanks are to Mr. Bandopadhyay of the NAM S&T Centre for detailing the arrangements for the workshop and Dr. V.P.Kharbhanda of the NAM S&T Centre for seeing through the compilation and printing process.

Editors

Introduction

Nanotechnology is the engineering, and related science, of the structures on the scale of molecules. It involves nearly all the scientific fields and specialties aimed at developing new materials and devices by precise manipulation of the structure of matter at the atomic- and molecular scales. The field of nanotechnology has found a variety of incredible applications with passive nanostructures including aerosols, colloids, coatings, nano-particle reinforced composites, nano-structured metals, polymers etc., and the active nanostructures embracing items such as targeted drugs, bio-devices, actuators, robots, sensors, molecular computers, surveillance devices and for use in space technology. Nanotechnology is hailed by many as the next industrial revolution, promising to change everything from the cars we drive to the clothes we wear, from the medical treatments our doctors can offer to our energy sources and workplaces, from new cancer therapies to pollution-eating compounds, from more durable consumer products to detectors for biohazards like anthrax, from novel foods to more efficient solar cells. Internationally, more than 100 foods have been manufactured, processed or packaged using nano-particles. The global nanotechnology market is expected to top $2.7 trillion by 2012.

Nature is already working at the nano-scale. But in so far as the manufactured nanostructures are concerned, these have special composition, reactivity and uniformity that may substantially increase the health and environmental vulnerability. Nano-particles are chemically more reactive than larger particles and can more easily get inside cells, even into the nucleus. Nano-tubes can behave like asbestos fibres and may possibly cause mesothelioma in living beings. One must recognise that free nano-particles do pose a hazard and this must be investigated from the beginning keeping in mind a balance between the benefits and risks from the use of nanotechnology materials and devices. The social and ethical issues concerning this fast emerging scientific field must be quickly addressed.

Although nanotechnology is in its infancy, this is the right time for the developing countries to invest in this new arena. Individual developing countries should take their own policy decision to identify the facet of advancements in nanotechnology which can address their specific economic, social and environmental needs.

In order to deliberate on the issues encompassing the growing significance of nanotechnology and its role in sustainable development, the Centre for Science and Technology of the Non-Aligned and Other Developing Countries (NAM S&T Centre) organised a 3-days international workshop on 'Nanotechnology: Present Status and Future Prospects in Developing Countries' at Kashan, Iran during 18-20 May 2009 jointly with the Iranian Research Organization for Science and Technology (IROST), Iranian Nanotechnology Initiative Council (INIC) and IOR-ARC Regional Centre for Science and Technology Transfer (RCSTT).

The Workshop was attended by 37 experts and senior professionals from 16 countries, including Armenia, Cuba, India, Indonesia, Iraq, Kenya, Libya, Malaysia, Mauritius, Myanmar, Nigeria, South Africa, Sri Lanka, Syria and Uganda, and the host country Iran. The overseas participants were from Armenia [Prof. Vladimir Aroutiounian, Head, Department of Physics of Semiconductors and Microelectronics, Yerevan State University, Yerevan]; Cuba [Dr. Ariel M. Felipe Gómez, Program Manager, Scientific Advisor Office, Council of State, Havana]; India [Dr. Harish Padh, Director, B. V. Patel Pharmaceutical Education and Research Development (PERD) Centre, and Project Director, National Institute of Pharmaceutical Education and Research (NIPER), Ahmedabad; Dr. Vinod Kumar Jain, Director, Amity Institute of Advance Research and Studies, Amity University, Noida; Dr. Uma Shanker Tandon, Scientist 'F', Council of Scientific and Industrial Research (CSIR), New Delhi]; Indonesia [Dr. Silvester Tursiloadi, Head, Division of Process and Catalysis Technology, Research Centre for Chemistry, Indonesian Institute of Sciences, Tangerang]; Iraq [Ms. Ndhal N. Hosain, Chief of Technical Observers, Central Organization for Standardization and Quality Control (COSQC), Baghdad]; Kenya [Prof. Bernard Odhiambo Aduda, Professor, Department of Physics, University of Nairobi, Nairobi; Dr. Erastus Gatika Gatebe, Lecturer, Department of Chemistry, Jomo Kenyatta University of Agriculture and Technology, Nairobi; Dr. Lutta W. Muhammad, Senior Research Officer, Kenya Agricultural Research Institute (KARI), Nairobi]; Libya [Dr. Nagib Ali Elmarzugi, Lecturer, Faculty of Pharmacy, Alfateh University, Researcher, Biotechnology Research Centre, Tripoli]; Malaysia [Mr. Radin Zulhazmi Bin Radin Abdul Halim, Principle Assistant Director, Industry Division, Ministry of Science, Technology and Innovation (MOSTI), Wilayah Persekutuan and Mr. Zailani Bin Safari, General Manager, Intelligence and Research Division, Malaysian Industry-Government Group for High Technology (MIGHT), Wilayah Persekutuan]; Mauritius [Mr. Hemraj Ramsurrun, Resource Officer, Rajiv Gandhi Science Centre, Port Louis]; Myanmar [Dr. Moe Zin Win, Researcher, Materials Science and Metallurgical Engineering Research Centre, Ministry of Science and Technology, Nay Pyi Taw]; Nigeria [Dr. (Mrs.) Zainab Hammanga, Deputy Director, Advance Materials Division, Raw Material Research and Development Council (RMRDC), Abuja]; South Africa [Ms. Nontombi Marule, Deputy Director, Emerging Research Areas, Department of Science and Technology, Pretoria]; Sri Lanka [Prof. Ajith P. de

Alwis, Professor, Sri Lanka Institute of Nanotechnology, University of Moratuwa, Malwana]; Syria [Dr. Nofal Ibrahim, Assistant Professor and Dr. Mhd. Ali Alsayed Ali, Researcher, Higher Institute for Applied Science and Technology (HIAST), Damascus;]; and Uganda [Mr. Henry Mugisha Bazira, Executive Director, Water Governance Institute, Kampala].

Workshop was conducted in nine technical sessions broadly categorised under the themes 'Characterization and Synthesis of Nano-Materials', 'Applications of Nanotechnology', 'Nanotechnology: Policy, Strategy and Market Development' and 'Nanotechnology: Status and Prospects in Developing Countries'. These sessions were co-chaired by Dr. Saeed Sarkar, Dr. M. Molanejad, Dr. Mohsen Jahanshahi, Dr. Safai, Dr. F. Davar, Dr. Bahram Khoshnevisan and Dr. Abdolhamid Bamoniri from Iran, and Dr. Abbas Sadri (ISESCO), Prof. Vladimir Aroutiounian (Armenia), Dr. Ariel M. Felipe Gómez (Cuba), Dr. Harish Padh and Dr. Vinod Kumar Jain (India), Prof. Bernard Odhiambo Aduda and Dr. Lutta W. Muhammad (Kenya), Dr. Nagib Ali Elmarzugi (Libya), Mr. Hemraj Ramsurrun (Mauritius), Dr. (Mrs.) Zainab Hammanga (Nigeria), Prof. Ajith P. de Alwis (Sri Lanka) and Dr. Nofal Ibrahim (Syria) from among overseas participants.

The foreign participants, who presented their papers, were Prof. Vladimir Aroutiounian (Armenia) on 'Nanotechnology–Present Status and Future Prospects in Armenia'; Dr. Ariel M. Felipe Gómez (Cuba) on 'The Cuban Approach towards Nano Biotechnology: A New Step on Scientific Development of the Country'; Dr. Harish Padh (India) on 'Drug Delivery and Drug Targeting using Therapeutic Nanoparticles'; Dr. Vinod Kumar Jain (India) on 'Applications of Micro and Nanotechnology in Biomedical Engineering' and 'MEMS and Nanotechnology Based Explosive Detector'; Dr. Uma Shanker Tandon (India) on 'Nano-Structures, Nano-Materials and Systems'; Dr. Silvester Tursiloadi (Indonesia) on 'Syntheses of Nano Size Anatase Type of Titania Aerogels by Addition of Silica'; Prof. Bernard Odhiambo Aduda (Kenya) on 'Status of Nanoscience and Nanotechnology in Kenya'; Dr. Lutta W. Muhammad (Kenya) on 'Harnessing Emerging Technologies for Kenya's Vision 2030 Development Strategy: Prospects for Leveraging Advances in Nano-Technology'; Dr. Erastus Gatika Gatebe (Kenya) on 'Development of Nanostructured Smart Delivery Systems for Pesticides and Fertilizers'; Dr. Nagib Ali Elmarzugi (Libya) on 'The Surface Study of Different Architectures of DMAEMA Polymer at Nano Scale for Transfer DNA in Gene Therapy'; Mr. Zailani Bin Safari (Malaysia) on 'Driving Nanotechnology Development in Malaysia'; Mr. Radin Zulhazmi Bin Radin Abdul Halim (Malaysia) on 'Nanotechnology: Present Status and Future Prospect in Malaysia; Mr. Hemraj Ramsurrun (Mauritius) on 'A Vision for Civic Engagement in Nanotechnology in Mauritius'; Dr. Moe Zin Win (Myanmar) on 'Present Status of Nanotechnology Research in Myanmar'; Dr. (Mrs.) Zainab Hammanga (Nigeria) on 'Nanotechnology–Present Status and Future Prospects in Nigeria'; Ms. Nontombi Marule (South Africa) on 'South African Nanotechnology Strategy'; Prof. Ajith P. de Alwis (Sri Lanka) on 'Sri Lanka Nanotechnology Initiative'; Dr. Nofal Ibrahim (Syria) on 'Growth and Characterization of PbS Nanocrystalline Thin Films Deposited on Glass Substrates by Chemical Bath Deposition', Dr. Mhd. Ali Alsayed Ali (Syria) on 'Characterization of Carbon Nanoparticles formed in Diluted Environments'; and Mr. Henry Mugisha

Bazira (Uganda) on 'Application of Nano-Science and Nanotechnology in Uganda's Economy'. Prof. Arun Kulshreshtha (NAM S&T Centre) made a presentation on 'South–South Cooperation through NAM S&T Centre'.

The participants from Iran, who made presentations during the workshop were Dr. Saeed Sarkar, Director, INIC on 'Iran 'Nano Initiative' Program, Targets and Achievement'; Dr. Mohsen Jahanshahi, Head, Nanobiotechnology Research Centre, Babol University of Technology on 'Nanofiltration Membranes: Preparation, Characterization and Industrial Application in Water / Waste Water Treatments'; and Dr. Bahram Khoshnevisan, Assistant Professor, Institute of Nanoscience and Nanotechnology, University of Kashan, Iran on 'Hydrogen Storage as a Fuel Resource in Future'

The Plenary Concluding Session was led by Dr. M. Molanejad (Director, International Cooperation, IROST), Dr. Mohsen Jahanshahi (Head, Nanobiotechnology Research Centre, Babol University of Technology) and Prof. Arun Kulshreshtha (Director, NAM S&T Centre) when a Kashan Declaration on 'Nanotechnology: Present Status and Future Prospects in Developing Countries' was adopted after considerable deliberations and debate.

The present publication is based on the deliberations of the Iran workshop and includes papers contributed by eminent specialists in the field. It comprises of twenty-one articles and review papers under four Chapters namely 'Characterisation and Synthesis on Nano-Materials', 'Application of Nanotechnology', 'Nanotechnology: Policy, Strategy and Market Development', and 'Nanotechnology: Status and Prospects in Developing Countries'.

I acknowledge with gratitude the deep involvement and determined efforts of Dr. Harish Padh, Director, B. V. Patel Pharmaceutical Education and Research Development (PERD) Centre, Ahmedabad, India; Prof. Bernard Odhiambo Aduda, Professor, Department of Physics, University of Nairobi, Kenya; and Prof. Ajith P. de Alwis, Sri Lanka Institute of Nanotechnology, Sri Lanka in editing this valuable publication. I am also indebted to the entire team of the NAM S&T Centre, particularly Mr. M. Bandyopadhyay, Dr. V.P. Kharbanda, Mrs. Manjari Manisha, Ms. Chitra Prasad and Mr. Pankaj Buttan in compiling the presented papers, liaising with the authors and editors and giving a shape to this volume.

I sincerely trust that this publication will serve as a valuable reference material not only for the developing countries but for other countries as well and will pave way for the promotion of South-South and North-South cooperation in the fast emerging area of nanotechnology.

Prof. Arun P. Kulshreshtha
Director, NAM S&T Centre

Contents

— Section II —
APPLICATIONS OF NANOTECHNOLOGY

— Section III —
NANOTECHNOLOGY: STATUS AND PROSPECTS IN DEVELOPING COUNTRIES

– Section I –

CHARACTERIZATION AND SYNTHESIS OF NANOMATERIAL

Chapter 1

Nanostructures, Nanomaterials and Systems

U.S. Tandon
Council of Scientific and Industrial Research,
2 Rafi Marg, New Delhi – 110 001, India
E-mail: tandonus@yahoo.com

ABSTRACT

The precision needed in the shape and dimensions of an ultimate nanostructure or system entail exacting restrictions on its engraving mechanisms. The generation of nano-structures producing miniaturized components such as lab-on-a-chip, DNA-array and NEMS calls for processes, which offer more flexibility than that available with the conventional surface milling. The shapes of components structured by classical processes are often governed by the crystalline properties of the substrate and chemical as well as physical parameters of the unit process. Reactive gas cold plasma has led to engraving of structures such as lines, holes, channels, grooves or pyramids of stipulated dimensions. Bottoms up approach like arc discharge, CVD, biomimetics and reverse micellar routes have given a spurt in the variety of nanomaterials being investigated. It has expanded the interest of researchers from nanoscience and technology towards the design of new nanomaterials with specific characteristics. Nano-biomaterials with better mechanical, thermal, chemical properties than found in known materials are being made. Nano-fibres, -particles, -phosphors and -composites on varied templates are being synthesized.

Keywords: *Biomimetic, Functionalize, Granule, Mushrooms, Nanocomposite, Nanoreactor, Stimuli, Tissue Engineering.*

Introduction

Spectacular developments in various unit processes of micro-nano-fabrication kept propelling the revolution in microelectronics for a period longer than two decades

(Tandon 1993). These unit processes later became the bases for unexpected growth of new progression and tools in mechanical, chemical, biological and other discipline of engineering. Macrostructures created with conventional engraving techniques invariably bore a footprint of the physics of the substrate and the chemistry of etchants. However, the adoption of iteratively optimized unit processes like micro-nano-machining facilitated the development and integration of representative devices such as MEMS, lab-on-a-chip, DNA-array and so forth. Miniaturized components such as the diaphragm, levers, grooves and gears for NEMS can thus be assembled (Tandon 2008) or co-fabricated to manufacture systems (Nakajima 2005) with reduced response time and power requirement. These technological advancements resulted through successive miniaturization, also known as the top-down elite approach. Its cutting edge infrastructure was rather expensive and select institutions could endeavor to miniaturize materials of scientific inquest. They used advanced lithographic tools, physico-chemical evaporation, AFM, and other techniques to downsize in 3-d, 2-d and 1-d space to fabricate nanostructures or devices and explore their characteristics. The applicability of such tiny devices is pushing ahead the evolution of fabrication sequence as well as their basic unit processes.

However, the realization of new phenomenon and a variety of useful applications of nanosized materials like CNT engaged the quest of scientists to evolve alternative avenues of realizing such structures and systems. Termed as bottoms-up approach, it included processes like CVD, laser evaporation, electro-polymerization etc. (Gong and Zhou 1995). Subsequently, the biological processes like cellular biosynthesis and biomimetics were also harnessed to develop dispersed nano-particles of metals, alloys, oxides etc. This approach of molecular nanotech improved the understanding to harness new properties governed by the phenomena occurring at these tiny dimensions. The design of new nanomaterials opened up avenues for these materials in IT, biotech, environment, chemical technology, space etc. Nano-sized materials of optimum shape are isolated, characterized and functionalized as per subsequent application. Under the influence of suitable stimuli, these nano-systems interface with tiny or complex systems of relevant arena. High strength nanocomposites, high permeability magnetic sensors, biomaterials, stimuli sensitive nano-carriers for drug delivery, hard coatings, catalysis, optical applications are being developed.

Creating the Structures and Systems by Chemical Machining

Nano-systems made by assembling or embedding sensors, logics and actuators have given way to the compact and reliable integrated systems where all the components are fabricated on the same substrate (Tandon 2004). It is crucial to identify suitable substrate, create structures of the stipulated complex shape and protect them while other components are being processed. Differential in machining rates along certain crystallographic planes vis-à-vis others could create microstructures like slant cavities, cantilevers and diaphragms and doping of substrate provided additional control. Sequential structuring processes like ultrasonic machining, EDM, plasma beams may not deliver either the needed precision or the throughput. While the surface processing will engrave the structures down to a certain depth, bulk processing would etch them through the entire thickness of the substrate. Machining the structures of nano-dimensions requires precision and flexibility in the process, both of which are available in reactive plasma. Representative nanostructures encountered in miniscule electronic devices or in chemical and biological micro-nano-analyzers may thus be created. The cold plasma system and its controls shown in Figure 1.1 were used to carve structures of decreasing dimensions and various shapes in a variety of substrates.

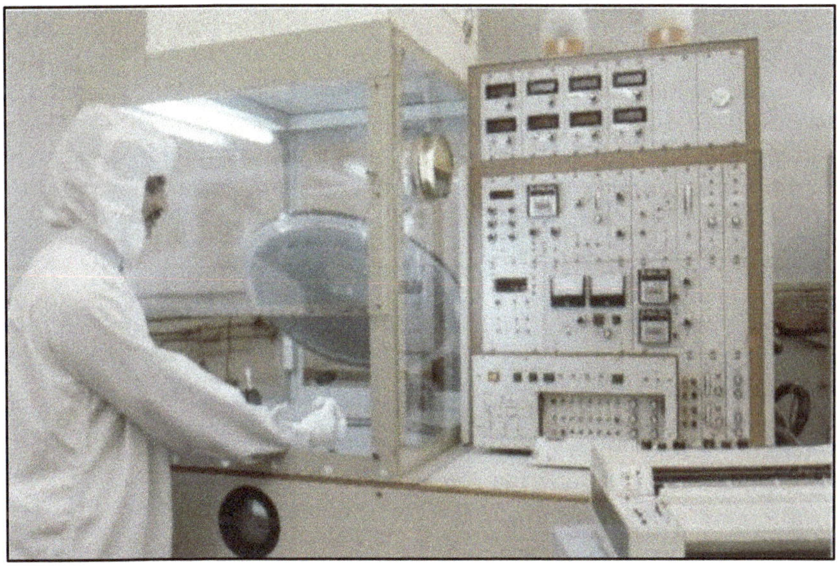

Figure 1.1: The Reactive Cold Plasma Chamber and its Control System

The full process of engraving the nanostructures may use several unit process steps like deposition, imprinting and etching sequentially or in parallel. Controlling the temperature of the substrate usefully limits the damage from radiation abundant within plasma. The combination of physical momentum and chemical reactivity of excited species deliver the potential to carve the structures and offer maneuverability to tailor their shapes (Lee and White, 1998). The ultimate structures, whether engraved or generated, are characterized with the help of their micrographs and also with the characteristics of the full structuring process. The shape, depth, distribution and profile of these structures vis-à-vis the set of process conditions *viz.* recipe is being analyzed to optimize and evolve new process sequence.

Plasma structuring processes developed for surface mode to make shallow devices are applicable to bulk machining to create tweezers or gears. For instance, when a conformal deposit of poly-silicon was made on a layer of sacrificial oxide (Tandon 2006) with a mold onto the substrate, the structure of a micro-bush could be engraved. Micro-nano-components of moderate aspect ratio are generally very thin. When created by surface process, these occupy a tiny footprint and are also resistant to mechanical shocks. For the fabrication of rugged components like gear trains that generate torque, high aspect ratio metallic patterns obtained by electroplating on thick resist are required (Conradie and Moore, 2002). Sacrificial layer on the substrate is used to create microstructures whose portions are hanging free.

Unique Process

Structuring with reactive plasma has been brought about from a sequence of sub-process steps at CEERI. The stipulated structures like channels, holes, trenches and mesa–designed in soft version and reproduced as opaque regions on transparent masks–were replicated onto a UV-sensitive polymer film purposely coated on the substrate. Unprotected regions of the substrate got exposed to react with suitable excited species and the volatile products were removed. Finally, the protecting polymer film was oxidized and the bared structures were analyzed for their dimensions, depth and profile. For a critical set of characteristics of ultimate microstructure, the recipe is often unique and its mapping on another plasma system may require further adjustment of parameters.

The set of parameters such as ambient pressure, chemical constituents and degree of ionization characterizing the plasma grossly affected the features of final structures, some of which are shown in Figures 1.2 and 1.3. Machining characteristics like anisotropy, rate and selectivity with bottom layer were also analyzed. Line-scan confirms its flat bottom onto which a micro- or nanosystems may be buried to protect it from the ambient. Analysis of iteratively generated structures indicates that rising ratio of carbon to other reactive components leads to deposition of films that effectively enhance the vertical structuring rate and thus the ultimate vertical dimensions of the structures (Pant and Tandon 1999).

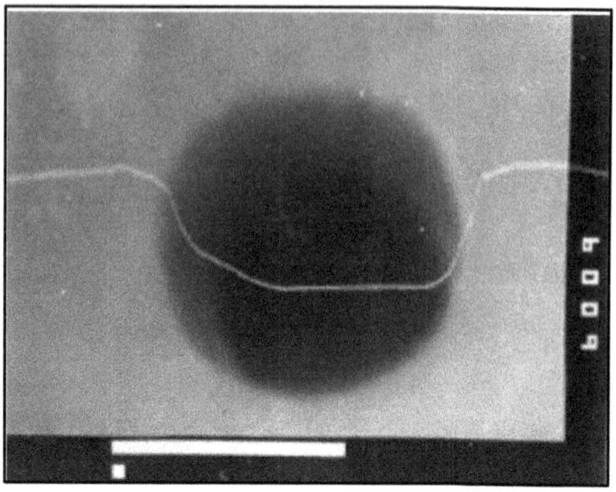

Figure 1.2: Micro-hole Twice as Deep as its Dia

Equi-spaced tapered-top channels 1000 nm wide at its bottom was engraved over its running length, which is three orders higher. Cross-sectional SEM (Figure 1.3) shows tapering at the top and like the deep hole in Figure 1.2, the aspect ratio of these structures can be increased many

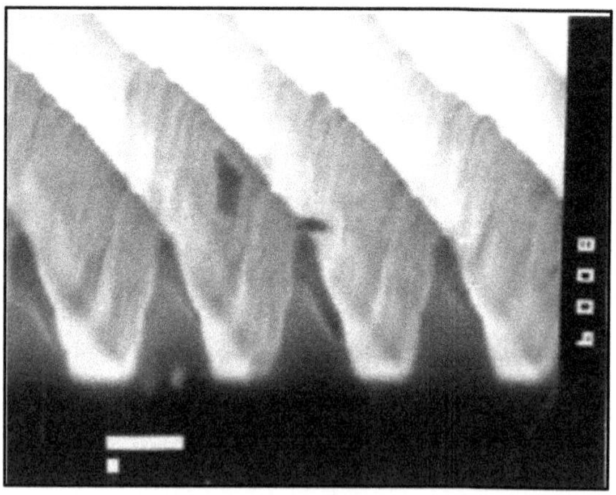

Figure 1.3: Tapered-top Channels for Fluidics

fold. Possibility of smooth micro-flow of liquid through these channels and through open bowl-shaped structures opens new vistas of their application in microfluidics (Tandon 2008). Requisite surface modification of such channels and capillaries is being undertaken at AMPRI by coating polymers like PDMA-g-PMMA to introduce bio-compatibility. Once converted to bio-sensors, these channels could separate serum proteins and peptide fragments by capillary electrophoresis or laser-induced fluorescence and also get integrated with lab-on-a-chip.

Diverse Structures

The excited species in cold reactive plasma offer maximum impact in their line-of-sight and this process of structuring can be automated. Several innovative recipes and their multi-step sequence were tried out to create structures of unconventional shapes. That is how a mushroom shaped structure (Figure 1.4) was created in a specific structural layer. Such a structure with plenty of lateral spread associated with isotropic process occupies wider area than the smaller tiny stencil through which it was created. Technology of generating such distinct shapes induces new applications and pushes up the demand of these structures although their dimensional spread has got enlarged. The capping top of mushroom structure may be sacrificed after it has served to protect the structural layer under it from the machining species.

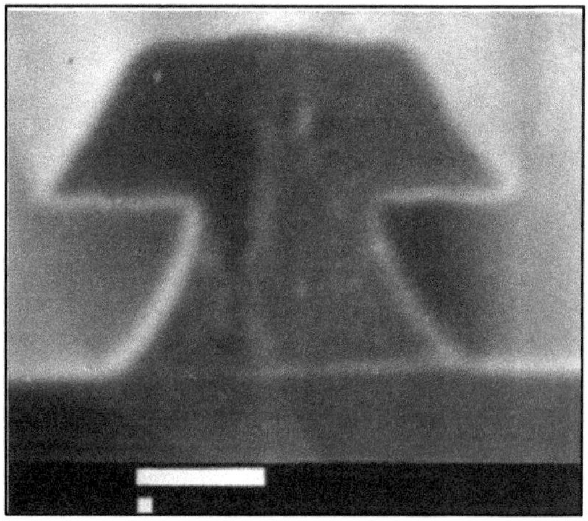

Figure 1.4: Mushroom-shaped Tiny Structure

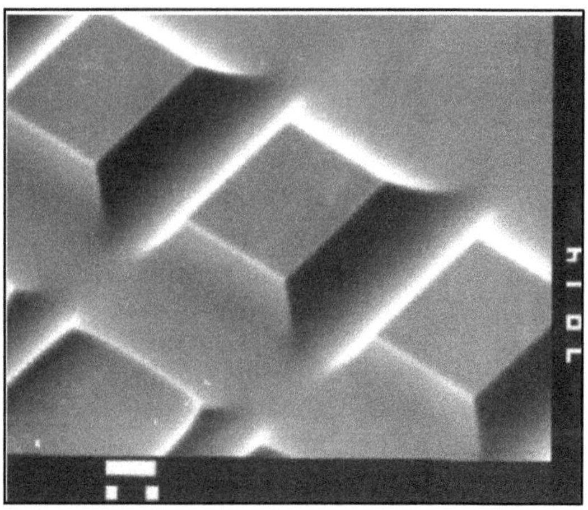

Figure 1.5: Pyramid-like Mesas

Another varied structuring sequence has delivered the structures having the shape of pyramid-like mesas with slant-wall profiles (Figure 1.5). The slant mesas may ideally host a matrix or linear array of sensing elements for specific tasks. The distance between the consecutive mesas enhances from their bottom to the top by almost two times of their thickness, which could result in new avenues for their applications. Such microstructures were generated through a planned deviation from their guiding micromask. It was done by restricting the rate of simultaneous deposition of polymer film and also ensuring that its removal is faster than deposition in the regions being engraved.

By resorting to process parameters that produce polymerizing components during the structuring, a deposit on the substrate was obtained and it was later pierced with ionic species available in a subsequent plasma process. The microstructures generated by machining through such punctures, which expand with micromachining, appear in Figure 1.6. The surface density and anisotropy of these sphere-like structures (Tandon and Pant 1992) could be varied through the any of the several process parameters applied in the two-step structuring process. The aspect ratio, lateral dimensions and the sharpness of these nano-granules can be tailored to specific desirable shape by iteratively optimizing the parameters for any substrate. The polymerizing and machining processes involve the ionized species of halogens and other reactive components, whose reaction with the substrate and surroundings needs to be carefully evaluated.

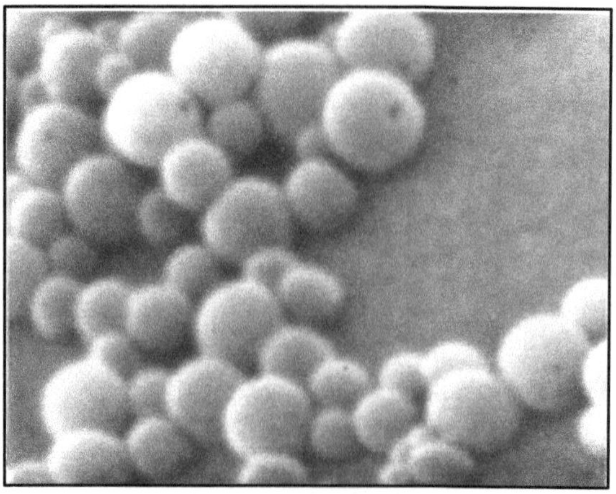

Figure 1.6: Nano-granules by Multi Step Process

Any increase in the aspect ratio of these structures like a trench or–whose linear dimension is several orders larger than its width–enhances the criticality in the profile of its sidewalls. The kink that spread its bottom and bow that notched it towards the top were partially inhibited through lateral deposition. Roughness in the sidewalls and the bowing appearing in Figure 1.7 arose from additional machining by high-energy species reflected from the bottom (Tandon 1997). Addition of chlorine in the reactive gas plasma marginally reduced the bow and improved the anisotropy of the sidewalls. A supplement of non-reactive machining eliminated the deposition of inhibitors and made the sidewalls somewhat smooth.

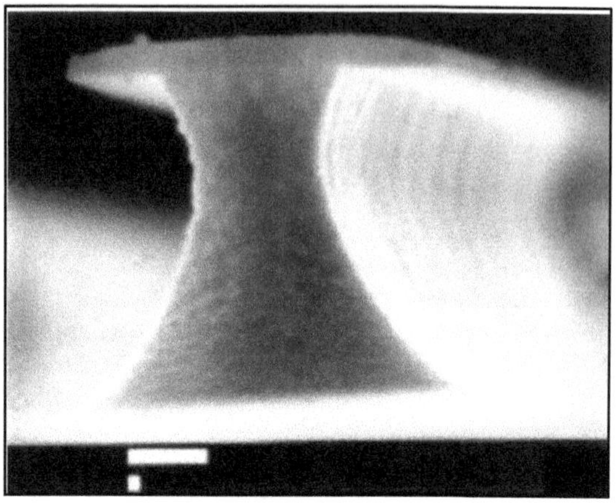

Figure 1.7: Uncapped Linear Microstructures

Creating Structures by Chemical Synthesis

While the top-down approach successively reduces the dimensions of a bulk structure to micro and nano-dimensions on a pre-selected substrate, another approach is to synthesize molecular material to nanostructures (Tandon 2003). Nanoparticles, -wires, -arrays, -tubes, -porous materials and -membranes are being synthesized on varied templates. They are then treated to become bio-compatible or functionalized to attach specific cargo or respond to stimuli and sometimes encapsulated for protection. Chemical, electrochemical, biomimetic, other biogenic and self-assembly approach are being adopted in CSIR laboratories. Some of these are executed in vacuum and others at ambient pressure. Nanostructures of organics, metals, alloys, oxides, biominerals, composites etc. are being developed for medicinal, electronic, catalytic, magnetic, pharmaceutical, structural and tribological applications. Nanophosphors and nanomaterials from the condensation of inert gases and complex oxide nanoparticles using reverse micellar routes have generated research interest. Metal nanoparticle composites in hybrid matrices, oriented arrays of complex oxides through low temperature roots, silica-Titania core shell particles and the use of biological templates like DNA and virus are among the new evolving projects.

Single and multi-walled carbon nanotubes as well as nanoshells and fullerenes were made at NPL using arc discharge at high temperature, moderate pressure and transition metal as catalyst (Figure 1.8).

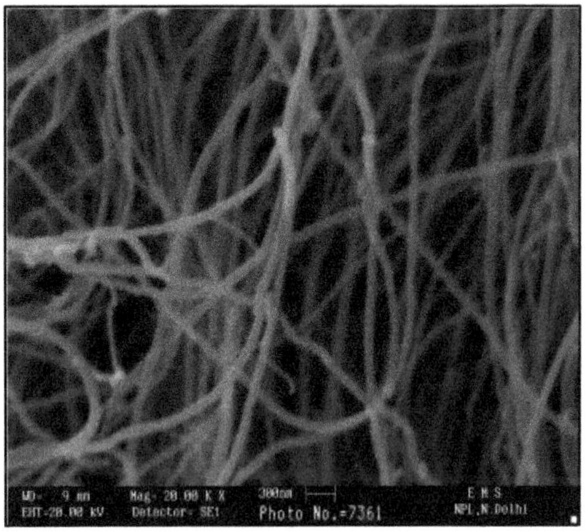

Figure 1.8: As Deposited Carbon Nanotube

Carbon Films

CVD with hydrocarbon catalyst rendered multiwall tubes whereas PECVD at 600°C delivered rose petal like carbon nano-sheets (Figure 1.9) on Ni/Si substrates in CH_4/Ar plasma (Srivastava, Vankar and Kumar 2005). CVD with hydrocarbon catalyst rendered multiwall tubes whereas PECVD at 600°C delivered rose petal like carbon nano-sheets (Figure 1.9) on Ni/Si substrates in CH_4/Ar plasma. Surface adsorption qualifies it as sensitive nano-sensor and porosity and large specific surface area make it ideal for biocomposite electrode. The strength, stiffness and fracture toughness of CNT-reinforced composite is finding applications in satellites,

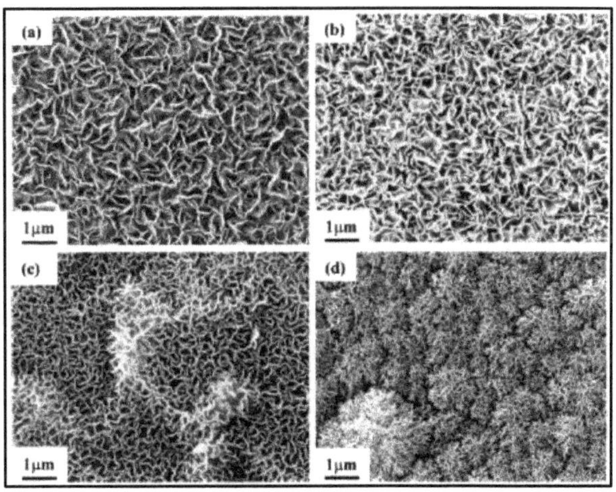

Figure 1.9: Petal like Nano-structured

aircrafts and bullet proof vests. The process of making organically capped CdSe nanoparticles having applications in LED and in fluorescence imaging has been optimized. CNTs, In_2O_3:Ag composite, $PbTiO_3$ and nano-structured ZnO film are being utilized to make flow sensors for chemical, biological fluids and pollutants. The nanopowder of Al-Mg alloy prepared by milling at cryogenic temperature gave light-weight, high strength, corrosion-resistant bulk, which finds applications in automobile and aerospace.

Biogenic Processes

Most of the physico-chemical processes used for making nanomaterials are hazardous, expensive and eco-unfriendly. Biogenic processes using fungi and actinomycetes helping the redox reactions thus got adopted as a welcome alternate from NCL (Ahmad 2003, 2005). Au, Ag and CdS nanoparticles made by intra- and extra-cellular biosynthesis using endophytic and plant pathogenic fungus respectively. These nanoparticles made from their requisite salts were characterized by XRD, UV-Vis, FTIR, TEM, SAED etc. They were found suitable for homogenous catalysis and could be immobilized in different matrices or in thin film form for optoelectronic and other electronic application. Similar extracellular synthesis of nanoparticles of Iron oxide, Zirconia, Titania and complex oxides like $BaTiO_3$ have been done. Fungus meditated bioleaching succeeded in creating amorphous SiO_2 nanoparticles (Figure 1.10).

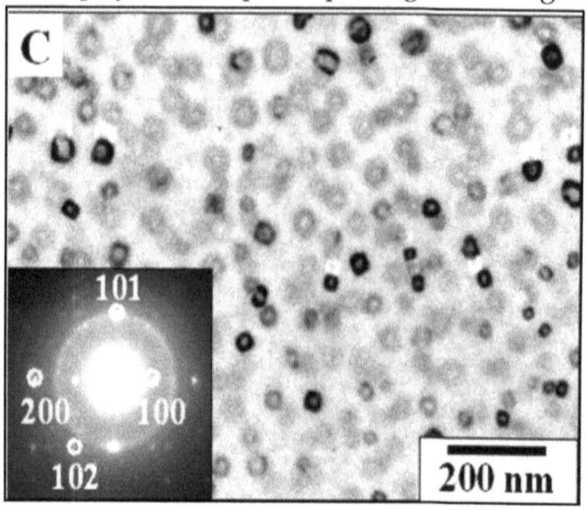

Figure 1.10: TEM of SiO_2 NP Calcined at 400°C

Biosynthesis of carbonate biominerals: $CaCO_3$ and $SrCO_3$–which have very large number of applications in various industries–have also been done. Extracellular synthesis of bimetallic 6-14 nm Au-Ag alloy nanoparticles (Figure 1.11) and biosynthesis of CdS nanoparticles have also been done. Protein shells surrounding the enzymes block electron transfer from an electrode to its active centre. But the rate of electron transfer from functionalized Au-nanoparticles to enzyme redox centre (5000/sec) is seven times faster than between glucose oxidase and its natural substrate.

Nanoparticles of hydroxyapatite (bone mineral) prepared at NML using biomimetic process at room temperature offer a great potential for drug delivery, tissue engineering, orthopedics, orthodontics and in osteo-inductive-conductive bio-ceramics (Figure 1.12). These nano-biomaterials are being characterized to meet mechanical, thermal, optical, or chemical requirements, which are more stringent than available with any single organic or inorganic material. The product has already been commercialized by an industry. By using the molecular cages of a globular

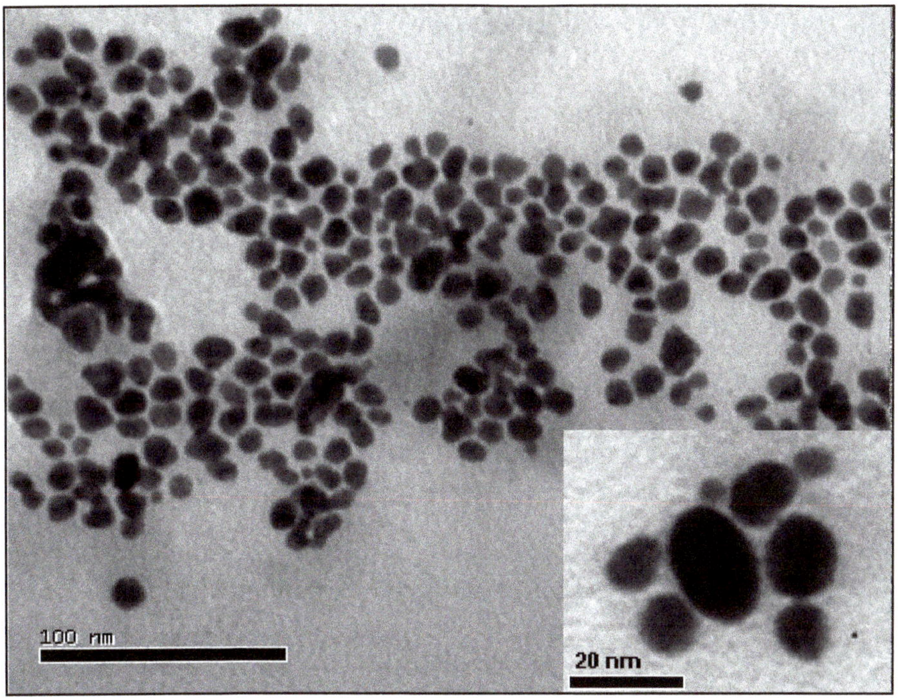

Figure 1.11: TEM of Au-Ag Alloy Nanoparticles

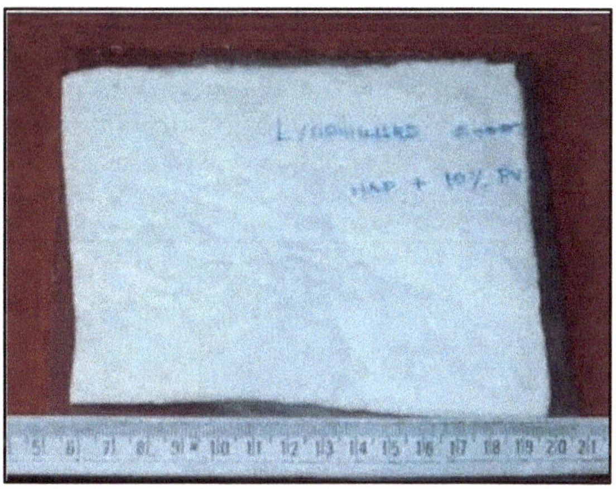

Figure 1.12: Biomimetically made Foam of HAp

protein named bovine serum albumin as nanoreactor, nanoparticles of super-paramagnetic Fe_3O_4, metallic Cu and semiconducting CdS have been synthesized with uniform morphology. Nanoparticles of carbide, boride and nitrides of Ti, Zr and Si have been prepared through Self-propagating High temperature Synthesis (SHS). Nano-ceramics like alumina, zirconia-toughened alumina, yttria-stabilized zirconia, aluminium nitride (Figure 1.13) are prepared by solution combustion, sol-gel and gas-phase synthesis.

Functionalization

Synthesis of photo-, electro-, and bio-active molecules with functional groups like–SH,–NH_2 and–COOH and anchoring them on water soluble metal nanoparticles is going on at NIIST. Similarly, supramolecular assemblies are being synthesized

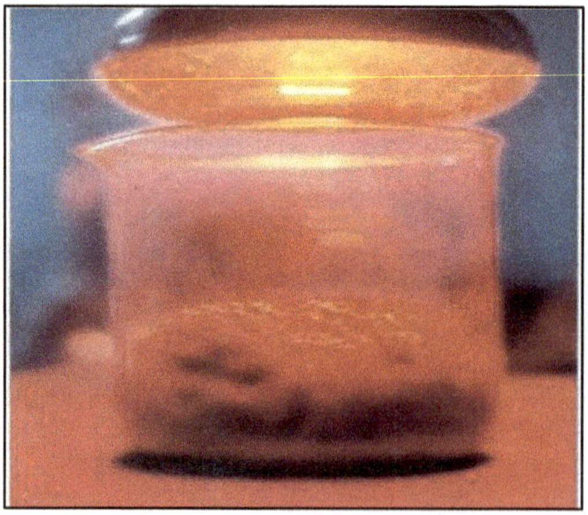

Figure 1.13: AlN Nanopowder 6 gm (250 ml)

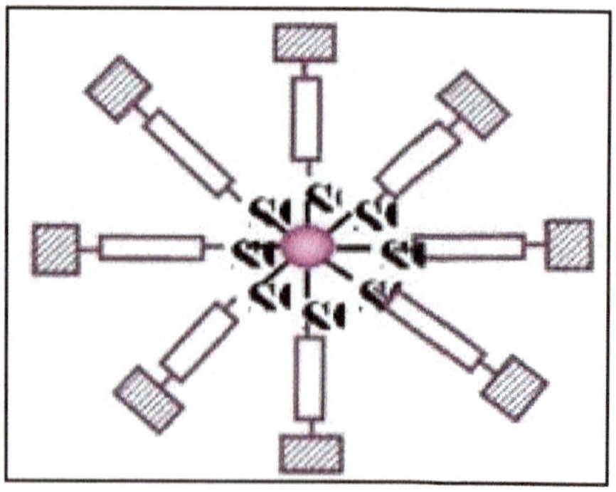

Figure 1.14: Photoresponsive Molecule on Au NP

(Figure 1.14) by attaching photo-responsive molecules to Au nanoparticle. These assemblies could bind or release biologically important molecules using external stimuli (Thomas and Kamat, 2003).

Through Plasmon coupling, Au nanorods are being dimerized and oligomerizerd to nanochains (Thomas *et al.*, 2004 and 2006) that detect Cysteine and Glutathione and protect the damage of DNA in the cells from oxidative stress (Figure 1.15). Nanosensors of phosphorescent nanomaterials are also being developed to detect (Ipe, *et al.*, 2006) alkali, alkaline earth and transition metal ions. Functionalized CdSe @ ZnS quantum dots and ZnS monolayers are being developed as optical sensors (Figure 1.16). Nano-sized photo catalytic titanium oxide is providing anti-algal and anti-bacterial coating on a variety of materials. Such transparent coatings decompose organic stains, algae or bacteria in presence of sun light. Intercalated layered clay

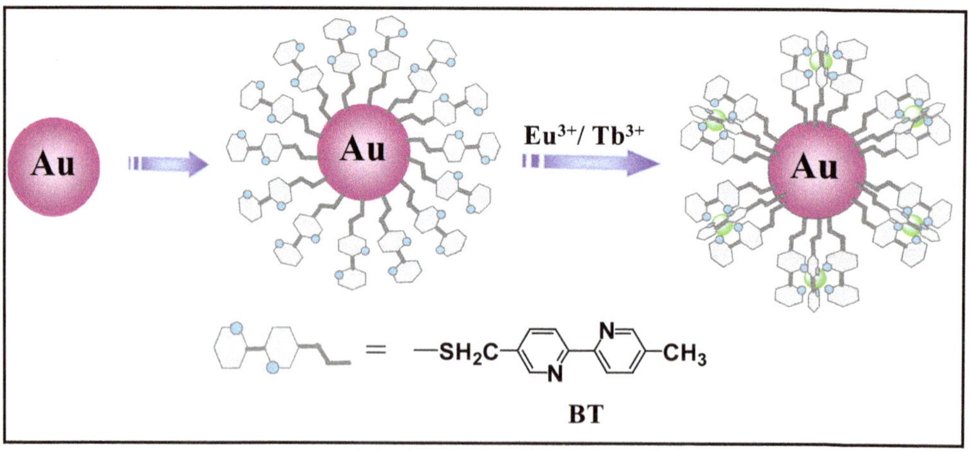

Figure 1.15: Phosphorescent Nanomaterials Sensor

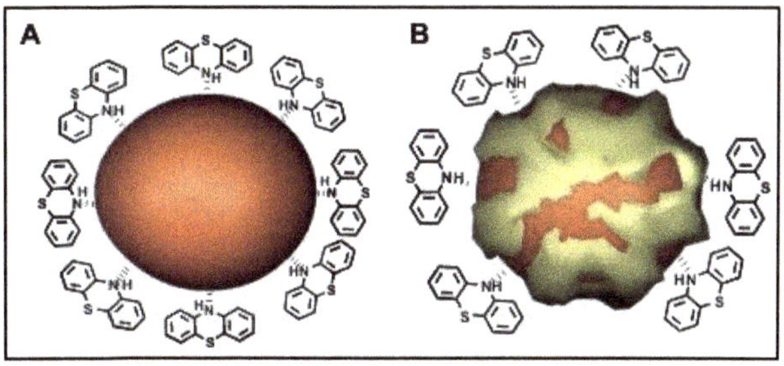

Figure 1.16: ZnS Monolayer as Optical Sensor

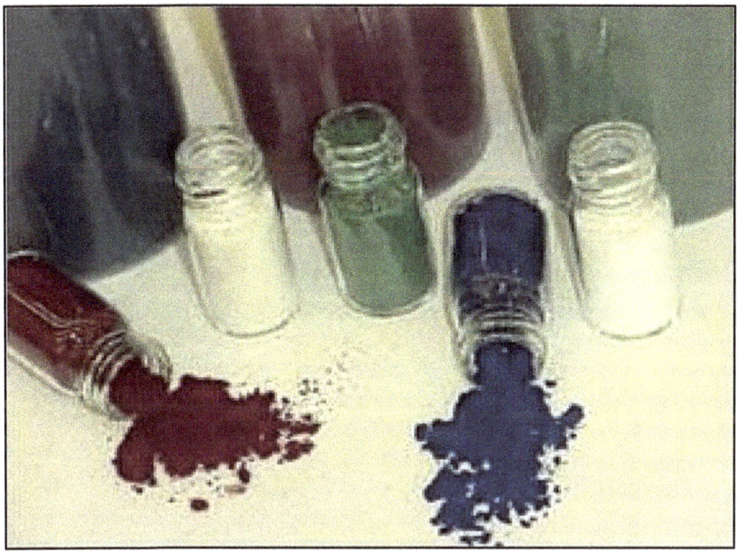

Figure 1.17: Eco-friendly Nano-pigments

with eco-friendly nano pigments (Figure 1.17) has been blended with polyethylene. Titania-Alumina nano-sized rare earth phosphate has been developed to give a contact angle of 104°.

By modifying the electrode surfaces with nano-reactors like metal nanoparticle-loaded dendrimer layers, nanoreactors were formed at CECRI (Raghu *et al.*, 2005) for oxidizing methanol to CO_2 (Figure 1.18). Novel effects of metal ion chelation on the properties of lipoic acid-capped Ag and Au nanoparticles were shown. Using 3-(3-thienyl) acrylic acid as capping agent, nearly mono-disperse gold nanoclusters were synthesized (Figure 1.19) and their application to single electron events (coulombic staircase) could be demonstrated.

Monosized nanostructured powder of Pt and Pt-alloys is being developed with the goal of integrating it in fuel cell configurations. It would offer highest power density as fuel cell catalyst. As a separate effort, Sn or Bi was filled in MWCNT and alloyed with Li to make novel anode for lithium-ion batteries (Prem Kumar *et al.*,

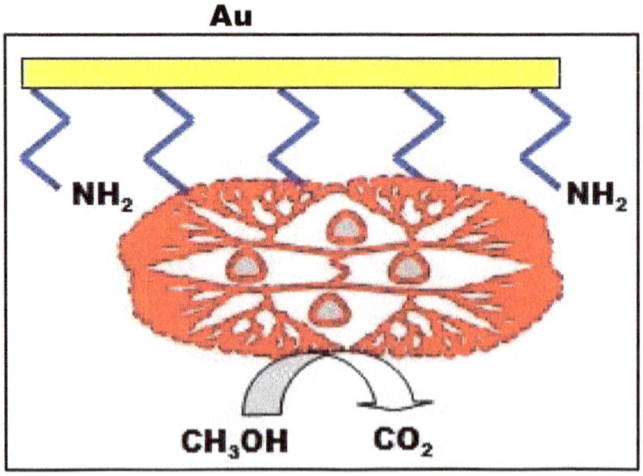

Figure 1.18: Self Assembled Dendrimer Nanoreactor

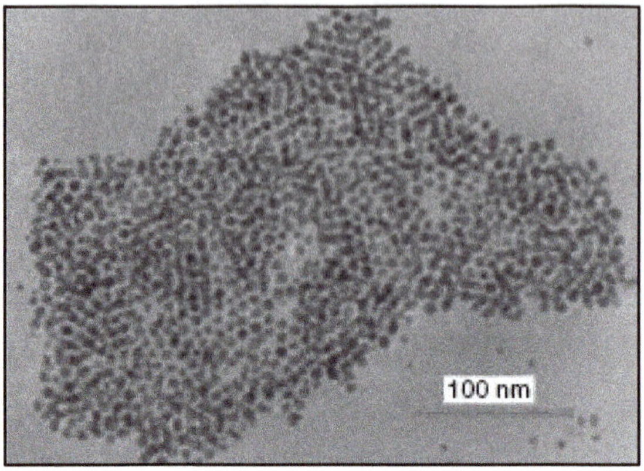

Figure 1.19: Capped Mono-disperse Au Clusters

2004). Its first-cycle insertion and de-insertion capacity *viz.* 2474 and 889mAhg^{-1} exceed the cumulative values from the constituents. By incorporating a conducting polymer Poly (ethylene-dioxy-thiophene) *viz.* PEDOT in nanoporous Au matrices, a nanocomposite was made (Mathiyarasu *et al.*, 2007). Its nanothin film on electrode surface executes catalysis and senses molecules of dopamine and uric acid.

Hybrid filter using nano-sized adsorbent in micro-ceramic membrane developed at CGCRI is being used to filter out As and Fe contaminants from ground water. Transparent coatings of inorganic-organic nanohybrid developed by CGCRI have enhanced the hardness and abrasion resistance of plastic sheets and ophthalmic lenses. Commercial production based on this technology is being undertaken by two industries. The hybrid has low/high refractive index layers which can be tuned to give antireflective as well as colourful coating. For SOFC, nanostructured doped LSCF cathodes, ZrO_2 core-Ni shell anode, and Gd-doped ceria electrolyte have been developed.

Numerous Ongoing Developments in Nanotechnology

The abundance of nanotech related R&D project is in drug delivery, catalysis, green energy, environment, biomaterials, biosensors, antimicrobial, development of toughened, light-weight, magnetic materials. Nevertheless, investigations are going on in various other streams of R&D within nanoscience and technology. These include detection of plant viruses, genetic modification of food, genes causing drug unresponsiveness; nano-filters for fluids and gases; nanoparticles from industrial wastes; Layered Double Hydroxides, functional nanoceramics, nanomedicine and toxicological aspects of nanotech. Their tiny dimension allows them to permeate biological membranes and the resulting toxic potential is being intensively studied and analyzed at IITR.

Conclusions

Microstructures like channels, holes or pyramids were machined with reactive gas plasma after they were imprinted on the substrate. The machining characteristics like slope, erosion and selectivity were optimized through plasma process parameters like power, chemistry and pressure. Distinct microstructures such as mushrooms, bows and spheres have been generated using unconventional recipes.

Fullerenes, CNTs, nanostructures of metals, alloys, salts, composites, foams have been prepared at low and ambient pressure, varying temperatures, from wastes and via biogenic modes. As core-cargo-shell, functionalized, capped and hydrophobic-hydrophillic modes their response to stimuli has generated wide variety of applications. Nanomaterials display exquisite optical, magnetic, hardness, catalytic, mechanical, biological and attachment properties, which make them a splendid material to start with and modify.

Acknowledgements

I thank my colleagues in CEERI Pilani who offered help in experiment and analysis. I am also grateful to all the scientists in different research laboratories of CSIR whose findings have been integrated. I am grateful to Prof. S.K. Brahmachari, DG CSIR and Dr. Naresh Kumar, Head RDPD for their precious encouragements.

References

Ahmad, A., 2003. *J. BioMed Nanotechnology*, 14: 824.

Ahmad, A., 2005. *J. BioMed. Nanotechnology*, 1: 47.

Conradie, E.H. and Moore, D.F., 2002. *J. Micromech. Microengg.*, 12: 368–374.

Gong and Zhou, 1999. *Proc of Intl Symp on Test and Measurement, ISTM*, pp. 23–26.

Ipe, B.I.K., Yoosaf, K. and Thomas, K.G., 2006. *J. Am. Chem. Soc.*, 128: 1907.Lee, S.S. and White, R.M., 1998. *J. Micromech. Microengg.*, 8: 230–238.

Mathiyarasu, J., Senthilkumar, S., Phani, K.L.N. and Yegnaraman, V., 2007. PEDOT–Au nanocomposite films for electrochemical sensing for dopamine and uric acid. *J. Nanosci. and Nanotechnology*, 7(6): 2206–2210.

Nakajima, N., 2005. *Asia Pacific Nanotech Weekly*, 3.

Pant, B.D. and Tandon, U.S., 1999. *Proc. International Conference on Smart Materials, Structures and Systems*. IISc, Bangalore, India, Allied Publishers.

Prem Kumar, T., Ramesh, R., Lin, Y.Y., and Ting-Kuo Fey, G., 2004. Tin-filled carbon nanotubes as lithium insertion anode materials for lithium batteries. *Electrochem. Comm.*, 6: 520.

Raghu, S., Berchmans, S., Phani, K.L.N. and Yegnaraman, V., 2005. Nanoparticulate platinum films on gold using dendrimer-based wet chemical method. *Pramana–J. Phys.*, 65(5): 821–830.

Srivastava, Vankar and Kumar, 2005. *TSF*, 492: 124.

Tandon, U.S. and Pant, B.D., 1992. Granulation of silicon surface through reactive ion etching. *J. Vacuum Sci. and Tech.*, B10: 2419–2421.

Tandon, U.S. and Khokle, W.S., 1993. *Patterning of Material Layers in Submicron Region*. John Wiley Inc., p. 96.

Tandon, U.S., 1997. Reactive plasma processes for the fabrication of nano-dimensional semiconductor devices. *Solid State Phenomena*, 55: 167–173.

Tandon, U.S., 2003. Strides in nanotechnology by India. *International Congress and Exhibition on Nanotechnology Nanotech2003 + Future*, Japan

Tandon, U.S., 2004. Micromanufacturing: Possibilities through micromachining. In: *Proc. of Indo-US Workshop on Advanced and Futuristic Manuf.*, IIT, Kanpur, India, p. 22.

Tandon, U.S., 2006. Engraving and generation of microstructures using reactive plasma. In: *First Japan-India Joint Seminar on Micro/Nano Manufacturing Science*, Tokyo, Japan

Tandon, U.S. and Sinha, A., 2008. Nanotechnology development in APO countries–India. In: *Proc. Study Meeting of Asian Productivity Organization*, Taipei.

Tandon, U.S., 2008. Microsensors, actuators and their applications. In: *Fourth IEEE Conf on Wireless Comm and sensor Networks*, Indian Inst. Information Technology, Allahabad.

Thomas, K.G. and. Kamat, P.V., 2003. *Acc. Chem. Res.*, p. 888.

Thomas, K.G. *et al.*, 2004. *J. Phys. Chem. B*, 108: 13066

Thomas, K.G. *et al.*, 2006. *J. Phys. Chem. B*, 110: 150

Abbreviations

AMPRI: Advanved Material Processing Research Institute, Bhopal

CECRI: Central Electrochemical Research Institute, Karaikudi

CEERI: Central Electronics Engineering Research Institute, Pilani

CGCRI: Central Glass and Ceramic Research Institute, Kolkata

CVD: Chemical Vapour Deposition

NCL: National Chemical Laboratory, Pune

NIIST: National Institute for Interdisciplinary Science and Technology, T'vanathapuram

NPL: National Physical Laboratory, New Delhi

PECVD: Plasma Enhanced Chemical Vapour Deposition

SHS: Self propagating High temperature Synthesis

SOFC: Solid Oxide Fuel Cell

Chapter 2

The Morphological Study of Different Architectures of (Dimethylamino) Ethyl Methacrylate Polymer at Nano Scale for Gene Delivery

Nagib A. Elmarzugi

Department of Industrial Pharmacy, Faculty of Pharmacy,
Alfateh University and Biotechnology Research Centre–Nanotechnology Research GP.,
P.O. Box 13100, Tripoli–Libya
E-mail: nelmarzugi@yahoo.com

ABSTRACT

The ability of transfer the DNA to the cell by a synthetic vector is one of the fundamental goals in biotechnology. Towards this end, the polymeric vector is considered a promising delivery vector; therefore the past several years have seen a considerable growth in the use of cationic polymers for gene delivery. This results from the realization that their unique architecture can translate into unusual chemical and physical properties and lead to a success in use as vectors for gene delivery. The physicochemical properties of many cationic polymers are often polar, highly functionalized, and soluble in water. They are dependant on their shape, as well as the presence of high positive charge density of protonized primary amine groups on their surface. Together with high solubility in water, these characteristics have led to the use of these polyamines to mediate efficient DNA transfer into numerous cell lines in vitro. The combination of Atomic Force Microscope AFM and Transmission Electron Microscope TEM can provide a range of morphological data. The basic surface study knowledge and information about the DMAEMA-based polymer systems individually at nano level, morphology, size, and width were analysed by both AFM and TEM. TEM image of DMAEMA homopolymer stained with a uranyl acetate stain. A large number of apparent individual molecules are observed; also small islands of aggregated materials are seen. The investigation of diameter of DMAEMA polymers show mean width ranging between 22.4 nm to 26.8 nm by AFM (Figure 2.1) and 16.9 nm to 24.75

nm by TEM (Figure 2.2). The diameter of the molecules imaged by TEM observed less, and many factors could affect such a dramatic decrease in the diameter, such as the dehydration during the sample preparation and staining process. AFM and TEM measurements show a globular morphology. The AFM data shows the diameter always to be larger than the height, indicating that the cationic polymers molecules are no longer globular or spherical but instead it seems dome-shaped when deposited on a mica surface, it is believed the compression of the tip influenced significant role in its formation as presented and the tip convolution could has additional effects on the formed complexes. The behaviour of the resulting polymer morphology was a compact globular structure. It is believed the interaction between surface groups along with short range Van der Waals forces are considered to be responsible for formation of the observed compact globular structures. Lastly, the observation was made to individual nanoscale molecules and both techniques displayed a similar structural morphology for the polymer, which is a nano-globular structure.

Keywords: *AFM (Atomic force microscope); TEM (Transmission electron microscope); DMAEMA: 2-(dimethylamino) ethyl methacrylate.*

Introduction

The past several years have seen a considerable growth in the use of cationic polymers for gene delivery. This results from the realization that their unique architecture can translate into unusual chemical and physical properties and lead to a success in use as vectors for gene delivery (Ducan, 2003)

Many materials are included in the characterisation process, DMAEMA homopolymer, DMAEMA-PEG copolymer, and DMAEMA-stat-PEG. The current study is aiming to provide an overview of the substrate requirements for macromolecular visualisation and imaging of the DMAEMA homopolymer, DMAEMA-PEG copolymer, DMAEMA-stat-PEG by Atomic Force Microscope (AFM) and Transmission Electron Microscope (TEM).

Moreover, to elucidate the effects of the drying process on samples by comparing dry to liquid data and to investigate the effects of different cations on imaging by AFM.

The polymeric structures of the polymers here have diverse morphology, so it is important when considering morphology that different complementary techniques are used. The physicochemical properties of many cationic polymers are often polar, highly functionalised, and soluble in water. They are dependant on their shape, as well as the presence of high positive charge density of protonized primary amine groups on their surface. Together with high solubility in water, these characteristics have led to the use of these polyamines to mediate efficient DNA transfer into numerous cell lines in vitro (Jackson *et al.*, 1998; Bielinska *et al.*, 1999).

TEM can be categorized into different classes according to the contrast scheme employed; the contrast best suited to TEM is Zernik phase contrast and differential interference contrast, since electron scattering by objects exerts only change in phase and not intensity of electron waves in general. This may be the fundamental cause of image contrast; always a major issue in TEM (Nagayama and Danev, 2003)

Grids are used to hold and support samples of polymer; copper grids are preferred for a use because they dissipate heat and electrostatic charges resulting from exposure to the electron beam. However, copper grids react with some reagents and certain buffer components. Therefore, in order to minimize unwanted reactions, provide sufficient support to the studied samples, obtain artefact free and gain images of sufficient contrast and resolution, support films with high transparency to electrons should be used (Hayat, 1986; Jiang *et al.*, 2001; Jung *et al.*, 2004). Fragile and very thin samples are unstable under the electron beam without such support films. For example, the examination of specimens such as cell fragments, bacteria and glycol methacrylate usually requires the use of support film. A film material should have a relatively low atomic number and density and should be very thin (10-20nm) to gain contrast and resolution, and to avoid attenuation of the intensity of important structural details in the image (Hayat, 1981; Jung *et al.*, 2004). In addition, the application of selective staining is necessary to increase the local electron scattering power of the specimen sufficiently so that an appreciable increase in image contrast will result. Its density should be higher than that of the embedding medium. Although, there is no electron staining reaction that is absolutely specific for biological polymer, Uranyl salts stain nucleic acids by ionic association with phosphate group (Hayat, 2000)

TEM gives the local changes and anisotropy of a sample visually. A TEM image contains much information on the morphology and structure of a sample. To some extent the analysis of the image could be separated into distinguished contrast and low distinguished contrast. Although, microscopy as a visual method is generally qualitative rather than quantitative analysis, a large number of dimensional details can be manipulated and analysed.

Imaging in liquid by AFM can eliminate unwanted drying and capillary forces by immersing tip and sample in liquid, also a fluid environment reduces or prevents the AFM tip and sample contamination. Furthermore, imaging in aqueous buffers may be critical for obtaining the morphology of bio-molecules in near physiological conditions. Operation in liquids also opens the possibility of studying biological interactions in situ (Hansma, 1996a:Garci'a and Pe'rez, 2002).

Different requirements are necessary to perform AFM imaging of polymer complexes experiments in biological conditions. The adsorption of colloidal particles, proteins, or other macromolecules at liquid-solid interface is of substantial importance in many biological processes. AFM studies typically use muscovite mica as the substrate of choice in which to image, this being primarily as a result of its atomically flat surface after cleavage along the basal plane of the potassium layer (Terashima, 2002).

Suitable substrates should combine atomic flatness and biocompatibility. Various substrates have been proposed for AFM, but mica remains up to now the most convenient surface for imaging bio molecules.

Prior to the invention of tapping mode AFM high resolution imaging of polymers was restricted to a few robust polymer surfaces, since forces used by other modes of

AFM often produced dramatic modifications of the polymer and copolymer surface morphology (Magonov and Reneker, 1997;Reiter *et al.*, 2000). Moreover, the visualisation by electron microscopy of polymer molecular morphology was also rare, except with high-resolution electron microscopy or cryo-TEM (Jackson *et al.*, 1998).

Experimental Methods

This focuses on the basic characteristics of the different polymer systems used in this study. The imaging was performed by AFM and TEM.

Results and Discussion

An image of a freshly cleaved mica surface is presented for a comparison purpose in Figure 2.1. The plain mica appeared by AFM as expected featureless and flat in appearance.

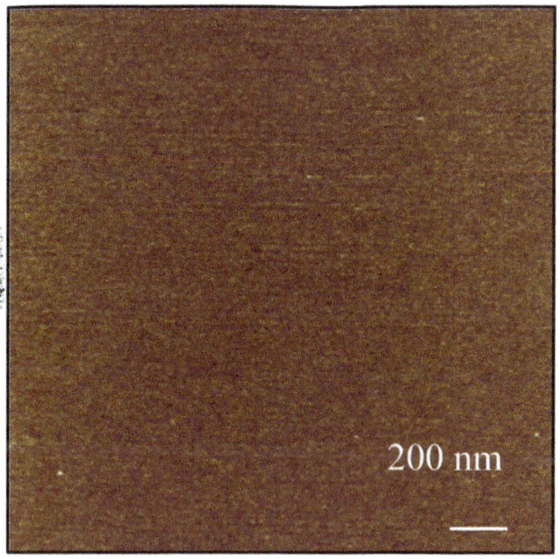

Figure 2.1: Image of Plain Mica by AFM, Z = 5 nm

Figure 2.2 shows an image of a TEM copper grid, including a support film (Pioloform), the image was taken at a high magnification of about 80,000x times. This image is provided to make the comparison clear between the substrate it self and individual molecules.

Figure 2.3 in liquid shows a topographical AFM image of DMAEMA homopolymer in 10 per cent PBS. The DMAEMA displays a spherical shape or compact globular morphology. Detailed

Figure 2.2: Image of TEM Substrate Copper Grids Containing the Supportive Film for a Comparison. Scale bar 500nm, Mag. 80 000x.

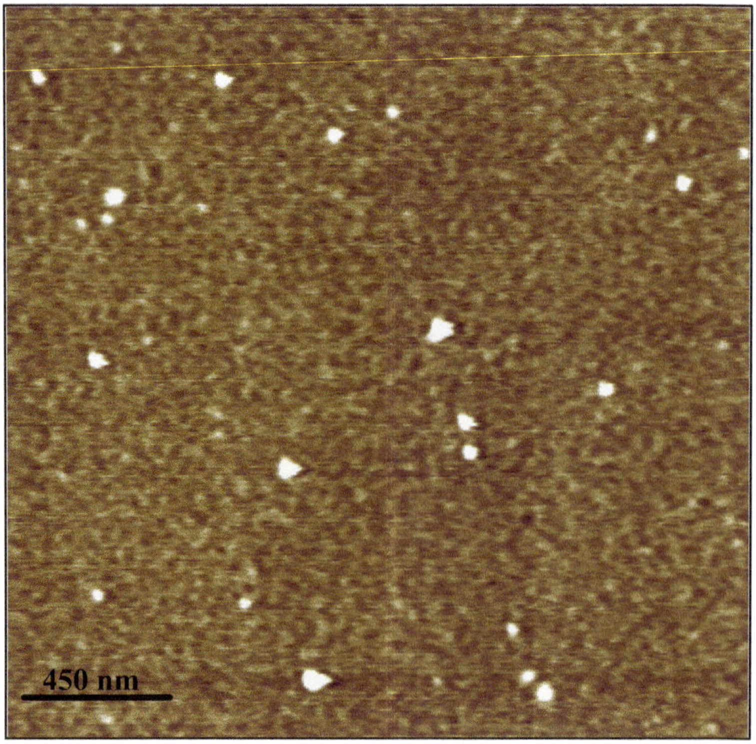

Figure 2.3: AFM Topography Image of DMAEMA in 10 per cent PBS (Liquid) on mica, Z= 3nm

cross sectional measurements on a number of such individual compact globular molecules give mean diameter of 22.5 ± 3.4 nm and mean height of 1.6 ± 0.5 nm.

In Figure 2.4 a dry image of DMAEMA polymer alone is shown. The particles again show a compact globular structure. The mean width of these particles was 25.5 ± 4 nm and a height of 1.5 ±0.7 nm. Hence, assuming the particles is sphere (by considering the measurements of the height and widths, although height measurements in AFM are always much to low. So we assumes width is height *i.e.* sphere) the mean volume of the particles is 637.8 ± 352.5 nm^3. It is believed the immobilisation of the bio molecules to the mica surface may cause an interruption in the folding process and the frequency and vertical velocity of the vibrating tip during the

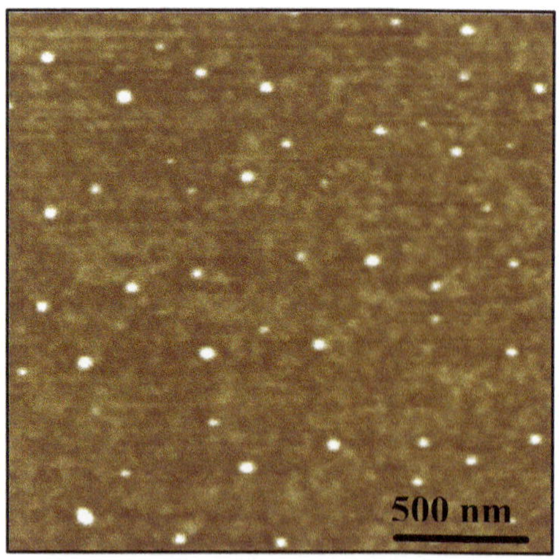

Figure 2.4: AFM Topography Image of DMAEMA in Air on Mica Z Range 9nm

AFM imaging are also most likely to be sufficiently high to cause changes in the observed morphologies.

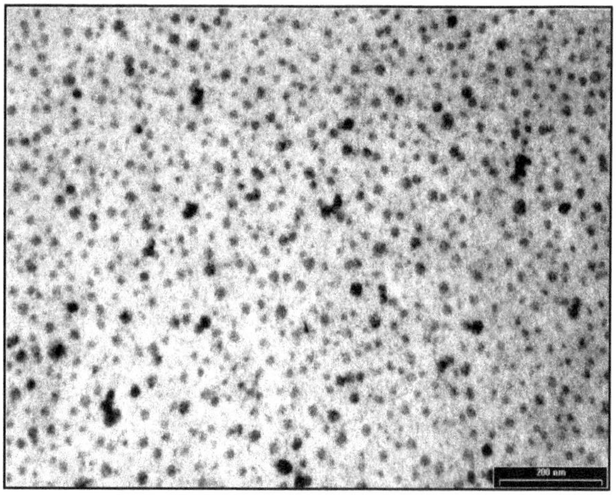

Figure 2.5: Image of DMAEMA-PEG in 10 per cent (bar = 200nm, 100 000x Mag.) by TEM

Figure 2.5 shows a TEM image of DMAEMA homopolymer stained with a uranyl acetate stain. A large number of apparent individual molecules are observed; also small islands of aggregated materials are seen. The particles were distributed all over the field of the grid. A circular structure was the predominant morphology, and mean width measurements were taken for a number of the molecules of 34.7 ± 10.6 nm.

The use of low electron energies in TEM so that only small interaction volumes of the primary electrons with specimen occurred allowed reproducible imaging of the polymer. Nevertheless, because of the extremely thin polymer layer in electron microscope investigation, the penetration depth of the electrons is still greater than the layer thickness, and part of the signal inevitably originates from the substrat (Suess *et al.*, 2004).

The data in Table 2.1 represents the average width of different polymeric systems used for investigation as a vector for gene delivery. DMAEMA, DMAEMA-PEG and DMAEMA-stat-PEG show slight difference in their mean widths. The DMAEMA-PEG copolymer gives the highest diameter.

Table 2.1: Mean Width (nm) of Different Polymer Systems by AFM

	DMAEMA	*DMAEMA-PEG*	*DMAEMA-stat-PEG*
Average	24.7 (N=12)	26.8 (N=9)	22.4 (N=10)
STDEV	±10.3	± 6.7	± 4.4

Table 2.2 shows the average height of the same polymeric systems by AFM. The DMAEMA homopolymer gives the greatest height.

Table 2.2: Mean Height (nm) of Different Polymer Systems by AFM

	DMAEMA	*DMAEMA-PEG*	*DMAEMA-stat-PEG*
Average	1.8 (N=9)	0.77 (N=9)	1.4 (N=10)
STDEV	±0.6	± 0.3	± 0.6

TEM mean width is presented in the Table 2.3; it demonstrates the highest apparent diameter for the DMAEMA homopolymer.

Table 2.3: Mean Width (nm) of Different Polymer Systems by TEM

	DMAEMA	*DMAEMA-PEG*	*DMAEMA-stat-PEG*
Average	34.75 (N=48)	17.8 (N=81)	16.9 (N=36)
STDEV	±10.9	±8.6	±7

The AFM data shows the diameter always to be larger than the height, indicating that the cationic polymers molecules, whether is PEGylated or non PEGylated are no longer globular or spherical but instead it seems dome-shaped when deposited on a mica surface, it is believed the compression of the tip influenced significant role in its formation as presented and the tip convolution could has additional effects on the formed complexes (Rivetti and Codeluppi, 2001).

In comparison to the DMAEMA, the observed increase in width of DMAEMA-PEG structures, at the same time displayed lowest height, suggests that the presence of PEG as A-B copolymer made slight widening in comparison with the other polymers.

The DMAEMA-polymer shows sufficient molecular rigidity in term of height and diameter onto mica surface to be visualized, despite the deposition on solid substrate could lead to different degrees of deformation as a result of the interplay between their inherent rigidity and interactions between the polymer molecules and mica surface.

The diameter of the molecules imaged by TEM decreased from DMAEMA homopolymer to DMAEMA-PEG and still observed less for DMAEMA-stat-PEG. Many factors could affect such a dramatic decrease in the diameter, such as the dehydration during the sample preparation and staining process.

Finally

The basic structural knowledge and information about the DMAEMA-based polymer systems individually at different Nano scale levels, morphology, size, and width were analysed by both AFM and TEM.

AFM and TEM are well-suited techniques to reveal important surface and morphological features that influence the performance of cationic polymers as step for gene delivery vectors. The combination of AFM and TEM provide a range of morphological data from a range of experimental conditions such as in dry as well as liquid. Individual polymers can be observed in different environments by both microscopic techniques.

The investigation of diameter of PEGylated and non PEGylated DMAEMA polymers show mean width ranging between 22.4 nm to 26.8 nm by AFM and 16.9 nm to 34.75 nm by TEM. AFM and TEM measurements show a globular morphology.

References

Bielinska, A.U., Chen, C., Johnson, J. and Baker, J.R., 2005. DNA complexing with polyamidoamine dendrimers: Implications for transfection. *Bioconjugate Chemistry*, 10: 843–850.

Duncan, R., 2003. The dawning era of polymer therapeutics. *Nature Reviews Drug Discovery*, 2: 347–360.

Garci'a, R. and Pe'rez, R.N., 2002. Dynamic atomic force microscopy methods. *Surface Science Reports*, 47: 197–301

Hayat, M.A., 2000. *Electron Microscope Biological Applications*. Cambridge University Press, USA.

Hayat, M.A., 1981. *Principles and Techniques of Electron Microscopy: Biological Applications*. Edward Arnold, London

Hansma, H.G. and Laney, D., 1996. DNA binding to mica correlates with cationic radius: Assay by atomic force microscopy. *Biophysics*, 70: 1933–1939

Jackson, C.L., Chanzy, H.D., Booy, F.P., Drake, B.J., Bauer, B.J., Armis, E.J. and Tomalia, D.A., 1998. Visualization of dendrimer molecules by transmission electron microscopy (TEM): Staining methods and cryo–TEM of vitrified, macromolecules, 31: 6259–6265.

Jung, H.J., Kittelson, D.B. and Zachariah, M.R., 2004. Kinetics and visualization of soot oxidation using transmission electron microscopy. *Combustion and Flame*, 136: 445–456.

Magonov, S.N. and Reneker, D.H., 1997. Characterization of polymer surfaces with atomic force microscopy. *Annual Review of Materials Science*, 27: 175–222.

Nagayama, K. and Danev, R., 2003. Image enhancement with phase plates in electron–phase microscopy. *Microscopy and Analysis*, 54: 17–19

Reiter, G., Castelein, G., Hoerner, P., Riess, G., Sommer, J.-U. and Floudas, G., 2000. Morphologies of diblock copolymer thin films before and after crystallization. *European Physical Journal E*, 2: 319–334.

Rivetti, C. and Codeluppi, S., 2001. Accurate length determination of DNA molecules visualized by atomic force microscopy: Evidence for a partial B– to A–form transition on mica. *Ultramicroscopy*, 87: 55–66.

Suess, C., Wenzl, F.P., Poelt, P., Scherf, U. and Leising, G., 2004. Ion-mediated organic electronics: An AFM and SEM study. *Microscopy and Analysis*, 59: 21–23.

Terashima, H., 2002. A direct measurement of the ion-exchange capacity of muscovite mica using a mettler ultramicrobalance. *Journal of Colloid and Interface Science*, 245: 81–85.

Tsukruk, V.V., 2006. Dendritic macromolecules at interfaces. *Advanced Materials*, 10: 253–257.

Chapter 3

Synthesis of Nano Size Anatase Type of Titania Aerogels by Addition of Silica

Silvester Tursiloadi
Research Center for Chemistry, Indonesian Institute of Sciences, Kawasan PUSPIPTEK, Serpong, Tangerang, Indonesia
E-mail: tursiloadi@yahoo.com

ABSTRACT

Anatase type titania exhibits high photocatalytic activity, and thus has attracted a great deal of attention in the field of photocatalysts for decomposition of environmental pollutants such as organic halides and for antifouling and antibacterial application. TiO_2-nanoparticles dispersed SiO_2 wet gels were prepared by hydrolysis of $Ti(OC_4H_9^n)_4$ and $Si(OC_2H_5)_4$ in a 2-propanol solution with acid catalyst. The solvent in the wet gels was supercritically extracted using CO_2 at 60°C and 22Mpa in one-step. Thermal evolution of the microstructure of the extracted gels (aerogels) was evaluated by XRD measurements, TEM and N_2 adsorption measurements. The as-extracted aerogel with a large specific surface area, more than $365 m^2g^{-1}$, contained anatase nanoparticles, about 5nm in diameter. The anatase phase was stable after calcinations at temperatures up to 1000°C, and BET specific surface area, total pore volume and average pore diameter did not change significantly after calcinations at temperature up to 800°C.

Keywords: *Ultrafine anatase, Sol-gel, CO_2 supercritical extraction.*

Introduction

Anatase type titania exhibits high photocatalytic activity, and thus has attracted a great deal of attention in the field of photocatalysts for decomposition of environmental pollutants such as organic halides (Weaver and Mills, 1997) and for antifouling and antibacterial application (Matsubara *et al.*, 1995; Ohko *et al.*, 1998). However, there are significant challenges in obtaining mesoporous titania with full anatase crystalline domains. On the other hand, at standard pressure (1bar) and normal temperatures, rutile is a thermodynamically more stable phase possessing smaller band gap energy (3.0 eV) than the anatase phase (3.2eV), and lower photocatalytic activity (Sclafani and Herrman, 1996; Kormann *et al.*, 1988). Anatase nanocrystals show higher photocatalytic activities than the bulk crystals because the nanocrystals facilitate the diffusion of excited electrons and holes toward the surface before their recombination (Anpo *et al.*, 1987). In addition, the anatase nanocrystals must be highly dispersed in the host matrix without aggregation to improve the activity. Mesoporous TiO_2-SiO_2 gel, a combination of highly adsorptive silica matrix and highly photo-active titania nanoparticles, will be an interesting material for photocatalytic applications.

The syntheses of ultrafine titania powders has been investigated using various techniques including hydrothermal treatment, oxidation of titanium powder, and vapor decomposition (Wang *et al.*, 2002; Cornell *et al.*, 1975; Mazdiyasni *et al.*, 1965). The sol-gel syntheses are widely used for preparation of transition metal oxides with nanoscale microstructures, and provide excellent chemical homogeneity and the possibility of deriving unique metastable phases at low reaction temperatures. The high porosity and the high specific surface area of material prepared by the sol-gel method make them very attractive from catalytic point of view. However, the titania powders prepared by the sol-gel method are usually amorphous, and the crystallization temperature is relatively high. The crystallization temperature can be reduced with the hydrothermal treatment, and depends on the temperature of hydrothermal treatment (Wang and Ying, 1999; Ovenstone and Yanagizawa, 1999). The heat treatment is a significant step with respect to device fabrication.

Excessive heating causes mesopore collapse, leading to the formation of dense materials, due to excessive crystal growth and sintering, which in turn will decrease device efficiency (Wang *et al.*, 2001; Mason, 1990). Additionally, the localized crystal domain limits the photoactivity of the resultant materials (Fujishima and Honda, 1972; Phillips, 1983; Gedanken *et al.*, 2001). Recently, the preparation of mesoporous titania which has highly porous and a fully crystalline phase domain has been scarcely explored (Srivastava *et al.*, 2002). Supercritical extraction techniques are recently used in material science to fabricate highly porous materials (Suh *et al.*, 1997). The supercritical extraction technique seems to be a good alternative to overcome the problems of low porosity and low specific surface area of sol-gel derived titania.

In this work, a new route to prepare mesoporous titania with highly porous and full anatase domains of TiO_2-nanoparticles dispersed SiO_2 aerogels were prepared by the sol-gel method and supercritical extraction, and the stability of anatase phase and the porous structure, *i.e.* pore size and specific surface area, has been discussed.

Experimental Procedure

TiO_2-SiO_2 wet gels were prepared by hydrolysis of titanium tetra-n-butoxide $Ti(OC_4H_9)_4$ (TNB) and tetraethylortosilicate $Si(OC_2H_5)_4$ (TEOS) in a 2-propanol solution with acid catalyst. The molar ratios used for the synthesis were [TNB+TEOS]: [H_2O]: solvent = 1: 13.4: 127 and [HNO_3]: [TNB+TEOS] = 0.06: 1. The ratios of [TiO_2]: [SiO_2] were 1: 0, 1: 2 and 1: 4 in mol. The TEOS was firstly dissolved in 2-propanol at room temperature, and a mixture of the catalyst solution, remaining 2-propanol, H_2O and HNO_3 was added, and then stirred for 30min. After that, TNB was added and then the catalyst solution was added under continuous stirring. The solution gelled around 3h after the addition of the catalyst solution. The gel time was defined as the time required after mixing for the vortex created by the stirring to disappear completely. The wet gel was placed in the flow of supercritical carbon dioxide in a supercritical extraction system at 60°C and 22Mpa for 4h. The supercritically extracted aerogels were calcined at 500, 600, 700, 800, 900 and 1000°C for 2h in air (Figure 3.1).

Changes in the microstructure of the aerogels during heating were evaluated by measurements of N_2 adsorption (Quantachrome, Autosorb). The specific surface area and pore volume of aerogels, before and after calcinations, were estimated by the BET and Barret-Joyner-Halenda (BJH) method (Barret *et al.*, 1951) using N_2 desorption curves. The pore and grain sizes of the samples were estimated from the images observed by a transmission electron microscope (TEM, Philips, TECNAI F20). Crystallization behaviors of the samples were detected by an X-ray diffractometer (Rigaku, RAD-C) after calcinations.

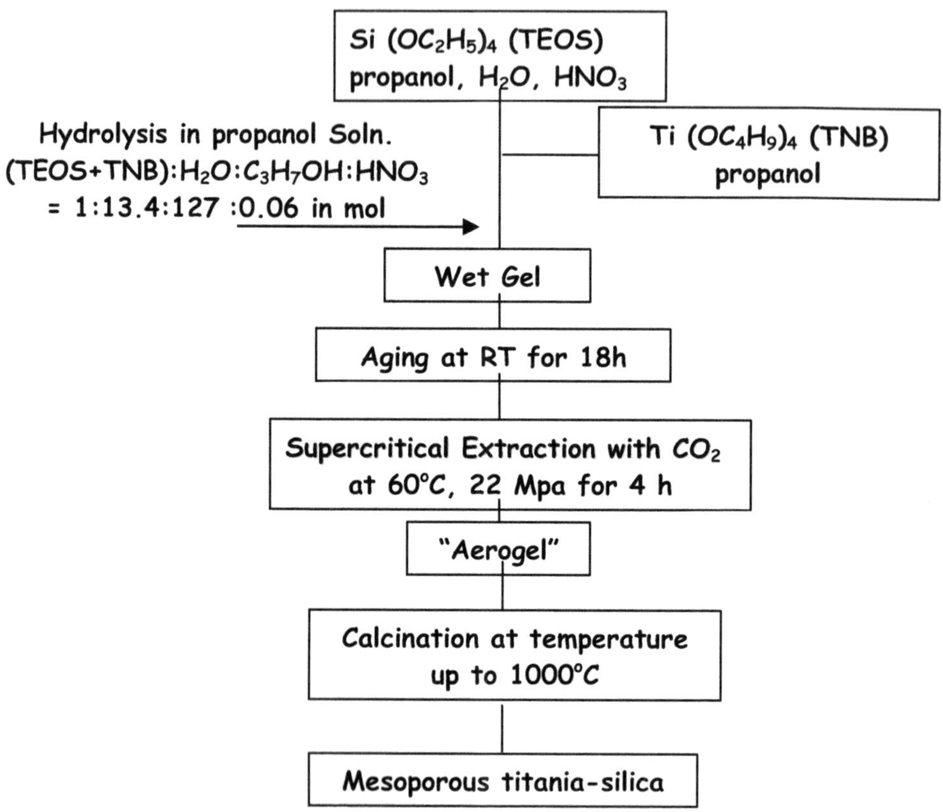

Figure 3.1: Flowchart of Sample Preparation

Results

TG and DTA profiles of TiO_2 and TiO_2-$2SiO_2$ aerogels are showed in Figure 3.2 and Figure 3.3. For the TiO_2 aerogel, weight loss around 5 per cent was mainly observed at around 80°C (Figure 3.2). Beyond 400°C the sample of TiO_2 aerogel practically lose no more weight. In the DTA curve, the TiO_2 aerogel showed endothermic peak around 80°C and weak triplet exothermic peaks around 150 and 250°C which accompanied weight loss around 8 per cent and 5 per cent respectively. The broad exothermic peak at around 310°C is followed by weight loss around 4 per cent also observed. For the TiO_2-$2SiO_2$ aerogel (Figure 3.3), weight loss around 13 per cent was mainly observed at around 100°C.

Beyond 400°C, the sample of TiO_2-$2SiO_2$ aerogel loses practically no more weight. In the DTA curve, the TiO_2-$2SiO_2$ aerogel showed endothermic peak around 100°C accompanied weight loss around 13 per cent and around 400°C weight loss around 5 per cent and no exothermic peak. Endothermic peaks appeared in almost the same temperature region for both samples namely at around 100°C as the evaporation of water and 2-propanol as hydrolyzing reagent and solvent, respectively.

The X-ray powder diffraction peaks of anatase were found for the as-extracted TiO_2 aerogel (Figure 3.4). The anatase phase was stable after calcination at 500°C. After calcination at 600°C, weak peaks of rutile were found along with the anatase peaks. After calcination at 700°C, the diffraction peaks of anatase disappeared and only the peaks of rutile were found. The X-ray diffraction peaks of anatase were found for the as-prepared TiO_2-$2SiO_2$ aerogel, too (Figure 3.5). The anatase phase in

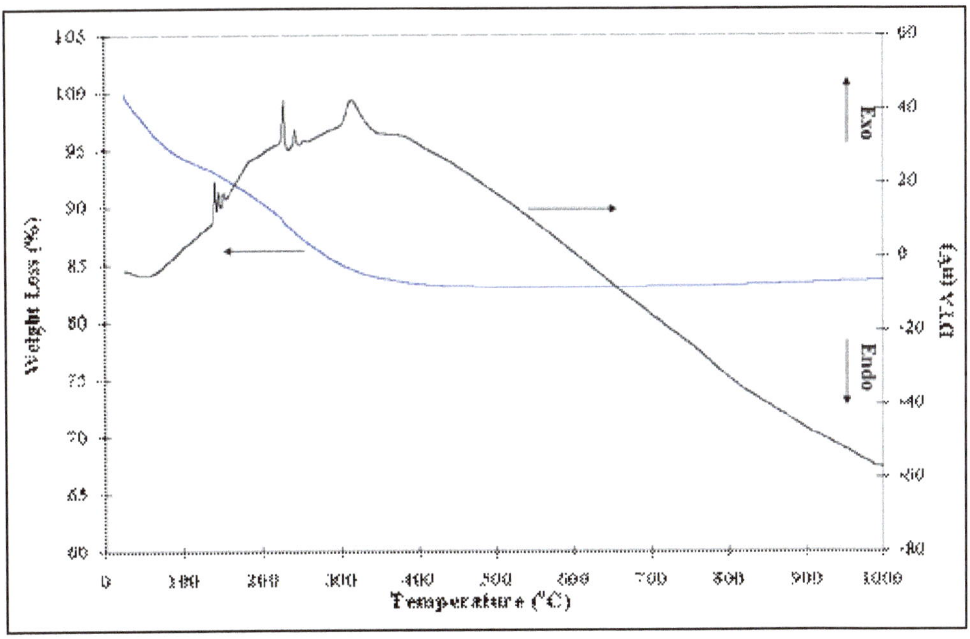

Figure 3.2: TG-DTA Profiles of TiO_2 Aerogel

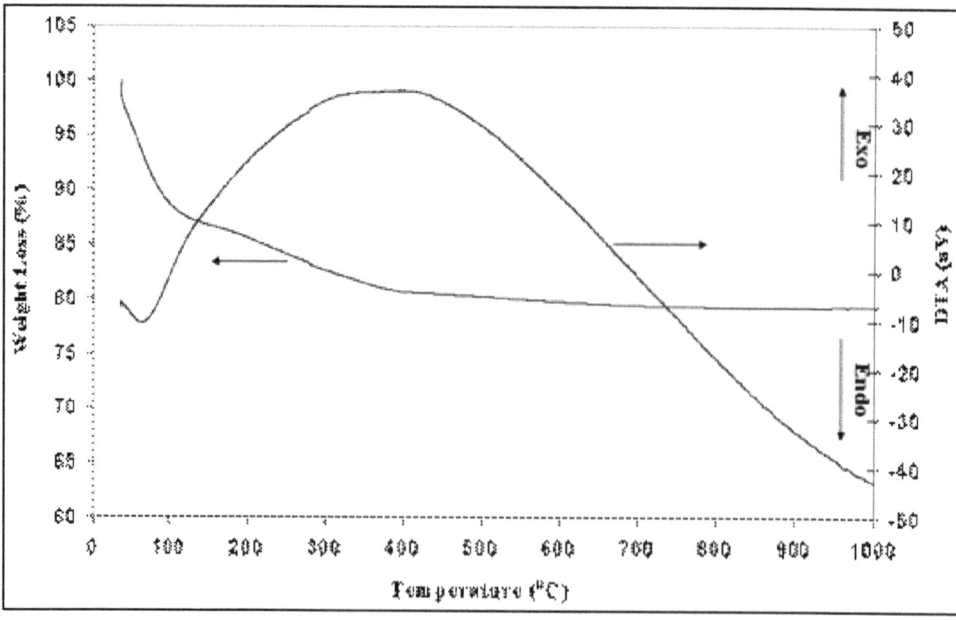

Figure 3.3: TG-DTA Profiles Aerogel of TiO_2-SiO_2

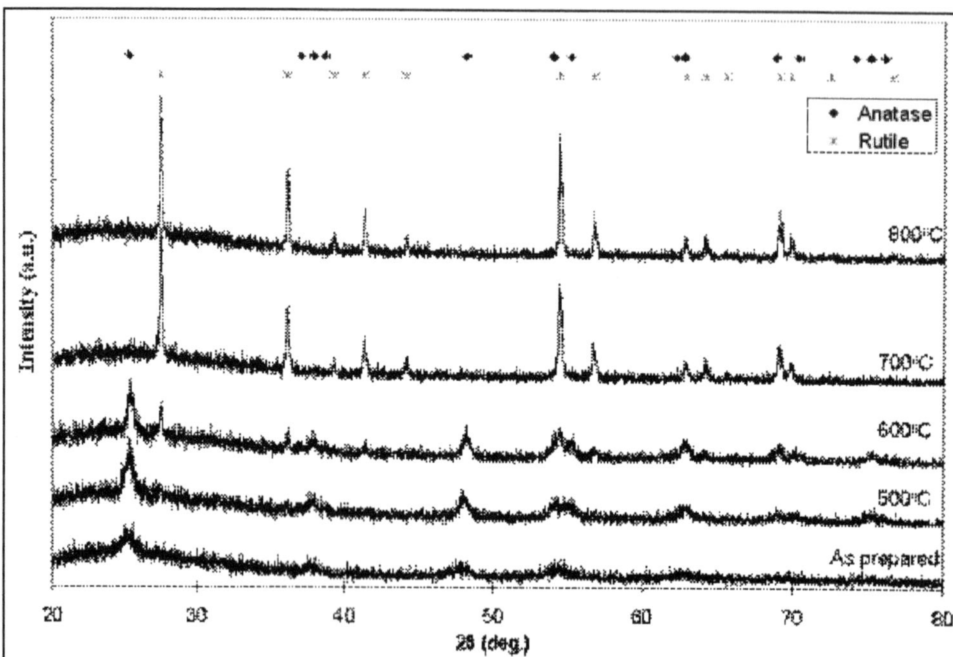

Figure 3.4: XRD Patterns of the TiO_2 Aerogels as-Prepared and After Calcination at Various Temperatures

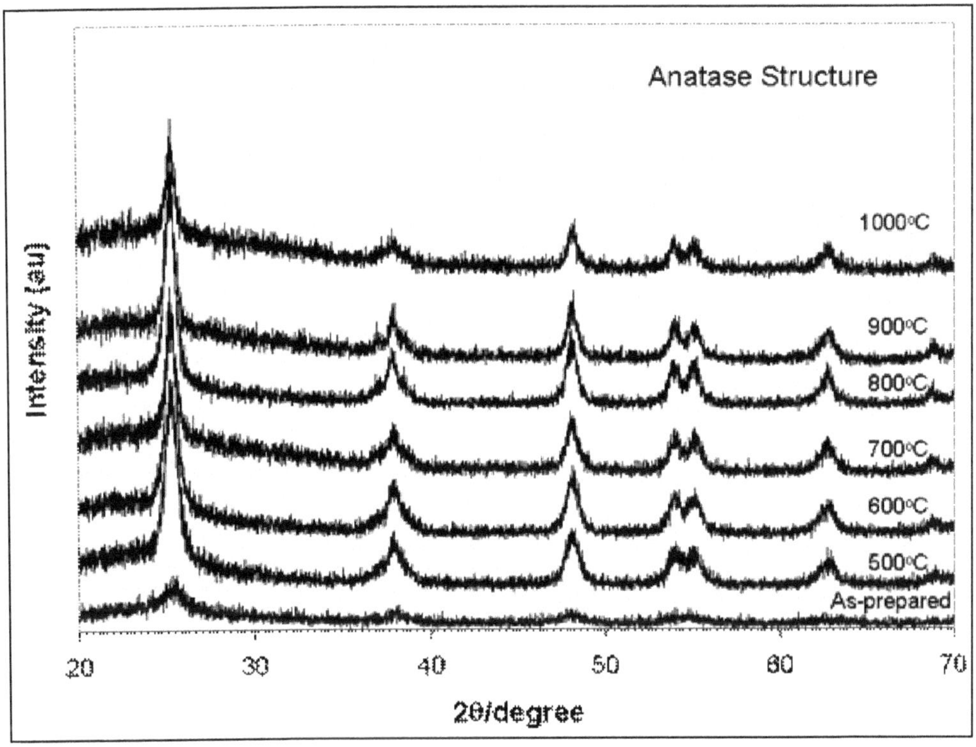

Figure 3.5: XRD Patterns of the TiO_2-$2SiO_2$ Aerogels as-Prepared and After Calcination at Various Temperatures

the TiO_2-$2SiO_2$ aerogel was stable after calcinations at temperatures up to 1000°C, and no peaks of rutile were observed (Figure 3.5). The as-extracted gel of TiO_2-$4SiO_2$ was amorphous by XRD (Figure 3.6). After calcination at 600°C, very weak peaks of anatase can be found, and hardly changed by calcinations at higher temperatures up to 900°C (Figure 3.6).

Figure 3.7 shows the TEM images and electron diffraction patterns for the as-prepared aerogels of TiO_2 and TiO_2-SiO_2. The crystalline structures were assigned to anatase from the electron diffraction patterns for TiO_2 and TiO_2-$2SiO_2$ aerogels. Many small anatase particles were observed throughout the as-prepared TiO_2 and TiO_2-$2SiO_2$ aerogels (Figures 3.7A and B). The particle size of anatase in the as-prepared TiO_2- and the TiO_2-$2SiO_2$ aerogels was about 5nm in diameter. The TEM images and electron diffraction pattern (Figure 3.8A) show that the anatase particle size in the TiO_2 aerogel increased about to 13nm after calcination at 500°C, and the anatase particle size in the TiO_2-$2SiO_2$ aerogel increased about to 10nm (Figure 3.8B). Amorphous particles were found in addition to the anatase particles for the TiO_2-$2SiO_2$ aerogel, but not for the TiO_2 aerogel. After calcination at 700°C, the particle size of the TiO_2 aerogel increased significantly, about 90nm in diameter, but the anatase particle size of the TiO_2-$2SiO_2$ aerogel did not changed significantly *i.e.* 10 nm (Figures

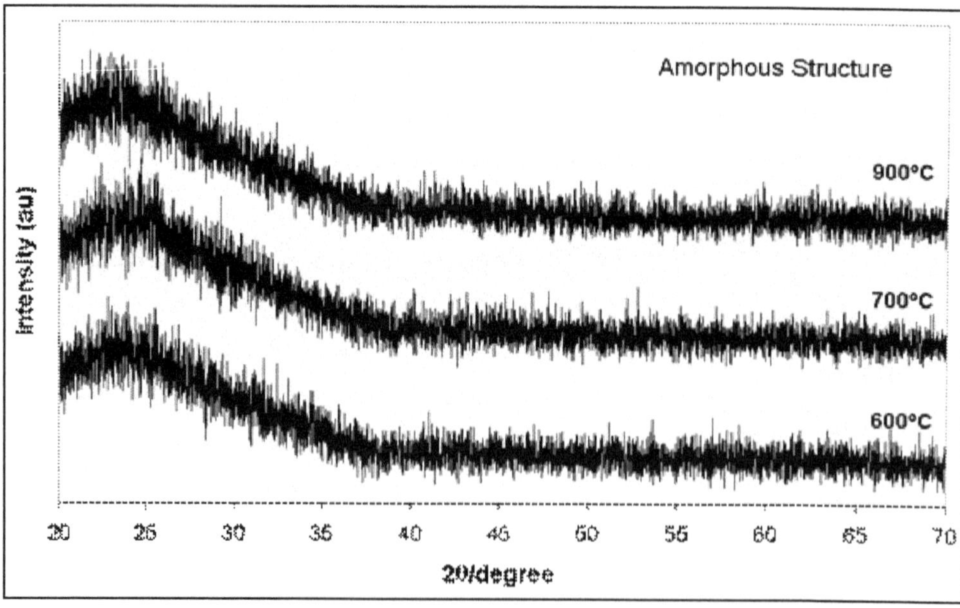

Figure 3.6: XRD Patterns of the TiO_2-4SiO_2 Aerogels After Calcination at Various Temperatures

3.9A and B). Amorphous particles were still found for the TiO_2-2SiO_2 aerogel in addition to the anatase particles.

Crystalline nanoparticles were hardly found for the TiO_2–4SiO_2 aerogels after calcinations at temperatures up to 700°C. However, anatase nanoparticles, about 5nm in diameter, were found for the TiO_2-4SiO_2 aerogel after calcination at 800°C (Figures 3.10B).

Table 3.1 and Figures 3.11–3.13 show the effect of calcination temperatures on the total pore volume, the specific surface area and the average pore diameter of the aerogels. The total pore volume of the as-prepared TiO_2 was 0.56cm^3g^{-1}, and after calcinations at 600°C, it became 0.35 cm^3g^{-1} and was larger than the total pore volume of the TiO_2-2SiO_2 aerogel, 0.25 and 0.28cm^3g^{-1}, respectively. After calcinations at temperatures higher than 600°C, the total pore volume of the TiO_2 aerogel decreased remarkably and became smaller than that of the TiO_2-2SiO_2 aerogel.

However, the pore volume for the TiO_2-2SiO_2 aerogels after calcinations at temperatures up to 800°C hardly changed. The specific surface area of the as-extracted TiO_2 aerogel, 195m^2g^{-1}, was smaller than that of the as-extracted TiO_2-2SiO_2 aerogel, 365m^2g^{-1}, and it drastically decreased after calcination at 500°C, 90m^2g^{-1}. On the contrary, changes in the specific surface area of the TiO_2-2SiO_2 aerogel by calcinations were not significant. After calcinations at 800°C, the specific surface area of the TiO_2-2SiO_2 aerogel was still large, 260m^2g^{-1}. Also, the changes in the average pore diameter of the TiO_2-2SiO_2 aerogel by calcinations, from 3.2nm as-extracted to 4.2nm after calcinations at 800°C, were not significant in comparison with the changes of the

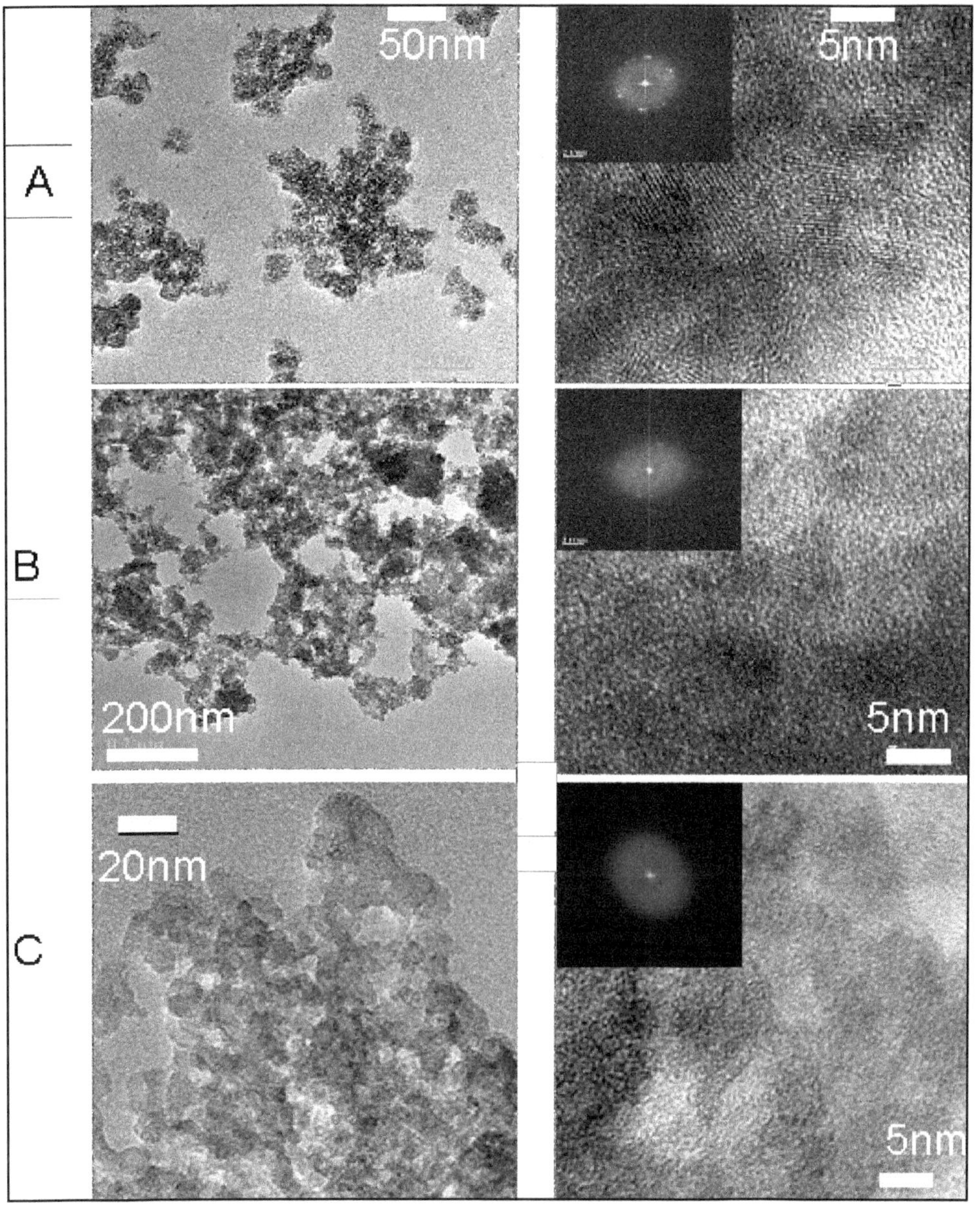

Figure 3.7: TEM Images and Electron Diffraction Patterns for the as-Extracted Aerogels; TiO_2 (A), TiO_2-$2SiO_2$ (B) and TiO_2-$4SiO_2$ (C)

TiO_2 aerogel, from 12.8nm as-extracted to 51.6nm after calcinations at 800°C. These results show that the porous structure of the TiO_2-$2SiO_2$ aerogel was thermally stable at temperatures up to 800°C in comparison with the TiO_2 aerogel. The as-extracted TiO_2-$4SiO_2$ aerogels had larger specific surface area, pore volume and pore size than other aerogels, and they were still large after calcination at 800°C.

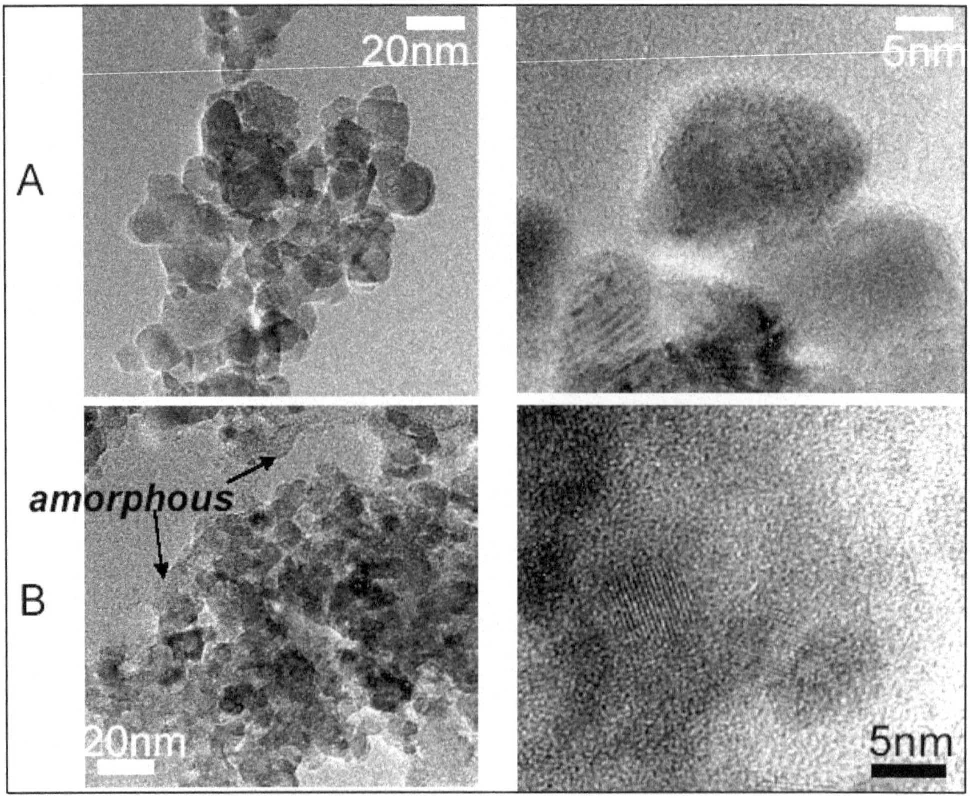

Figure 3.8: TEM Images for the Aerogels after Calcination at 500°C; TiO_2 (A) and TiO_2-$2SiO_2$ (B)

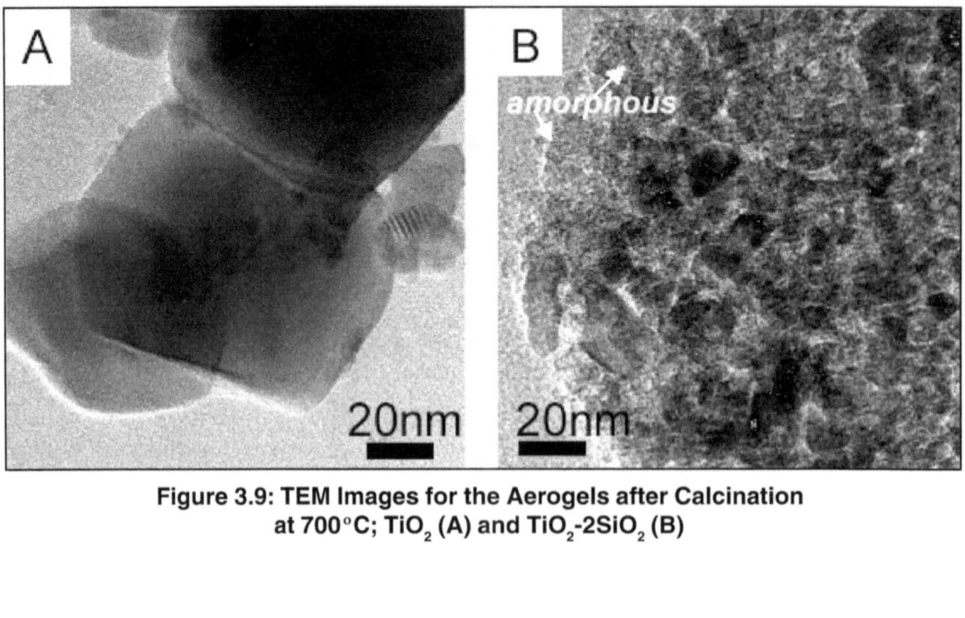

Figure 3.9: TEM Images for the Aerogels after Calcination at 700°C; TiO_2 (A) and TiO_2-$2SiO_2$ (B)

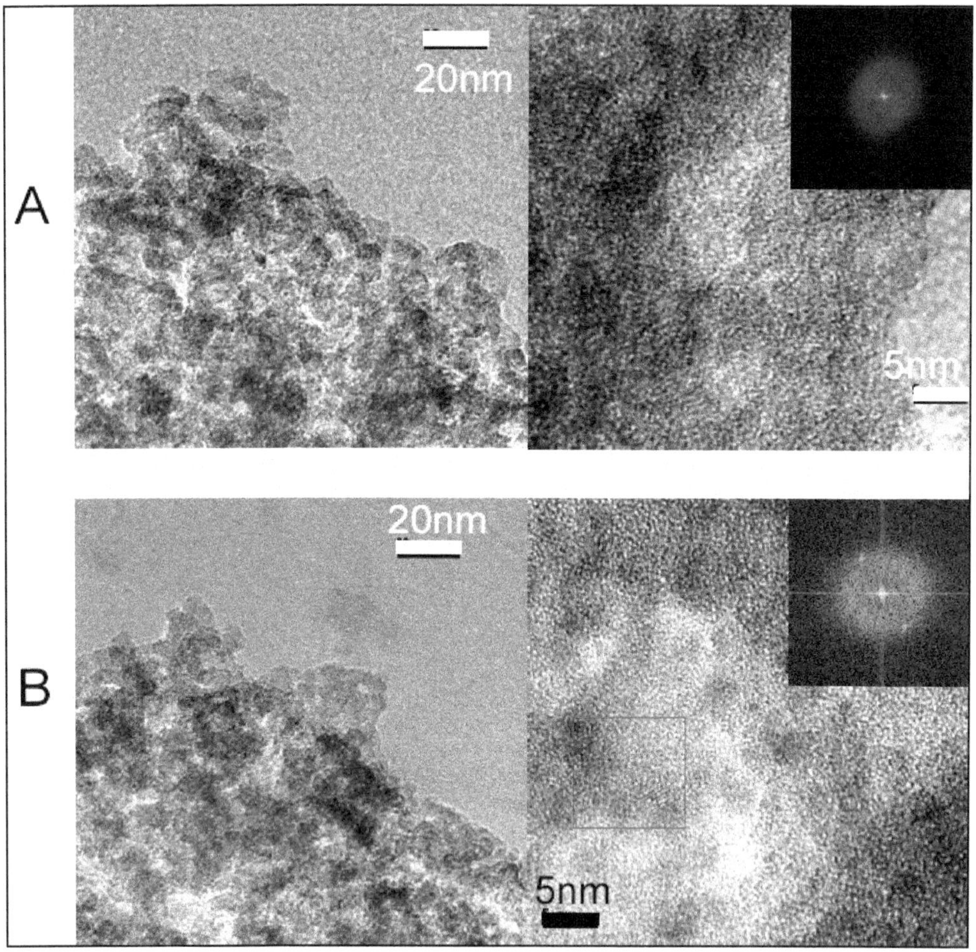

Figure 3.10: TEM Images and Electron Diffraction Patterns for TiO_2-$4SiO_2$ Calcined at 700°C (A) and at 800°C (B)

Figure 3.11 shows the effect of calcination temperature on the total pore volume of the TiO_2 and TiO_2-$2SiO_2$ aerogels. The total pore volume of the as-prepared TiO_2 and after calcinations at 600°C, was larger than the total pore volume of the TiO_2-$2SiO_2$. After calcinations at temperatures higher than 600°C, the total pore volume of the TiO_2 aerogel decreased remarkably and became smaller than that of the TiO_2-$2SiO_2$ aerogel. However, after calcinations at temperatures up to 800°C, the pore volume for the TiO_2-$2SiO_2$ aerogels did not change significantly.

Figure 3.12 shows the effect of calcination temperature on specific surface area of the TiO_2 and TiO_2-$2SiO_2$ aerogels. The specific surface area of TiO_2 as-prepared is $195m^2/g$ and after calcined at 500 °C is $90m^2/g$, smaller than that of specific surface area of TiO_2-$2SiO_2$, namely 365 and $338m^2/g$, respectively. The specific surface area of TiO_2 at all temperature calcinations is far smaller than that of specific surface area of TiO_2-$2SiO_2$. After calcination at 500 °C and more, the value of specific surface area of TiO_2 decreases drastically, but does not change significantly for TiO_2-$2SiO_2$.

Table 3.1: Specific Surface Area, Cumulative Pore Volume, and Average Pore Diameter of the TiO_2 and TiO_2-SiO_2 Aerogels after Calcination at Various Temperatures[a]

	as-extracted	*500°C*	*600°C*	*700°C*	*800°C*
TiO_2					
Surface area (m^2g^{-1})	195	90	58	19	5
Pore volume (cm^3g^{-1})	0.555	0.487	0.345	0.164	0.053
Average pore diameter (nm)	12.8	18.7	20.1	36.5	51.6
TiO_2-SiO_2 1: 2					
Surface area (m^2g^{-1})	365	338	309	296	260
Pore volume (cm^3g^{-1})	0.249	0.299	0.282	0.274	0.253
Average pore diameter (nm)	3.2	3.8	3.9	3.9	4.2
TiO_2-SiO_2 1: 4					
Surface area (m^2g^{-1})	566	435	392	329	273
Pore volume (cm^3g^{-1})	1.136	0.808	0.836	0.736	0.746
Average pore diameter (nm)	8.1	7.6	8.7	9.2	11.4

a: The accuracy of N_2 adsorption measurements was 0.1 per cent, and the reproducibility of these values for each sample was within 10 per cent.

The specific surface area of TiO_2-$2SiO_2$ was stable on high temperature, but not for TiO_2. Figure 3.13 shows the effect of calcination temperature on average pore size diameter of TiO_2 aerogel and TiO_2-$2SiO_2$ aerogel. The average pore size diameter of TiO_2 as-prepared is 12.8 nm and after calcined at 600 °C is 23.8 nm, larger than that average pore size diameter of TiO_2-$2SiO_2$, namely 2.8 and 3.7 nm, respectively. After calcination at 700 °C for 2h, average pore size diameter of TiO_2 increased drastically namely 36.5 nm, but the average pore size diameter of TiO_2-$2SiO_2$ did not change significantly, namely 3.8 nm. The average pore size diameter of TiO_2 as-prepared and after calcined up to 800 °C is the larger than that of TiO_2-$2SiO_2$. The average pore diameter of TiO_2-$2SiO_2$ was stable at high temperature, but not for TiO_2.

Discussion

Exothermic peaks are hardly found for the TiO_2-$2SiO_2$ aerogel. It may be caused by the solvent (alcohol) that hardly remained in the TiO_2-$2SiO_2$ aerogel. The solvent of the TiO_2-$2SiO_2$ may be removed by the supercritical extraction. The temperature needed to eliminate residual alcoxy group of TiO_2 aerogel were found up to 400 °C that is indicated that a lot of residual alcoxy group still remain.

By the supercritical CO_2-extraction in one-step, highly porous TiO_2-SiO_2 gels with a large specific surface area have been obtained. Aerogels are usually prepared

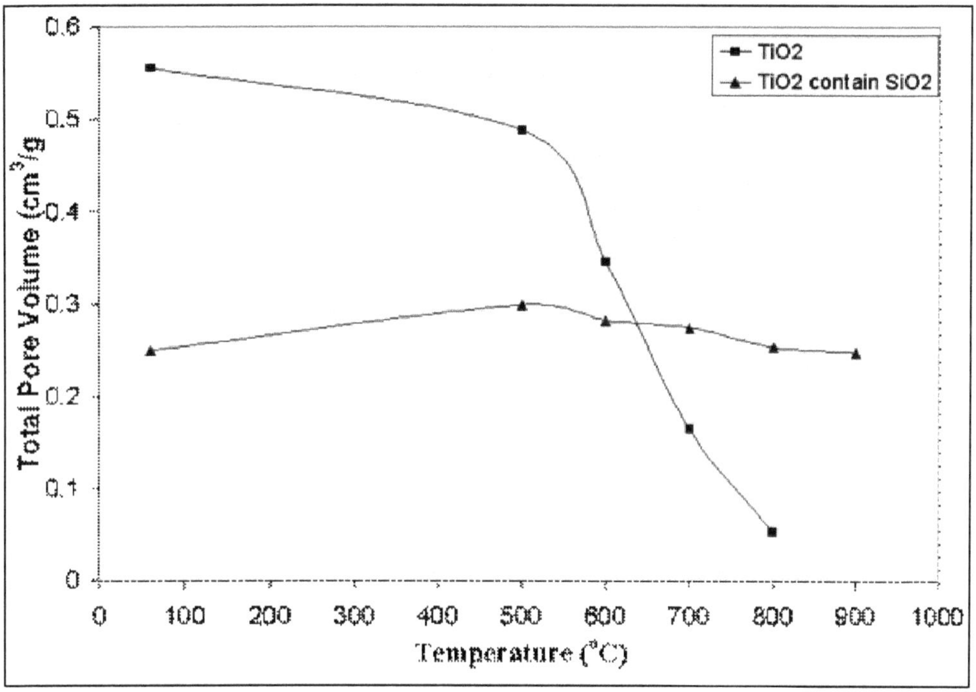

Figure 3.11: Cumulative Pore Volume of the TiO_2 and TiO_2-$2SiO_2$ Aerogels as-Prepared and after Calcination at Various Temperatures

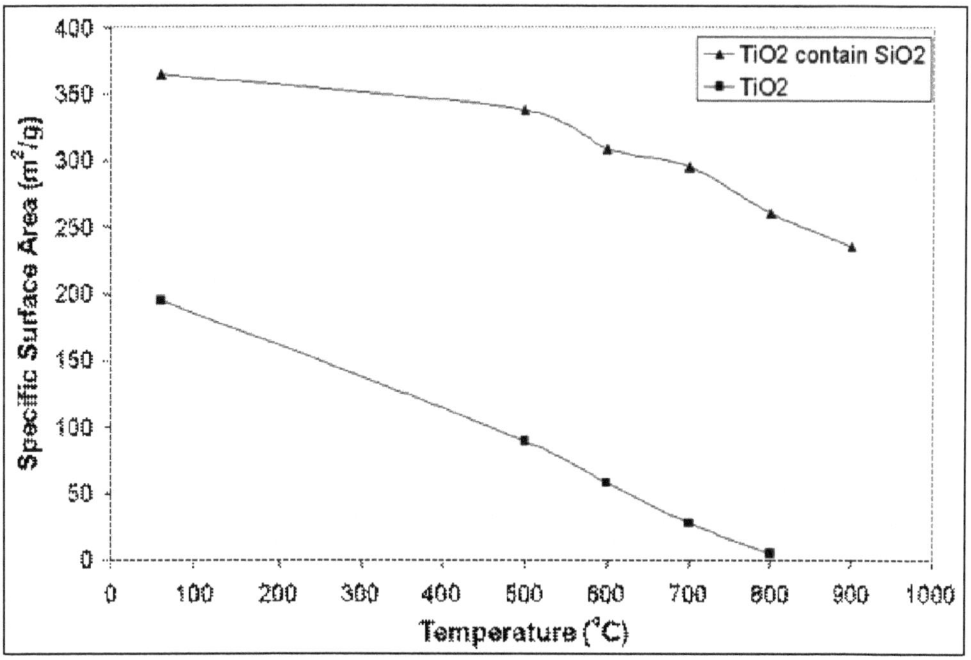

Figure 3.12: Specific Surface Area of the TiO_2 and TiO_2-$2SiO_2$ Aerogels as-Prepared and After Calcination at Various Temperatures

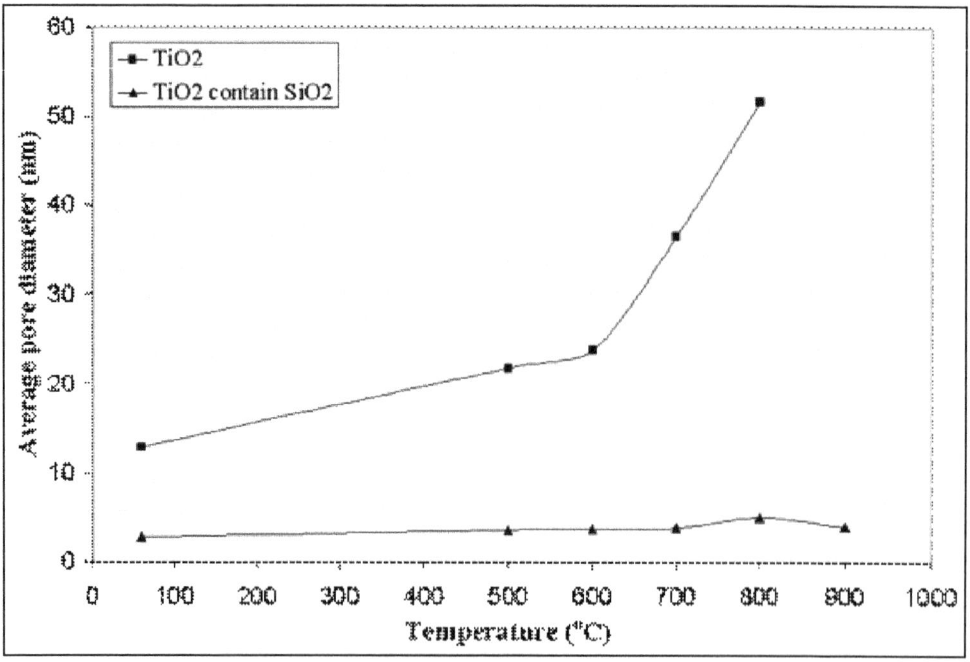

Figure 3.13: Average Pore Diameter of the TiO_2 and TiO_2-$2SiO_2$ Aerogels as-Prepared and after Calcination at Various Temperatures

by the two-step drying process, *i.e.* substitution of solvent in wet gels with liquid CO_2 and then supercritical extraction.

The substitution of solvent usually needs a long time and complex operations, and therefore, the one-step extraction method will be the better alternative. The effect of CO_2 supercritical extraction on the crystallization can be seen in Figures 3.4 and 3.5. Under the high pressure during the CO_2 supercritical extraction (22Mpa), crystalline anatase nanoparticles were formed for both of the TiO_2 and the TiO_2-$2SiO_2$ aerogel. The as-prepared TiO_2-$4SiO_2$ aerogel is amorphous by XRD (Figure 3.6), probably due to the low concentration of crystalline phase. After calcinations at 800°C, the electron diffraction pattern for small particles in the TiO_2-$4SiO_2$ aerogel shows the anatase structure.

Anatase phase cannot be found if drying is performed under an ambient pressure. In general, the structure of anatase is formed at temperatures higher than 400°C (Wang and Ying, 1999; Imai *et al.*, 1997). However, it was reported previously that anatase phase can be found for TiO_2 gel film after hydrothermal treatment at temperatures in the range between 80°C and 180°C (Imai *et al.*, 1997).

Water vapor was considered to enhance the crystallization of TiO_2 gel through cleavage of Ti-O-Ti bond by hydrolysis (Imai *et al.*, 1997). Under high pressure during the supercritical extraction, residual H_2O may enhance the deposition of anatase. On the other hand, solubility of substances into supercritical fluids is usually high, and it can be considered that anatase nanoparticles are formed in supercritical CO_2 by the solution-deposition mechanism from amorphous TiO_2. However, the solubility of TiO_2 in supercritical CO_2 is not known, and the further discussion should be needed.

The relationship with anatase nucleation and role of SiO_2 were show in Figures 3.14a and b. The anatase nanoparticles in the TiO_2 aerogel, about 5nm in diameter (Figure 3.7A), grow up to ca. 13nm after calcinations at 500°C (Figure 3.8A), the diffraction peaks of rutile are found at 600°C (Figure 3.4), and the particles of rutile grow up to ca. 90nm after calcinations at 700°C (Figures 3.9A and 3.14a). On the contrary, grain growth of anatase nanoparticles in the TiO_2-$2SiO_2$ aerogels was hardly observed at temperatures up to 700°C (Figures 3.8B and 3.9B). The grain growth is restricted with addition of SiO_2 (Figure 3.14b). Anatase nanoparticles in the TiO_2-$2SiO_2$ aerogel are surrounded with SiO_2 nanoparticles and highly dispersed, thus the grain growth rate is low (Figure 3.14b). Also, a large amount of SiO_2 will retard the crystallization of TiO_2 and the phase transition from anatase to rutile, probably due to the small particle size of anatase (Figure 3.14b). After calcinations at 600°C, diffraction peaks of rutile were found for the TiO_2 aerogel, but not for the TiO_2-$2SiO_2$ aerogels. The shift of crystallization- and phase transition temperatures of small particles and thin films from those of bulk materials, usually higher temperatures, is often reported.

Crystallization or phase transformations take place through nucleation and growth processes. The Gibb's free energy is difference between the amorphous and crystalline phases or between two crystalline phases (Kasuga *et al.*, 1999; Kumar *et al.*, 1993). In the phase transformation of nanocrystalline anatase particles of TiO_2 shows in Figure 3.14a. Interface nucleation step A, nucleation of rutile starts within the interface between two contacting anatase particles, and the nucleation rate is determined by the probability of contact of two anatase particles (Zhang and Banfield, 1999). Nuclei growth step B, as for any nucleation event, the formation of a stable

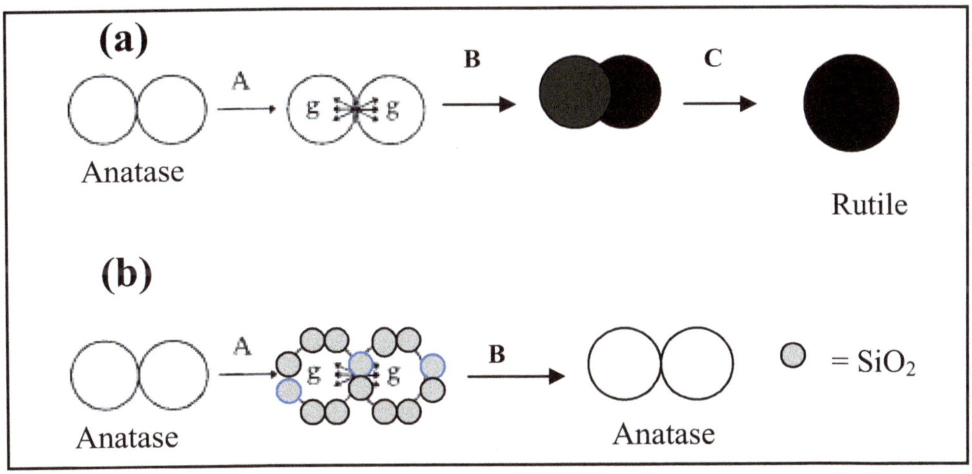

Figure 3.14: Diagram Showing Steps Involved in the Phase Transformation of Nanocrystalline Anatase Particles TiO_2 (a) and TiO_2-$2SiO_2$ (b). The *g* is the nuclei growth rate.

nucleus is dependent upon interfacial energy. Fluctuation of a rutile nucleus within an anatase particle of around critical size for rutile would almost certainly result in rapid transformation of the entire particle (Figure 3.14a), removing the interfacial energy contribution. The particle size in diameter of anatase phase of TiO_2 did not change after calcination up to 600°C, namely 13 nm (step A and B). Thus, the critical nucleus size of TiO_2 is 13 nm (Table 3.1 and Figure 3.8A). Formation of a rutile particle steps C, involves growth at the particle-particle neck from constituents evidently provided by surface diffusion. The growth of rutile may happen when a rutile crystallite comes into contact with an anatase crystallite forming a larger rutile crystallite or when two rutile nuclei merge together.

In the phase transformation of nanocrystalline anatase particles of TiO_2-$2SiO_2$ shows in Figure 3.14b. Interface nucleation step A, nucleation of rutile starts within the interface between two contacting anatase particles, and the nucleation rate is determined by the probability of contact of two anatase particles. Nuclei growth step B, is restricted with addition of SiO_2 and critical size for rutile did not achieve. Anatase nanoparticles in the TiO_2-$2SiO_2$ aerogel are surrounded with SiO_2 nanoparticles and highly dispersed, thus the nuclei growth rate is low. Also, a large amount of SiO_2 will retard the nuclei growth of TiO_2 and the phase transition from anatase to rutile, probably due to the small particle size of anatase *i.e.* 10 nm (Figure 3.9).

The above calcination experiments TiO_2-$2SiO_2$ indicate that no anatase transform to rutile. Author consider this to be a consequence of the dependence of thermodynamic stability on particle size, noting that TiO_2-$2SiO_2$ and TiO_2 growth rates differ, causing changes in relative stability during thermal treatments. Figures 3.8 and 3.9 illustrate the nanocrystalline anatase particles (nm) after calcination at 500 and 700°C. It is seen that, at particle size of anatase c.a. 13 nm and c.a.10nm for TiO_2 and TiO_2-$2SiO_2$ respectively, anatase is the most stable phase; for particles sizes greater than 13 nm for TiO_2, rutile is the most stable phase. For TiO_2-$2SiO_2$, the particle size is stable in small size namely c.a.10 nm after calcination at 700°C or more, anatase is the most stable. The thermodynamic of the size effect on the thermal stability of TiO_2 anatase nanoparticles synthesized by MOCVD was reported by W. Li *et al.* (Li *et al.*, 2004). 12nm anatase had the lowest transformation starting temperature and thermal stability as compared with 17 and 23 nm samples (Li *et al.*, 2004). Anatase is most thermodynamically stable at sizes less than 11nm, brookite is most stable at sizes between 11 and 35 nm, and rutile is most stable at size greater than 35 nm (Zhang and Banfield, 2000).

The specific surface areas of the aerogels after calcinations at various temperatures are shown in Table 3.1. With increasing calcination temperatures, the specific surface areas of the TiO_2 aerogel decrease drastically, but did not change significantly for TiO_2-SiO_2 aerogels. This is due to the restriction of the grain growth with addition of SiO_2.

Conclusions

The technique of CO_2 supercritical extraction makes it possible to produce highly porous anatase with small particle size. With addition of a large amount of SiO_2, the grain growth rate of crystalline TiO_2, anatase, is restricted, and anatase phase is thermally stabilized. The rutile phase was not found after calcinations at temperatures up to 1000°C. The transformation sequence among the anatase and rutile is size dependent, because the energies of the anatase and rutile are sufficiently close that they can be reversed by small differences in surface energy. If particle sizes of the TiO_2-$2SiO_2$ and TiO_2 phases are equal or smaller than 13nm anatase is most thermodynamically stable. The rutile is most stable at sizes greater than 13 nm. The specific surface area, the pore volume and the average pore size of TiO_2-SiO_2 aerogels do not change significantly after calcinations at temperatures up to 800°C.

References

Anpo, M., Shima, T., Kodama, S. and Kubokawa., Y., 1987. *J. Phys. Chem.*, 91: 4305.

Barret, E.P., Joyner, L.G. and Halenda, P.H., 1951. *J. Am. Chem. Soc.*, 73: 373.

Cornell, R.M., Posner, A.M. and Quirk, J.P., 1975. *J. Colloid Interface Sci.*, 53: 6.

Fujishima, A. and Honda, K., 1972. *Nature*, 238: 37.

Gedanken, A., Tang, X., Wang, Y., Perkas, N., Koltypin, Y., Landau, M.V., Vradman, L. and Herskowitz, M., 2001. *Chem. Eur. J.*, 7: 4546.

Imai, H., Morimoto, H., Tominaga, A. and Hirashima, H., 1997. *J. Sol-Gel Sci. and Tech.*, 10: 45.

Kormann, C., Bahnemann, D.W. and Hoffmann, M.R., 1988. *J. Phys. Chem.*, 92: 5196.

Kasuga, T., Hiramatsu, M., Hoson, A., Sekino, T. and Niihara, K., 1999. *Adv. Mater* and 11: 1307.

Kumar, K.N.P., Keizer, K., Burggraaf, A.J., Okubo, T. and Nagamoto, H., 1993. *J. Mater. Chem.*, 3.

Li, W., Ni, C., Lin, H., Huang, C.P. and Ismat Shah, S., 2004. *J. Applied Physics*, 96(11): 663.

Matsubara, H., Tanaka, M., Koyama, S., Hashimoto, K. and Fujishima, A., 1995. *Chem. Lett.*, pp. 767.

Mazdiyasni, K.S., Lynch, C.T. and Smith, J.S., 1965. *J. Am. Ceram. Soc.*, 48: 372.

Mason (Ed.), 1990. *Sonochemistry: The Uses of Ultrasound in Chemistry*. Royal Society of Chemistry, Cambridge.

Ovenstone, J. and Yanagizawa, K., 1999. *Chem. Mater.*, 11: 2770.

Ohko, Y., Tryk, D.A., Hashimoto, K. and Fujishima, A., 1998. *J. Phys. Chem. B*, 102: 2699.

Phillips, R., 1983. *Sources and Applications of Ultraviolet Radiation*. Academic Press, New York, pp. 124.

Sclafani, A. and Herrman, J.M., 1996. *J. Mater. Res.*, 100: 13655.

Srivastava, D.N., Perkas, N., Gedanken, A. and Felner, I., 2002. *J. Phys. Chem. B*, 106: 1878.

Srivastava, D.N., Chappel, S., Palchik, O., Zaban, A. and Gedanken, A., 2002. *Langmuir*, 18: 4160.

Suh, D.J., Park, T.J., Kim, J.H. and Kim, K.L., 1997. *Chem. Mater.*, 9(9): 1903.

Weaver, S. and Mills, G., 1997. *J. Phys. Chem. B*, 101: 3769

Wang, C., Deng, Z.X., Zhang, G., Fan, S. and Li, Y., 2002. *Powder Technol.*, 125: 39.

Wang, C.C. and Ying, J.Y., 1999. *Chem. Mater.*, 11: 3113.

Wang, T.Y.Q., Yin, L.X., Palchik, O., Hacohen, Y.R., Koltypin, Y. and Gedanken, A., 2001. *Langmuir*, 17: 4131.

Zhang, H. and Banfield, D.F., 1999. *American Mineralogist*, 84: 528.

Zhang, H. and Banfield, J.F., 2000. *J. Phys. Chem. B*, 104: 3481.

Chapter 4

Development of Nanostructured Smart Delivery Systems for Pesticides and Fertilizers

Erastus Gatebe and Harrison Wanyika

Department of Chemistry,
Jomo Kenyatta University of Agriculture and Technology,
Nairobi, Kenya. P.O Box 62000-00200, Nairobi
E-mail: erastusgatebe@yahoo.com

ABSTRACT

Agriculture forms the backbone of third world economies. The Kenyan agricultural sector have been experiencing major challenges including an increasing risk of disease, high production costs due to the high cost of fertilizer and pesticides, threats to agricultural production from changing weather patterns and a growing demand for healthy, safe food. Nanotechnology, which is the manipulation of matter at the scale of atoms and molecules, is rapidly converging with biotechnology and information technology in changing the way food is produced and systems in agriculture that are in use. Owing to the failure of biotechnology to address the issue of environmental and health concern as well as control and ownership, nanotechnology will address the challenges raised in the field of biotechnology. In our current study, we intend to show how nanotechnology can be used to improve food security in Kenya and how the technology can be used to improve on agriculture without compromising the safety and health of the people of Kenya. The expected outcome of the research is reduced crop production costs by decreasing the quantities of fertilizer and pesticides used in agricultural production and by treating only the affected part of a crop at an early stage of stress.

Keywords: *Nanotechnology, Biotechnology, Smart delivery systems, Nanotubes, Food security, Agriculture.*

Introduction

General Introduction

In the recent past, various players in the field of nanotechnology have been exploring the possibility of using the new science to sustain mankind as resources become scarce and demand increases. Although biotechnology had seen the revolutionization of traditional food production practices, various challenges have been encountered. Key among them is ethical issues, health as well as environmental ones.

In a study carried out by Helmut Kaiser, the argument was that nanoscience will be used to design by shaping molecules and atoms in order to get the desired output. In this study, the prediction has been that nanotechnology in biotechnology and nano Bioinformation will be the driving force in this field. In these areas, the key ingredient is production of less waste and utilizing less energy in production especially for food manufacturers.

As noted by various authors (Gatebe, 2007; Heinze, 2004; Borm *et al.*, 2006; www.nano.gov; Davies, 2006; www.wilsoncenter.org/nano; www.icfi.com; Maynard, 2006) nanoscale materials present regulatory challenges similar to those posed by products using new emerging technologies. These materials often have chemical, physical, or biological properties that are dissimilar from those of their larger ones. These differences may include altered magnetic properties, altered electrical or optical activity, increased structural integrity, or altered chemical or biological activity (ICF International, 2006). Due to these properties, nanoscale materials have great potential for use in a vast array of products (www.icfi.com).

Currently, many reviews have been published that focuses on the state of knowledge in terms of behavior, the interaction with biological systems (both for advantageous and toxicity outcomes), and potential environmental disposition of nanoscale materials. These publications focuses on the risks associated with these nanoscale materials, an array of these can be found in the peer-reviewed public literature, which touches on biological systems interactions with nanomaterials mainly highlights possible risks as a result. The emphasize on these publications focuses on the need for research on such areas as characterization of materials in the nanoscale range; methods for identifying hazards; understanding biological response to nanoscale materials; and characterizing nanoscale material exposure and transport (in humans and the environment)(Maynard, 2006).

In the developing world and especially Kenya, food security has been a concern and with the changing environment, urgent measures need to be taken to arrest the problem. Currently, food scarcity and drought has been declared a national disaster with the government appealing for donation to feed the over 10 million persons being affected.

Although this study intend to unlock the potential of nanoscience in addressing the food situation, greater concern especially the legislation of nanoparticles and nanotubes will be explored since guidelines must be adhered to. Our project also proposes to come up with mechanisms, which can be used to check the safety and usefulness of this technology while still enhancing its productivity.

In sub-Saharan Africa the biggest challenges is how to feed the ever-increasing population. With the projected deficit of food and desertification to increase ten fold in the next ten years. Urgent need and measures are required to stop or arrest the drift occasioned also by global warming and adverse weather change. In particular, the increasing desertification is posing a serious threat to the people of this region who rely on natural erratic rainfall patterns.

In order to achieve our objectives, we intend to come up with a new system that facilitates ingredient intake (fertilizers) and pesticides in order to prevent or avoid food loss. In this regard, various analytical techniques will be utilized to study the effects, efficacy and mode of application in vitro. Based on the data obtained, a pilot study will be established at JKUAT to demonstrate the working principles of the new technology. Data obtained will then be disseminated to the farmers through workshops, seminars and fliers.

This proposed project is expected to open new lines/ thinking in the field of nanoscience and secure both environmental and health issues that might arise.

Problem Statement

Agriculture in Kenya has become unprofitable due to the abnormally high agricultural production costs as a result of the high costs of farm inputs especially fertilizers and pesticides.

The diminishing land size as the population increases necessitate the overhaul of the current agricultural production practices.

Justification

Nanotechnology has the potential to develop smart agricultural systems that would minimize input (*i.e.* fertilizers, pesticides, herbicides etc) and maximize output (*i.e.* crop yields) by delivering chemicals in a controlled and targeted manner.

Hypothesis

Nanotechnology could be used to improve the efficiency of fertilizers and pesticides and consequently lower the costs of agricultural production.

Objectives

General Objective

To develop nanostructured smart delivery systems for plant fertilizers and pesticides.

Specific Objectives

1. To synthesize polydiacetylenes nanotubes.
2. To characterize the polydiacetylenes nanotubes.
3. To carry out in vitro tests to determine the ability of the nanotubes to act as a smart delivery systems for enhancing the ability of plants to absorb nutrients from fertilizers.

4. To carry out in vitro tests to determine the ability of the nanotubes to act as smart delivery systems for pesticides.
5. Demonstrate the working principles of the new technology in situ.

Methodology

Design, Synthesis and Characterization of Self Assembled Polydiacetylene Nanotubes

Methods used in this section are modifications of previously established protocols.

General Methods

Self-assembled nanotubes from dicarboxylic PDA (DCPDA) will be synthesized using layer-by-layer (LBL) techniques inside a template. Alumina template with 100 nm pore diameter will be used for the synthesis of DCPDA nanotubes. The Zr-carboxylate will be used for interlayer linking chemistry between the DCPDA layers because Zr-carboxylate is robust and bond formation kinetics is fast (Gatebe, 2007).

Preparation of DCPDA Monomer

The diacetylene monomer 10, 12- docosadiyndioic acid (95 per cent) are purchased and of the highest quality. The monomer is prepared by dissolving 0.0144 g in 40 ml of chloroform to make 1 mM DCPDA. The solution is then filtered through 0.45 µm filter paper to discard any polymer formed due to thermal or photo polymerization of DCPDA. The solution is covered with aluminium foil to avoid photo polymerization and is prepared freshly for each deposition cycle.

Layer by Layer Deposition of DCPDA

Glass slides and gold substrates will were cleaned for organic contaminants by dipping in an oxidizing solution of piranha solution prepared by mixing 3:1 concentrated sulphuric acid with hydrogen peroxide (30 per cent) for 20 minutes. The glass substrates is then rinsed with nanopure water and hydrolyzed with 2M HCl for 5 minutes to functionalize them with silanol group. They are then rinsed with nanopure water and dried under argon gas. The hydrolyzed substrates is then primed in a solution containing 95 ml ethanol, 1 ml acetic acid and 4 ml (3-amino propyl) triethoxysilane for 1 hour followed by rinsing with ethanol and then baking for 10 hours at 90°C. The oxidized glasses is then phosphorylated for 20 minutes using 0.2 M $POCl_3$ and 0.2 M 2, 4, 6-collidine in dry acetonitrile under stream of argon. The substrates are then rinsed with acetonitrile and water to generate phosphoric acid on the surface. The zirconation is performed using 5 mM solution of $ZrOCl_2$ (50:50 ethanol/water) for 20 minutes.

The cleaned gold substrates were functionalized by soaking the substrates in 1 mM 16-mercaptohexadecanoic acid (MHAD) in chloroform for 18 hours at room temperature. The zirconation of the carboxylic acid monolayer is performed as described above. All the substrates are incubated in 1 mM DCPDA. Multilayer zirconation followed by DCPDA adsorption cycles with fresh solution as the number of layers desired is repeated. The photo polymerization of the PDA monomers is performed after every deposition step using 254 nm wavelengths for 2 minutes (4.5 mW at 1 cm).

UV-VIS Spectroscopy

The UV–VIS Spectrophotometry is being utilized to study the kinetics of the new nanoscale based plant ingredients.

The measurement of absorption spectra for the multi-layer DCPDA films was accomplished with a Perkin Elmer Lambda 25 UV-VIS instrument. The substrates were put in a quartz cuvette and absorbance measured from 300-800 nm ranges. The background was corrected using cleaned non-functionalized glass substrate. The band pass was set at 5 nm and scanning speed of 500nm/sec.

Fluorescence Spectroscopy

Fluorescence spectrophotometer was used to study the emission characteristics of the materials.

The emission spectroscopic measurements were performed using Perkin-Elmer LS 55 with excitation wavelength of 470 nm. The substrates were then held in a clear quartz fluorescence cuvette. The band pass was set at 5 nm and scanning speed of 500nm/sec.

Surface Characterization

Fourier Transform Reflectance Absorption Infrared Spectroscopy (FT-RAIRS)

External reflection FTIR spectra of the multilayer on gold substrates was measured with Nicolet Nexus 670 FT-IR fitted with nitrogen cooled MCT-B detector and a Velma II variable-angle seculars-reflectance accessory operating at a beam incidence angle of 80° with respect to surface normal. The samples will be averaged at 1000 scans at 2 cm^{-1} resolution against a background of bare gold substrate. The spectrometer was purged continuously with dry nitrogen gas to minimize water vapor in the sample.

Fluorescence and Electron Microscopies

The nanotubes were first visualized after dissolving the alumina template in 2 M NaOH using optical fluorescence microscope (Leica DMRI) equipped with a CCD camera. The topography and morphology of the nanotubes was determined using scanning electron microscope (SEM) and transmission electron microscope (TEM). Micro Rheometer will be used to carry out the characterization of the liquid based products.

The samples for SEM analysis were coated with 5 nm gold/palladium alloy to make them conductive.

In Vitro Tests for the Synthesized Nanotubes as Delivery Systems for Pesticides and Fertilizers

The synthesized nanoscale plant ingredients will be incorporated in pesticides and fertilizers and in vitro tests will be done with plants grown in green houses to determine the efficacy of the novel ingredients as smart delivery systems.

Results and Discussion

Our preliminary data shows that using the dicarboxylic acid, the multiple layers of DCPDA has been synthesized. The multilayers are being characterized for the biodegration assays in order to ensure that environmental pollutants are not introduced into the soil. These data are to be used for validation studies. Self-assembled nanotubes from dicarboxylic PDA (DCPDA) were synthesized using layer-by-layer (LBL) techniques inside a template. Alumina templates with 100 nm pore diameter were used for the synthesis of DCPDA nanotubes. The Zr-carboxylate was used for interlayer linking chemistry between the DCPDA layers because Zr-carboxylate is robust and bond formation kinetics is fast (Erastus Gatebe *et al.*, 2008) (Figure 4.1).

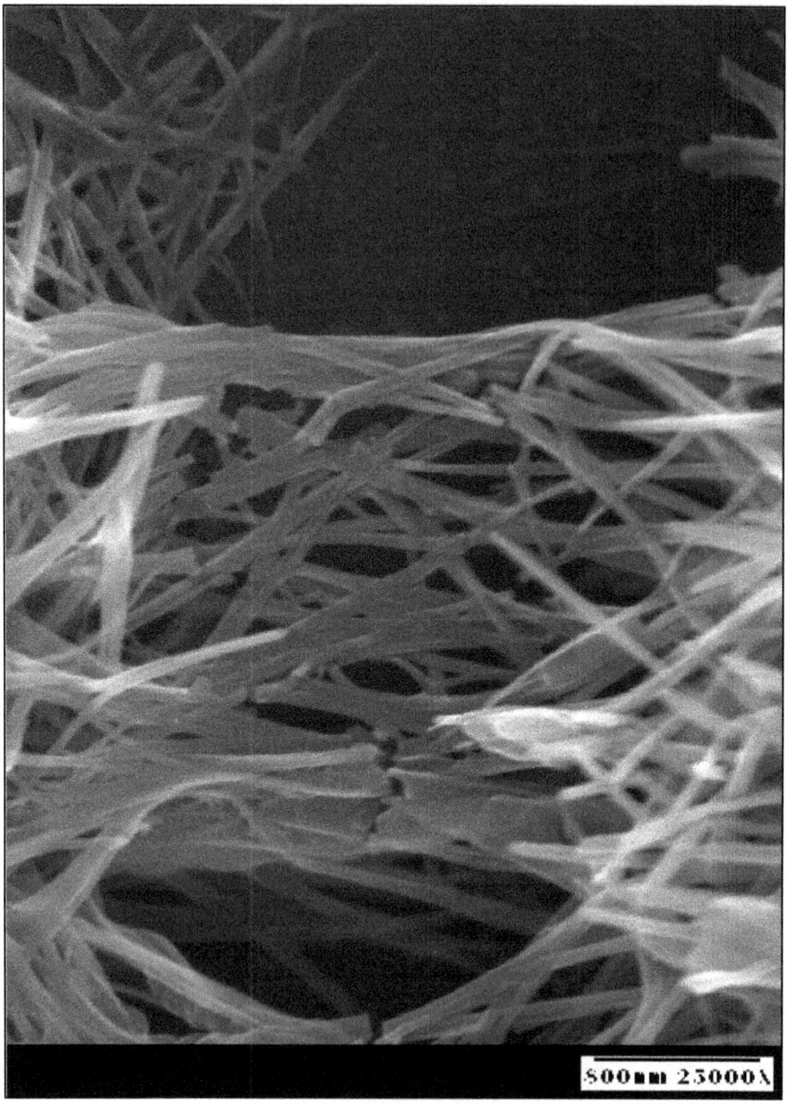

Figure 4.1: Shows the Nanotubes of the DCPDA

Conclusion

Bottom-up approach is one of the methods for synthesizing tailored materials with nanoscale features. Our work has utilized self-assembling to make nanometer size DCPDA nanotubes using a template-assisted technique. The zirconium-carboxylate Chemistry and LBL techniques were used to make nanotubes of desired inner diameter. Using the proposed method, higher hierarchical structures with complex features can be made with multifunctionality.

Expected Outcomes

1. Reduce crop production costs by decreasing the quantities of fertilizer and pesticides used in agricultural production and by treating only the affected part of a crop at an early stage of stress.
2. Optimizing the yield per acreage.
3. Ensuring the country move from food importer to exporter and also increasing the per capita income

References

Borm, P.J.A., *et al.*, 2006. *Australian Safety and Compensation Council. A Review of the Potential Occupational Health and Safety. Implications of Nanotechnology.* July 2006. ISBN 0 642 32609 6; the potential risks of nanomaterials: a review carried out for ECETOC. *Particle and Fiber Toxicology*, 3: 11.

Davies, J.C., 2006. *Managing the effects of Nanotechnology*. Woodrow Wilson Institute for Scholars. Available at www.wilsoncenter.org/nano.

Erastus Gatebe, Hayley Herron, Rashid Zakeri, R., Pradeep, Samir Aoudi, Punit Kohli, 2008. Synthesis and characterization of polydiacetylene films and nanotubes, Journal of American Chemical Society (JACS).

Gatebe, E.G., 2007. Fluorescence resonance energy transfer in conjugated organic nanostructures. *Ph.D. Thesis*, pp. 67–72.

Heinze, T., 2004. Nanoscience and nanotechnology in Europe: Analysis of publications and patent applications including comparisons with the United States. *Nanotechnology Law and Business*, 1(4): 10.

ICF International, 2006. *Characterizing the Environmental, Health, and Safety Implications of Nanotechnology: Where should the federal government go from here.*

National Nanotechnology Initiative (NNI), 2006. *Environmental, Health, and Safety Research Needs for Engineered Nanoscale Materials*, or, www.nano.gov.

Maynard, A.D. and Davies, J.C., 2006. *Nanotechnology: A Research Strategy for Addressing Risk*. Woodrow Wilson Institute for Scholars www.icfi.com

Miller, J.C. and Serrato, R.M.J. Kudhahl, 2005. *The Handbook of Nanotechnology: Business, Policy and Intellectual Property Law*. John Wiley and Sons Inc.

Chapter 5

Characterization of Carbon Nanoparticles (CNPs) Formed in Diluted Environment

Alsayed Ali

Higher Institute of Applied Sciences and Technology (HIAST), Syria

E-mail: masayed@hiast.edu.sy

ABSTRACT

This is part of a study on the formation of carbon nanoparticles from the nucleation of gas phase PAH molecules produced in low temperature reactors. We have assessed atomic force microscopy for complementary ex-situ characterization of the morphology of these particles. The feasibility of ex-situ characterization by atomic force microscopy (AFM) was first performed by the analysis of soot particles formed in a diffusion flame and deposited onto a substrate. It was found that the mica surfaces are appropriate as substrates because they have a negligible roughness and are extremely simple to prepare. The AFM was operated under tapping mode preferably over contact mode which may move the nanoparticles on the surface. This phase was followed by preliminary experiments of particles formation in the chamber of the CRESU reactor dedicated to the study of reaction rates of condensable species.. This chemical reactor is designed to generate dense uniform flows using a Laval nozzle, over a wide range of temperatures (23-500K) and containing large amounts of PAH vapors. Atomic force microscopy was used to establish the particle size distribution and morphology of these particles deposited onto mica substrates. These results combined with time of flight mass spectrometry results (particle mass distribution) will, in future, help in the understanding of the influence of the nature of the precursors, the flow temperature, the presence of ions, etc. on the chemical and physical structures of the nanoparticles.

Keywords: *Carbon nanoparticles, Soot, Atomic force microscopy, Diffusion flame, CRESU, PAHs.*

Introduction

Formation Mechanisms of Nanoparticles

Dust particles span a continuous size distribution from large molecules–such as polycyclic aromatic hydrocarbons (PAHs)–to micrometer-sized particles to nanometer-sized particles, and influence many processes in the evolution of the interstellar medium such as the energy balance through the photoelectric effect, the ionization balance through recombination with electrons and ions, the radiative budget and the chemical composition of molecular clouds.

The mechanisms of nucleation and growth of carbonaceous dust nanoparticles from hydrocarbon molecular precursors in the circumstellar shell of carbon-rich stars are not well understood. Some carbon particle formation models attribute a key role (of what?) to physical condensation of medium-sized PAH molecules (such as pyrene) (Frenklach, 2002). L. Biennier and coworkers (2006) have started kinetics studies on the nucleation of particles from gas phase PAH molecules. An interesting aspect of this research is the characterization of such particles. Our objective, which is a part of a more global project, is to assess ex-situ characterization methods to determine the morphology of these particles.

Particle Deposition on Surfaces

Atomically smooth surfaces have been studied in great detail over several decades using ultra-high vacuum environment, scanning probe microscopy and a variety of spectroscopic techniques. Nanoparticles or molecules can be deposited on surfaces to be characterized in ex-situ methods or to control the reaction products or growing on surfaces. Surfaces, in general, can reconstruct, meaning that the atomic structure is different on surfaces due to the smaller number of chemical bonds of surface atoms. Surfaces have atomic steps and defects, and surface atoms react preferentially with these defects and step edges. The presence of nanoparticles influences the surface property and can change the surface reconstruction. When depositing nanoparticles or molecules on surfaces, we need to have an idea about their two dimensional packing or the position and orientation of individual molecules or nanoparticles.

Molecular transport is easier on surfaces onto which the molecules are weakly bound. Although the molecules consist of large number of atoms they can take a random walk very much like the diffusion of atoms on a surface. Isolated medium-sized PAH molecules deposited on a surface will stick only if the surface is maintained at low temperature. Above a certain temperature which will depend on the vapour pressure of the species, they are expected to evaporate. It has been estimated that the Van der Waals forces that hold together large PAH clusters are not small (binding energy of about 0.25 eV per molecule for pyrene clusters). Therefore, large PAH clusters may be deposited onto the surface at relatively high temperature (500K) and remain intact. It is however, to be kept in mind that the molecules could still be moved by the probe. Indeed the probe itself is commonly used to pick up and place atoms and molecules or nanoparticles.

Ex situ AFM Characterization Mode

Pico Pulse Molecular Imaging Atomic Force Microscope (AFM) (and Digital Instruments Nanoscope AFM during the first phases of the project) is used to observe the particulate deposits. This instrument can be run in two different operational modes, contact-mode and tapping-mode. We recall that the better known contact-mode technique consists of scanning performed with the tip directly touching the surface to be analyzed; it is widely used in all cases in which the observed surfaces are resistant to micro-mechanical solicitations.

The tapping-mode technique consists of oscillating the probe in oscillation close to the sample surface, permitting the tip to touch the surface just slightly. The oscillating frequency is chosen close to the mechanical resonance frequency of the holding cantilever. The major advantage of running in tapping-mode is that strong tip sample interactions are avoided. The contact force is reduced to less than 100 pN, which allows observation of samples that are not strongly bonded to the substrate surface. Therefore, accidental removal or shifting of the sample on the substrate surface is prevented. In addition, tapping-mode attenuates the undesirable convolution effects that two touching surfaces always generate.

It is also worth mentioning that in tapping-mode the phase shift detection feature allows investigations of the viscoelastic properties of the samples. The phase shift detection consists of measuring the phase shift between the input signal that governs the oscillation and the deflection/oscillation output signal of the cantilever. Because the feedback acts independently from the modulation signal both topography and phase detection can be acquired at the same time.

Because of its many attractive features, the tapping- mode technique is preferred over the contact mode to observe the deposits. Early attempts were made with the AFM in the contact mode. The AFM was equipped with a Nanoprobe silicon probes type FM, characterized by a force constant of 1.2 to 3.5 N/m, to minimize probe-sample convolution effects. The images recorded are shown in Figure 5.1 which shows that the contact mode is not appropriate because the cantilever becomes rapidly covered by the particles that are loosely bound to the surface causing the image to be non-interpretable, besides the difficulties in AFM contact mode parameters choosing in the case of small force of surface–particles.

Carbon Particles Formation in Diffusion Flame

In order to assess the analysis method, we first had to be assured of the presence of nanoparticles on the substrate. We turned to combustion systems that are well known to generate large amounts of particles.

The production of the carbon particles generated in a diffusion flame chamber is well studied and has been already validated by d'Alessio and co-workers (Barone *et al.*, 2003). Therefore, it seems logical to start our investigation by forming, depositing and analysing the carbon soot particles generated in a diffusion flame chamber to validate our approach. Although, the aim here is not to understand how particles are formed in flames, we should recall that the current consensus is that PAH molecules are involved in soot particles formation. We can distinguish several opposing scenarios

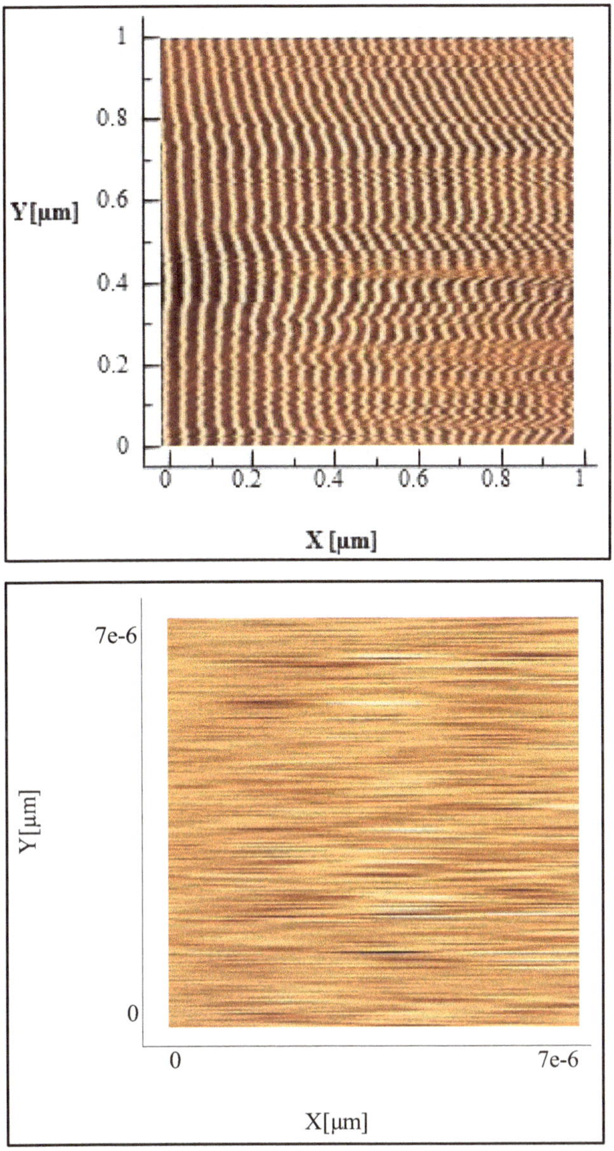

Figure 5.1: Illustration of Some Problems Encountered with AFM Operating in the Contact Mode

especially concerning the molecular to particle transition. In the dominant one (Barone, A.C.) the fuel is broken down in small carbon chains and radicals by pyrolysis. They then react to form the first aromatic rings which subsequently grow by hydrogen abstraction acetylene addition (HACA mechanism). Once the aromatics have achieved a critical size, physical condensation triggers the formation of PAH clusters. These Van der Waals clusters then grow by surface reaction with acetylene and by coagulation.

Another scenario has been proposed by Reilly and co-workers. They suggest that large PAH droplets are first formed in the flames and that sudden dehydrogenation starts simultaneously at numerous sites on the surface to produce soot nuclei. Nucleation sites grow then in size by integration of surrounding PAHs. Finally, soot nuclei begin to interact and line up end to end in a fractal pattern due to van der Waals interactions. As these aligned nuclei grow, they fuse (Reilly, 2000).

More recently, Siegmann *et al.* (2002) proposed that soot particles first originate from chemical condensation of one to two rings structures and aliphatic bridges forming nuclei. PAHs are then synthesized on the particles using their surface as templates and evaporate into the gas-phase when synthesis is completed.

Atomic Force Microscopy of Soot Particles

Soot particles were generated by a diffusion flame of ethylene C_2H_4 (100 slm Air, 45 slm C_2H_4). The flame was characterized by a radius of 1.1 cm, a height of 4 cm and a velocity of 40 cm/s. The particles formed in the flame were collected on a substrate of mica ($K_2O.3Al_2O_3.6SiO_2.2H_2O$). In our case, there is small surface segregation due to the high smoothness of mica surfaces. Because of the low reactivity of mica surfaces, it is expected that physisorption dominates over chemisorption.

The substrate of mica was cleaned by taking off a sheet by an adhesive tape just before the experiment. We recall that Mica has a roughness of the atomic order. The sample was analysed by AFM operated in tapping mode (see Table 5.1 for characteristics). The nanoparticles were collected at different heights above the burner (H_{AB}): 22mm, 35mm, and 61.5 mm across the whole transverse section of the flame and in less than 1 second.

Table 5.1: The AFM Cantilever Mechanical Properties in Tapping Mode

Parameters	*Symbol*	*Range*	*Unit*
Material		n^+-Silicon	
Resistivity		0.01-0.025	Ωcm
Thickness	T	2.5 to 4.0	μm
Length	L	225	μm
Width	W	20 to 30	μm
Force constant		1.2 to 3.5	N/m
Tip height	H	10-15	μm

Different images were obtained for each sampling position; those reported in Figures 5.2–5.4 are representative of the sampled material. Measurements at high magnifications obviously result in much more detailed structure, enhancing the visibility of the smaller diameter particulate fraction, but electromechanical noise and the limits of feedback electronics put a limit to this procedure. On the other hand, statistics that are more reliable are obtained when larger samples are considered. Thus, a convenient compromise was found by performing the measurements on surfaces of 1000 x1000 nm.

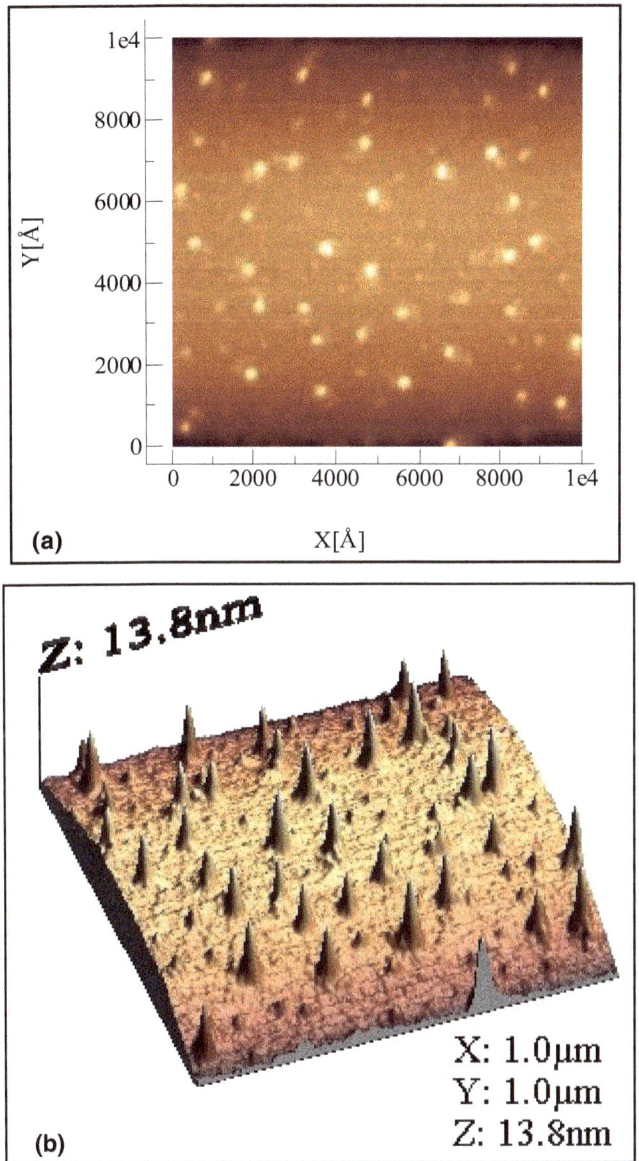

Figure 5.2: AFM 2D (a) and 3D (b) Images of Carbon Particles Collected on a Substrate of Mica at a Height Above the Burner H_{AB} = 22 mm of a C_2H_4/Air Diffusion Flame

Analysis of Carbon Soot Particles Generated in a Diffusion Flame

The different parameters (surface, volume and height) of carbonaceous particles were evaluated using a homemade software using LabView (see Figure 5.5). This program allows virtual separation of the particles from the substrate by subtracting a threshold plane from the given image data. This operation requires a very accurate

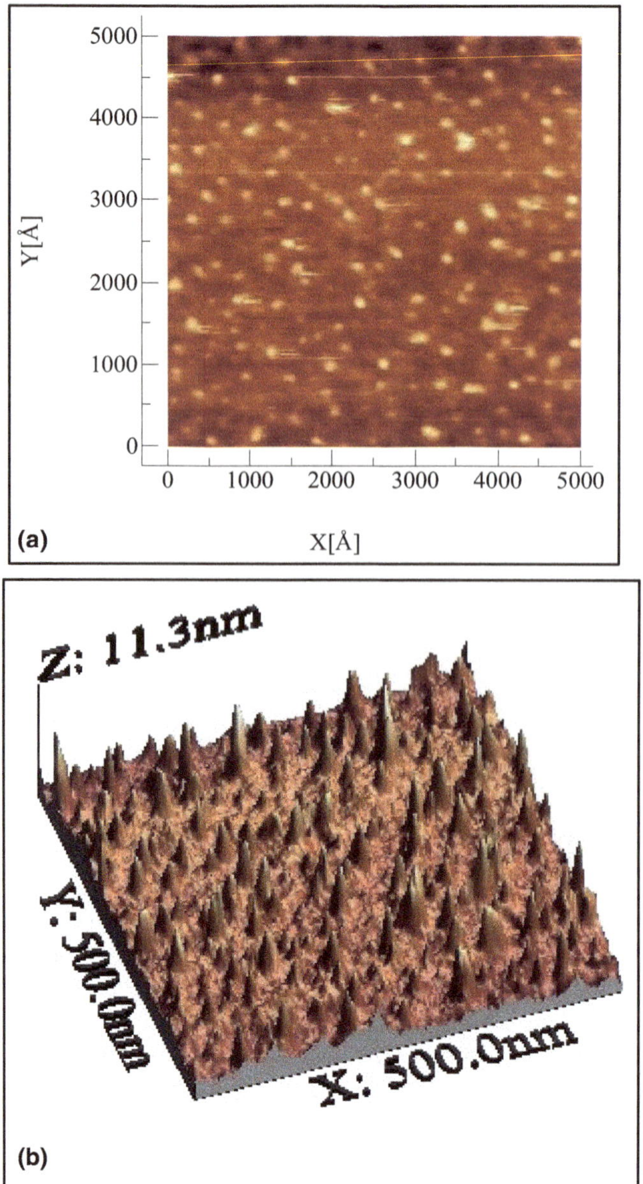

Figure 5.3: AFM 2D (a) and 3D (b) Images of Carbon Particles Collected on a Substrate of Mica at a Height Above the Burner H_{AB} = 35 mm of a C_2H_4/Air Diffusion Flame

execution, since a wrong choice of the threshold value could significantly affect further volume estimations, with an error of up to about 20 per cent.

An equivalent apparent diameter (ED) was calculated from the measured particle volume, assuming spherical particles. Among the various parameters obtained from

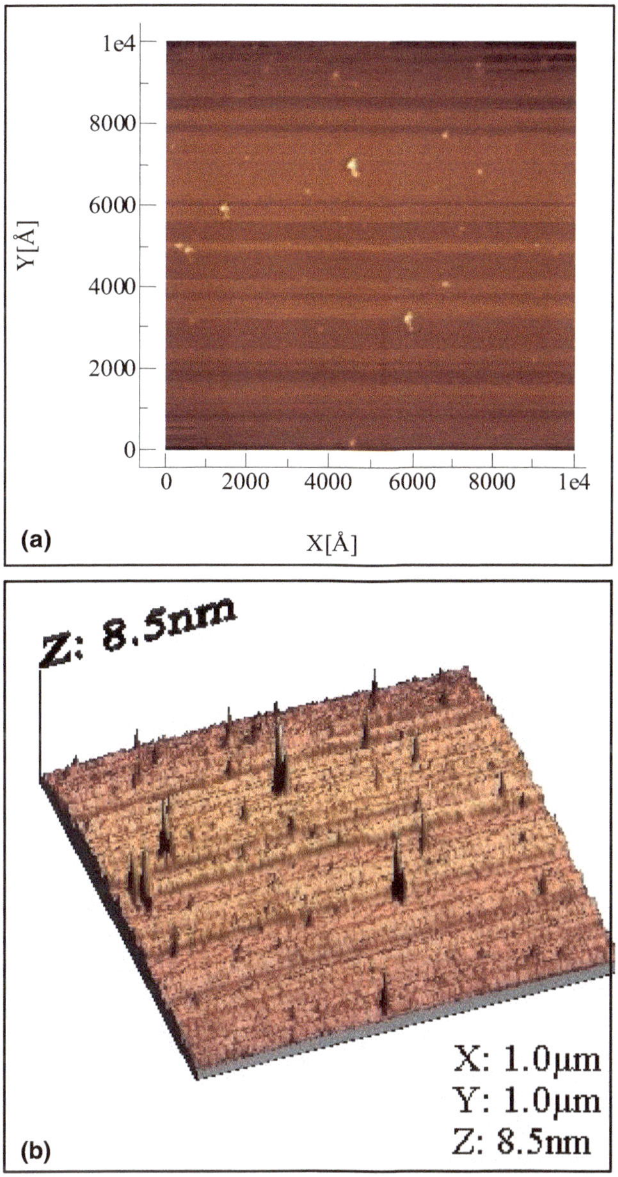

Figure 5.4: AFM 2D (a) and 3D (b) Images of Carbon Particles Collected on a Substrate of Mica at a Height Above the Burner H_{AB} = 61.5 mm of a C_2H_4/Air Diffusion Flame

the measurements, one is particularly significant in characterizing flame generated inception particles, and it is the particle volume represented by frequency of different (ED) nanoparticles (as presented on Figures 5.6a,b,c), and the volume distribution function *P*(R) calculated from equation (1).

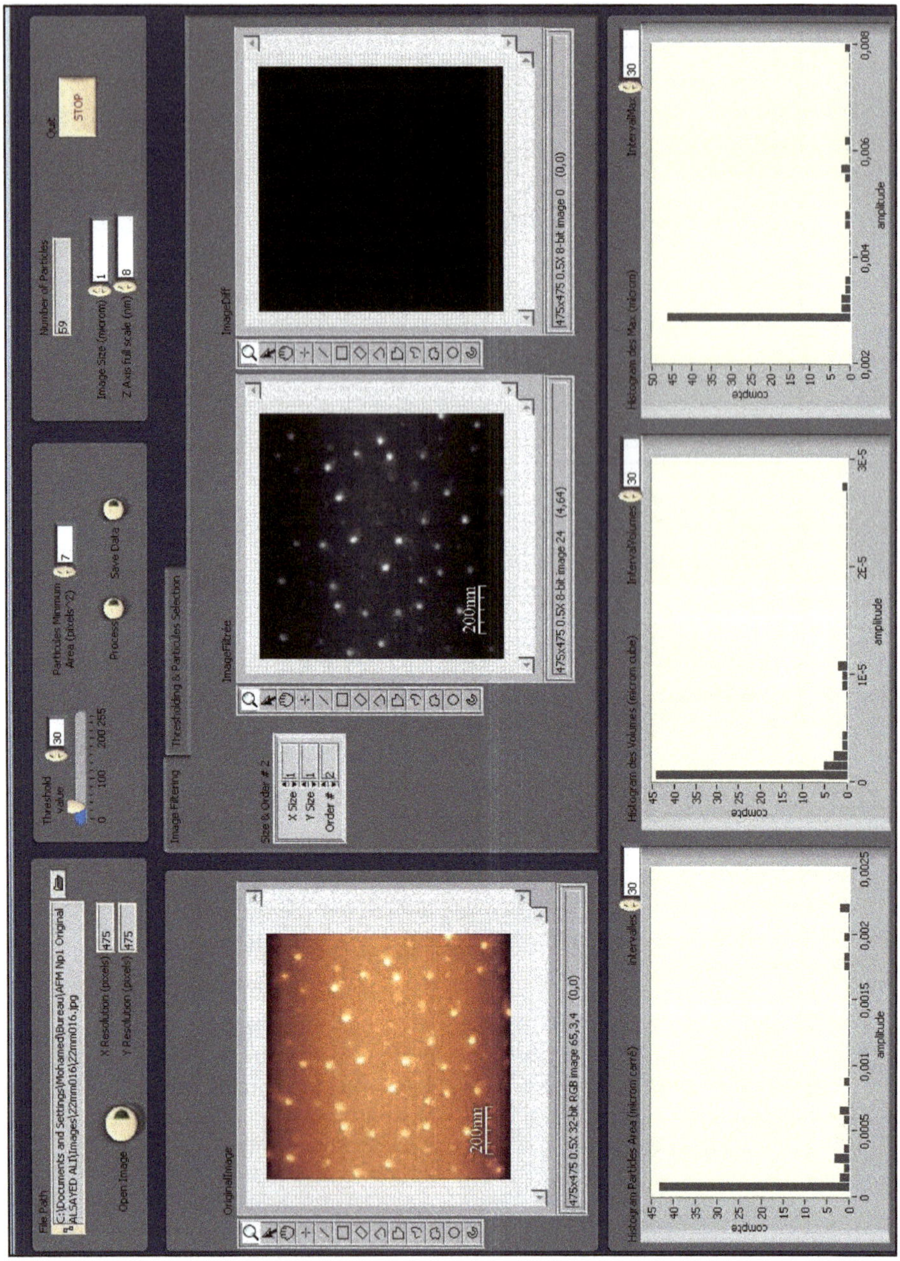

Figure 5.5: The Developed Software Using LabView Used to Analyse the AFM Images to Determine the Histograms of the Area, Volume and Height of the Particles

Figure 5.6: Particle Size Distribution vs Particle Diameter

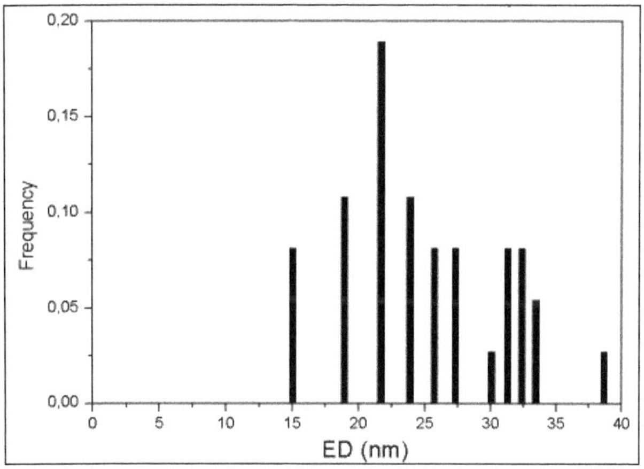

(a) H_{AB} = 22 mm

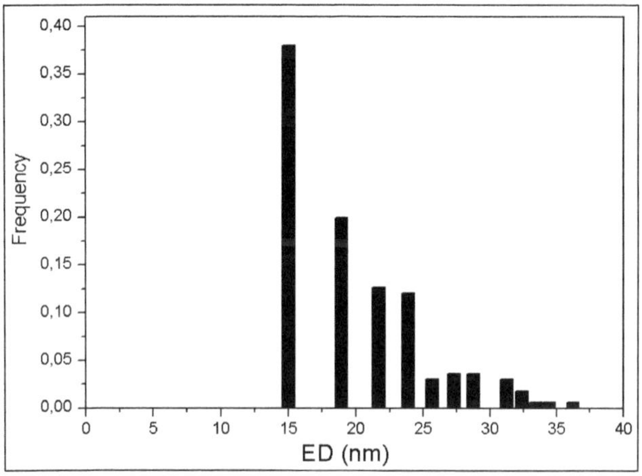

(b) H_{AB} = 35 mm

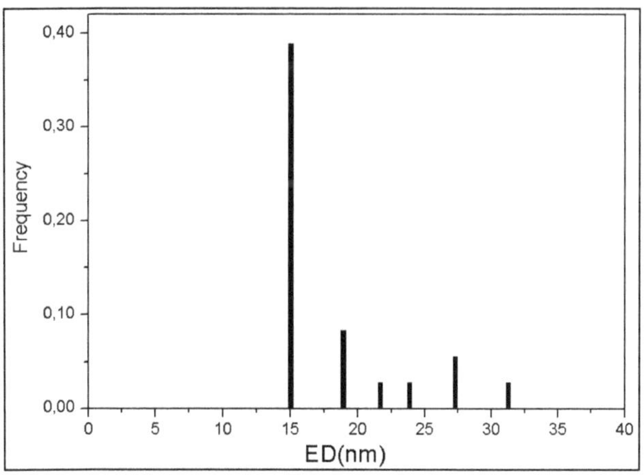

(c) H_{AB} = 61,5 mm

$$P(R) = \frac{4\pi}{3} R^3 N(R) \tag{1}$$

where, $N(R)$ is the relative number of particles of radius $R = ED/2$ in the system. This $P(R)$ is used in the calculation for a polydisperse system of solid spheres (as shown in Figures 5.7–a,b,c).

Previous studies (Barone *et al.*, 2003) have shown that the results of AFM measurements are in good agreement with measurements performed with the UV light scattering technique on the same flames, and our results confirm the same provided that::

1. The AFM is run in tapping mode to avoid moving the particles that are weakly bound to the surface of the substrate;
2. Mica substrate is preferred because of its low roughness which is necessary to distinguish very small structures from the background noise; its elevated mechanical and chemical resistance to high temperatures; and the possibility to obtain extremely clean surfaces (cleavable sample).

Our results displayed in Figures 5.2–5.4 show that we observe a slightly bimodal distribution of the particle size. As seen in Figure 5.10, the size of the nanoparticles formed in our diffusion flame is comparable to findings of previous studies performed by AFM or Small Angel X-ray spectroscopy (SAXS) (Mitchell *et al.*, 2006) as shown in Figures 5.8 and 5.9. The functions of particle size distribution are reported for different sampling conditions. We find that the soot particle size distribution measured in our experiment is somewhat narrower than in SAXS measurements.

Having had our approach using AFM validated by the measurements of carbon soot particles generated in a diffusion flame, we are now in position to start the collection of nanoparticles formed in the Cinétique de Réaction en Ecoulement Supersonique Uniforme (CRESU) reactor.

Carbon Particle Formation in a CRESU Reactor

The principal objective of this work was the characterization of the nanoparticles formed in the supersonic uniform CRESU flow from molecular precursors. Once PAH molecules are in the flow, they may condense to give rise to PAH cluster depending on their concentration and on the flow temperature. Mass spectrometry characterization of the species present in the supersonic expansion reveals a sudden nucleation of PAH molecules after a certain threshold is passed. Biennier and coworkers have determined a value of 80×10^{12} cm^{-3} for the critical density for the onset of condensation for the pyrene molecule at 60 K (Sabbah *et al.*, to be published). In our series of experiments, large quantities of pyrene were injected into the flow to make sure that large clusters were generated (shift of the size distribution towards higher sizes).

The experimental parameters of the CRESU reactor are presented in Table 5.2.

Figure 5.7: Plot of Volume Distribution vs Particle Radius

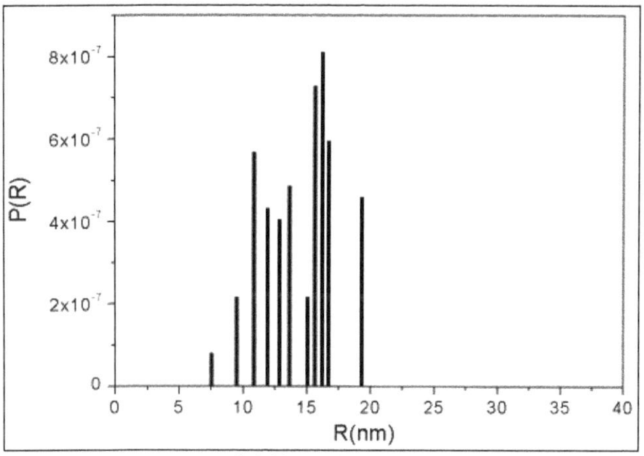

(a) H_{AB} = 22 mm

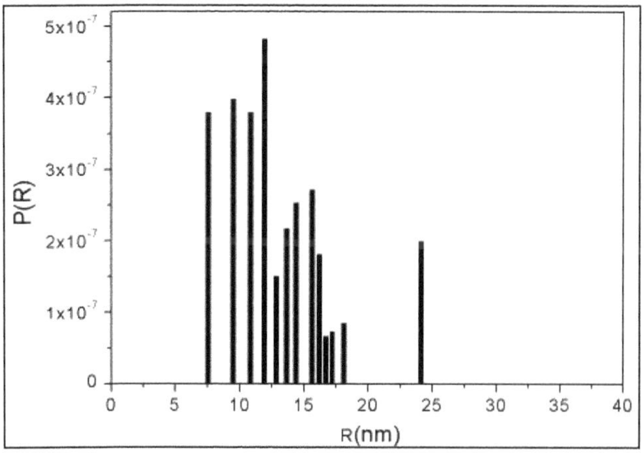

(b) H_{AB} = 35 mm

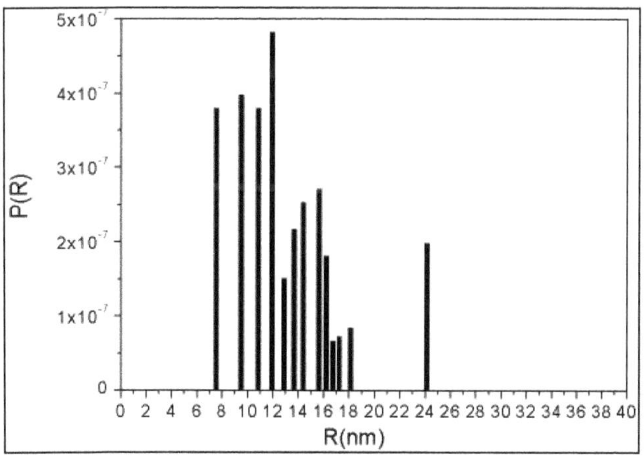

(c) H_{AB} = 61,5 mm

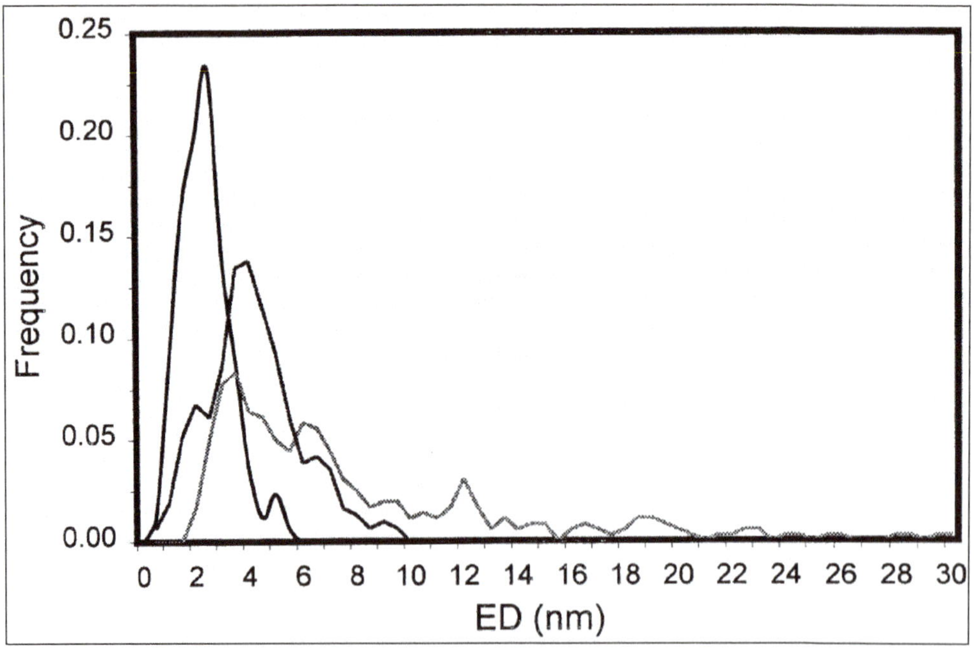

Figure 5.8: Particle Size Distribution in the C_2H_4/Air Flame, C/O ~ 0.77, for H_{AB} Equal to 4 mm (Black line), 6 mm (Dark gray line), 8 mm (Bright gray line), and Respectively. Measurements were performed by AFM.

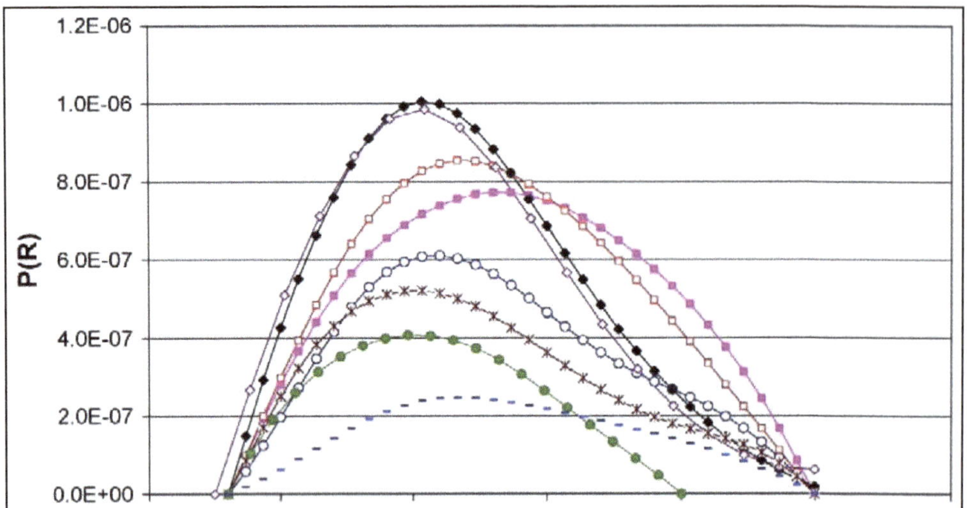

Figure 5.9: Plots of Volume Distribution vs Particle Radius for Heights Above the Burner H_{AB} of 10 mm (•); 15 mm (o); 20 mm (■); 25 mm (9); 30 mm (♦); 35 mm (◊); 40 mm (x); and 45 mm (–) determined using the GNOM package Measurements Performed by SAXS (Afer Mitchell JBA *et al.,* 2006).

Table 5.2: The General Conditions of the Experiment of Nanoparticle Formation in the Supersonic Flow and Deposition on the Substrate

Parameters	*Symbol*	*Value*	*Unit*
Temperature in Injection room	T_{res}	470	K
Pressure in Injection room	P_{res}	52	mbar
Supersonic temperature	T_{jet}	60	K
Carrier gas	He		
Chamber pressure	P_{jet}	0.317	mbar
Density of gas used to form the supersonic injection		4.01×10^{16}	molec.cm^{-3}
PAH density in the supersonic jet		1×10^{14}	molec.cm^{-3}
Hydrodynamic time		529	µs

Once the substrate was cleaved, it was fixed on the substrate holder and inserted into the reactor. We note that because the inserted plate generates a shockwave, it was expected that the surface temperature would quickly rise to a temperature close to its environment which corresponds to the reservoir temperature.

The first substrate exposed to a pure He flow served as the reference sample. The subsequent samples were collected for different exposure time (amount of particles) and PAH flow concentration (particle size distribution). After particle collection, the samples were removed from the reactor, slowly re-pressurised under argon

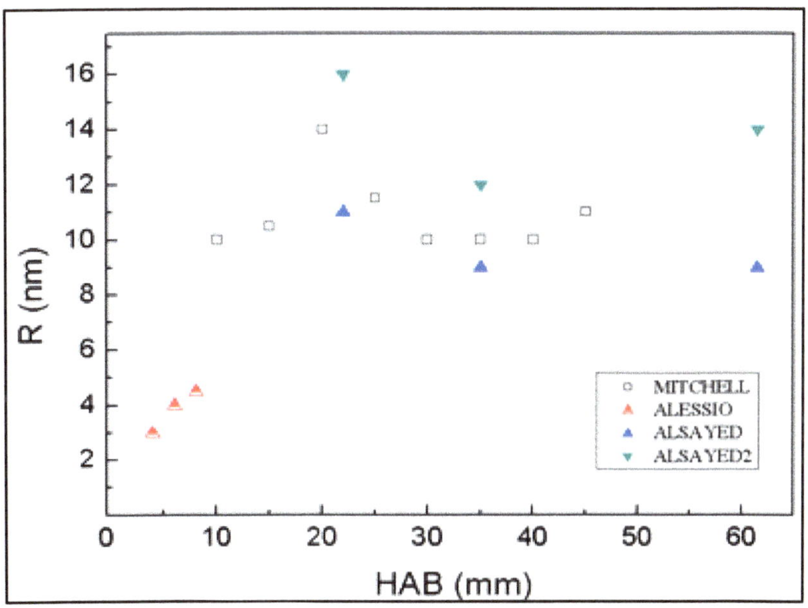

Figure 5.10: Average Size of Soot Particles Henerated in a C_2H_4/Air Diffusion Flame Analysed by AFM (•) and by SAXS (9). In our work, we found a slightly bimodal distribution.

atmosphere, transferred to plastic boxes and analysed within a few days by AFM operating at room pressure. Considering the temperature at which the samples were maintained, oxidation should play no major role.

Experimental Results

In Figures 5.11 and 5.12 are represented the first successful images of nanoparticles formed and collected in a CRESU reactor under the experimental conditions mentioned in Table 5.2.

Analysis of the Carbon Particles Generated in a CRESU Reactor

The first obvious observation that we can make is that we did not produce uniform layers. Uniform layers are expected to be generated through one by one molecule deposition. Initial surface defects are then amplified to produce steps, holes and hills. This is very different from the landscape that we image. Moreover, the continuous surface heating by the shockwave favours the evaporation of the relatively volatile pyrene monomers over the clusters shifting the size distribution towards the higher sizes. There is unfortunately no information available in the literature on the stability of the large clusters at low or high temperatures but considering the calculated binding energy for small pyrene clusters (Frenklach, M., Mitchell, J.B.A.,)only large ones will survive at the temperature of 500 K.

The second interesting observation is that we measure a rather uniform size distribution of clusters centred on 20 nm (see Figure 5.13). Even if we assume very porous particles, the average particles should contain more than a thousand of pyrene monomers. We see from the histogram of nanoparticle volume versus diameter (Figure

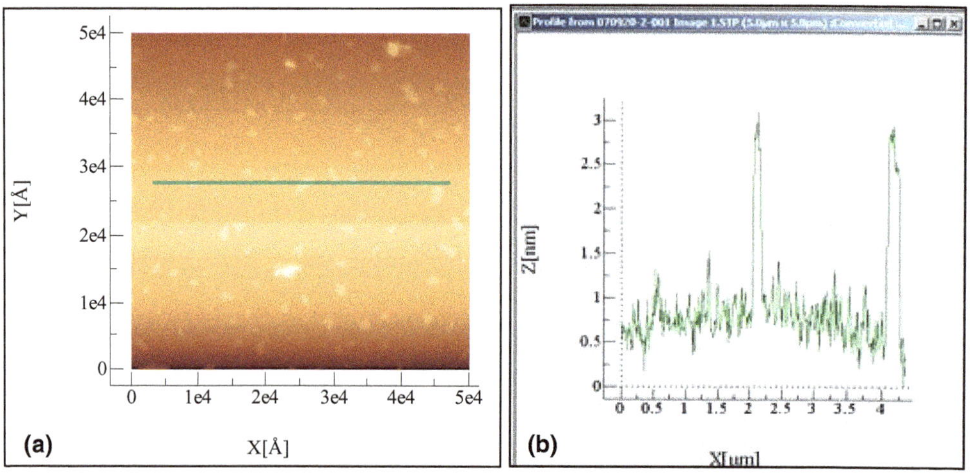

Figure 5.11: (a) AFM 2D image of the sample of reference. (There are no PAH molecules introduced in the CRESU chamber of reaction. There is some evidence of residual PAHs in the experimental chamber). (b) Cross section of the AFM 2D image (a) along the line presented on (a). The typical height of the surface defects or impurities remains below a few nanometers.

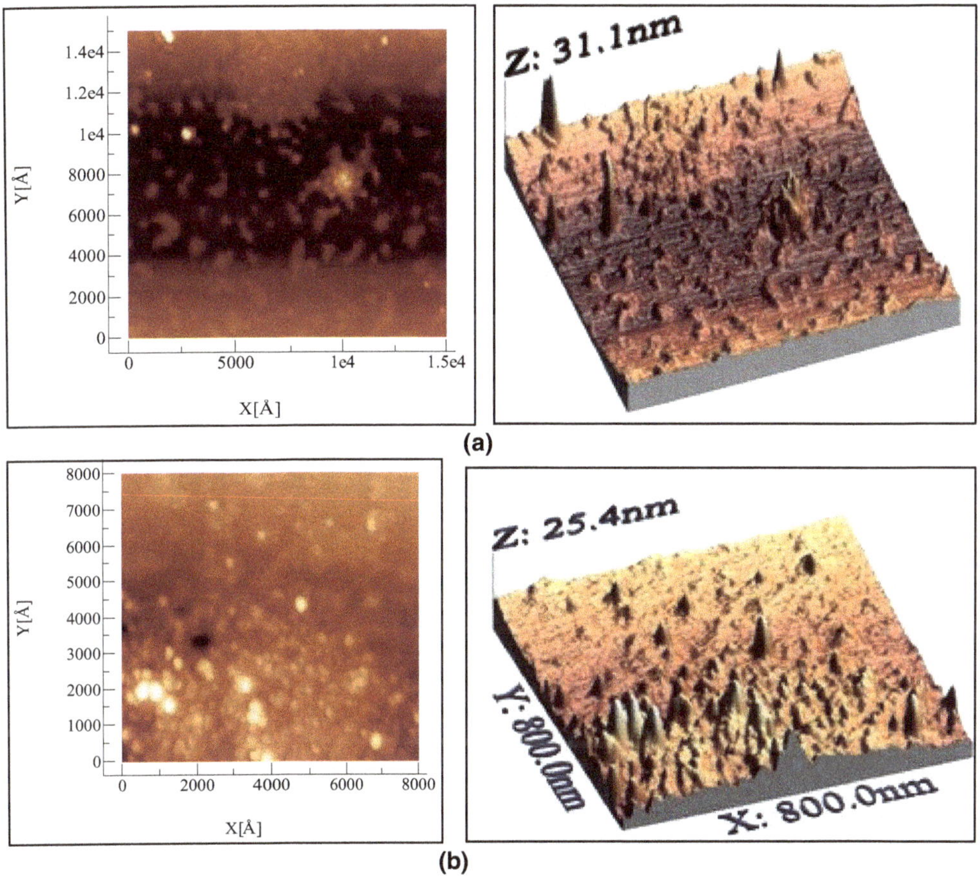

Figure 5.12: AFM 2D and 3D Images of Carbon Particles Generated in CRESU Flow Reactor in which the PAH Initial Density was 10^{14} Molecule.cm^{-3}. The substrate of mica is exposed during 15 min: (a) image dimension are of 1.5μm x 1.5 μm. (b) Image dimension are of 800 nm x 800 nm second location on the substrate.

5.13), that the nanoparticles are of the same size order and have a volume distribution comparable to those formed in flames. Earlier attempts made in contact mode AFM showed that the particles were easily moved by the tip, suggesting that the particles are weakly bound to the surface. It seems however, unlikely that these clusters are directly formed on the surface. This conclusion is supported by the fact that the structures observed are tridimensional (*i.e.*, not flattened).

The third thing we see is that some of these clusters have themselves coagulated hence creating giant structures over 100 nm in diameter but which are rather difficult to characterize in more depth. This observation constitutes a marked difference from the measurements obtained in the flame. These observations taken together support a scenario of gas-phase formation of the clusters.

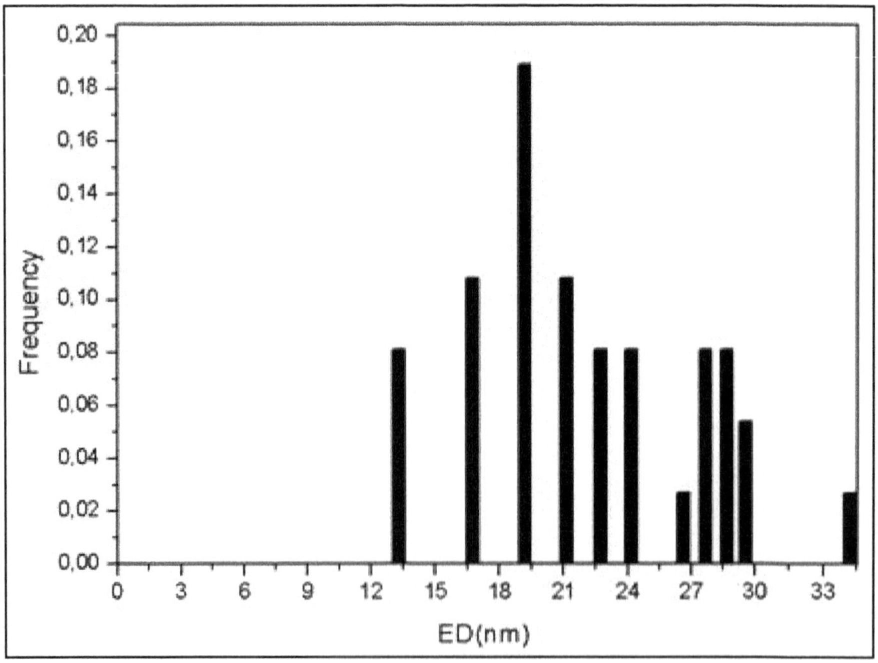

Figure 5.13: Particle Size Distribution vs. Particle Diameter (Analysis of images displayed on Figure 5.12)

Conclusion and Future Prospects

The experimental results can be summarized below:

1. We have shown that carbon nanoparticles are generated in the diffusion flame with a measured size distribution comparable to previous studies. This study confirms the validity of ex-situ characterization by AFM operated in tapping-mode.
2. Preliminary results have shown that large carbon clusters are generated in the CRESU reactor from gas-phase pyrene molecules at low temperature.
3. Considering the short reaction time and the large size achieved, carbon particles generated into the CRESU flow are most likely very porous.
4. The combination of AFM images of the particles (morphology and size distribution) with the recording of high resolution mass spectra of species generated in the reactor (mass distribution) may in future lead to the determination of the porosity of the particles. These results could contribute to determining the influence of the different parameters (temperature, kind of precursor and experimental conditions–pressure gas, and measured porosity) on the particle formation, agglomeration and deposition.

Acknowledgements

I would like to thank Dr. Biennier. L. Dr. Rowe B. and Ohana J. of the Experimental Astrochemistry Team of IPR, at University of RENNES I, in RENNES, France, for the collaboration and giving me the opportunity of doing this work with them. Many thanks also to all members of the Experimental Astrochemistry Team, many appreciations to the Higher Institute for Applied Sciences and Technology for the financial support.

References

Barone, A.C., D'Alessio., A. and D'Anna., A., 2003. *Combustion and Flame,* 132: 181–187.

Frenklach, M., 2002. *Phys. Chem. Chem. Phys.,* 4: 2028–2037.

Mitchell, J.B.A., Courbe, J, Florescu-Mitchell AI, di Stasio, S. and Weiss, T.J., 2006. *Appl. Phys.,* 100: 12.

Mitchell, J.B.A., Le Garrec, Florescu-Mitchell and di Stasio, S.J., 2006. *Combustion and Flame,* 145: 80–87.

Reilly, P.T.A., 2000. *Combustion and Flame,* 122: 90–104.

Sabbah, H., Rowe, B., Sims, I. and Biennier, L. (*In Press*).

Siegmann, K., 2002. *Journal of Electron Spectroscopy and Related Phenomena,* 126: 191–202.

Chapter 6

Growth and Characterization of PbS Nanocrystalline Thin Films Deposited on Glass Substrates by Chemical Bath Deposition

Iyad Ali, Nofal ibrahim*, Louay Mansour and Abdel Karim Sharba

Higher Institute of Applied Science and Technology (HIAST),
P.O.BOX 31983, Damascus, Syria
**E-mail: nofal.ibrahim2006@gmail.com*

ABSTRACT

In this paper we present the preparation of PbS nanocrystalline thin films using Chemical Bath Deposition (CBD) technique. We have performed this work in order to study the photoconductivity of PbS semi-conductor thin films. The details of the preparation method are described. Thickness of deposited films has been determined using mechanical and optical methods. From the optical absorption measurements we have determined the band gap values. Using the first approximation parabolic bands model and the values of band gaps, we have determined the size of PbS nanocrystallites. Also, we have investigated the electrical and photoelectrical behaviors of the PbS films. Our study shows that the size of PbS thin films nanocrystallites affects the photoconductive properties of the material. Furthermore, investigations show that there are two different sizes of grains located in two different layers, the first one, with grain's size of about 20 nm, concerns the part of PbS deposited directly on the glass substrate and the second layer, with grain's size of about 80 nm, concerns the PbS deposited on the first layer.

***Keywords**: PbS, Nanocrystalline thin Films, Chemical bath Deposition, Photoconductivity.*

Introduction

Photoconductive lead sulfide (PbS) was one of the earliest and most successful semiconductor-based photoconductors. Even today it is one of the most sensitive uncooled detectors in the 1 to 3 μm spectral region. It has high quantum efficiency, fast time response, and can operate at room temperature (Blount, 1973; Espevik, 1971). The spectral response can be extended into infrared up to 5μm by cooling the detector due to the positive temperature coefficient of the energy gap (Johnson, 1984). PbS nanocrystaline thin films were deeply studied in the past and a variety of its physical properties have been reported in the literature (Steigerwald, 1989; Wang, 1989; Chakravorty, 1992; Qadri, 1996; Thielsch, 1998). Nevertheless, it still has now a great experimental interest. Investigations confirm its ability to become a versatile technological platform for the creation of better optoelectronic devices. They show a strong quantum confinement effect when the nanocrystallite size becomes equal to the dimension of Bohr exciton (Qadri, 1996; Thielsch, 1998).

Various techniques are available for the deposition of polycrystalline thin films of PbS. Chemical Bath Deposition (CBD) is an inexpensive method and does not require sophisticated vacuum equipment (Pintilie, 1995; Davis, 1966; Nair, 1989). Many PbS characteristics can be tailored over a wide range by adjusting the chemistry of the deposition such as the reaction time, temperature, pH, heat treatment and the film thickness. These characteristics include detectivity, responsivity, resistance and spectral response range (Moss, 1973; Träger, 2007; Johnson, 1984).

In the present study, we report the optical, electrical, and photoconductive properties of PbS nanocrystaline thin films grown on glass substrates using chemical bath deposition method. The remainder of this paper is organized as follows. Experimental details are given in Section with particular emphasis on the compounds involved in the reactions and the deposition parameters. In next Section, we present the optical and photoconductive properties of the grown films. In particular, the presence of two layers with two different sizes of crystallites was demonstrated by optical and electrical measurements. Finally, the parameters from our measurements are compared with the literature and conclusions that can be drawn on the basis of the experimental data and calculations are mentioned.

Experimental Details

Samples Preparation

The PbS thin film deposition was performed on glass substrates by CBD method. Before the deposition, the substrates were very carefully washed using oxidant mixing ($K_2Cr_2O_7$:H_2SO_4; 1:10), HNO_3, and were completely rinsed with distilled water. The cleaning of the substrate surface is extremely important for the quality of the thin film that will be obtained later. The substrates were mounted vertically and immersed into the chemical bath.

We have performed 4 different sets of deposition. The chemical bath contained an aqueous solution of the following constituents mixed in the sequence: $Pb(NO_3)_2$ 0.011M, NaOH 0.109M, $CS(NH_2)_2$ 0.044 and H_2O. The reaction process is considered as follows:

$$Pb(NO_3)_2 + 2NaOH \rightarrow Pb(OH)_2 + 2NaNO_3 \quad \text{(a)}$$

$$Pb(OH)_2 + 4NaOH \rightarrow Na_4Pb(OH)_6 \quad \text{(b)}$$

$$Na_4Pb(OH)_6 \rightarrow 4\,Na^+ + HPbO_2^- + 3OH^- + H_2O \quad \text{(c)}$$

$$CS(NH_2)_2 + OH^- \rightarrow CH_2N_2 + H_2O + SH^- \quad \text{(d)}$$

$$HPbO_2^- + SH^- \rightarrow PbS + 2OH^- \quad \text{(e)}$$

The precipitation took place in the presence of a reducing agent (hydroxylamine hydrochloride, $NH_2OH.HCl$ 0.018M) and of Bi ions ($Bi(NO_3)_2$). The reducing agent and the Bi salt are introduced in the bath in order to control the nucleation/growth process during the deposition of the PbS layer (Pentia, 2001b; Nascu, 1995). It is known that the presence of small amount of Bi ions lead to enhance the grain size (Pentia, 2001a). The composition of the chemical bath and the solution temperature for each set are given in Table 6.1. We have varied the reaction parameters in sets #3 and #4 in order to study the effect of Bi ions concentration and the temperature on the grain size.

Table 6.1: Composition of the Chemical Bath and the Solution Temperature for Each Set

Sample	*Compounds Involved in the Reaction*	*Solution Temperature (°C)*
#1	Essential compounds: $Pb(NO_3)_2$, $CS(NH_2)_2$, NaOH, H_2O	30
#2	Essential compounds + ($NH_2OH.HCl$ 0.018M, $Bi(NO_3)_2$ 5.27×10^{-5} M)	30
#3	Essential compounds + ($NH_2OH.HCl$ 0.018M, $Bi(NO_3)_2$ 1.05×10^{-4} M)	30
#4	Idem #3	24

In order to obtain a reasonable thickness, the reaction was repeated four times for each sample, the deposition duration in each time is 20 minutes. The starting pH of the reaction (pH 13) was kept constant for all depositions. After the reaction, the samples were annealed at 100 °C for 7 hours to improve the sensitivity of the films and stabilize the parameter.

Characterization Methods

The thickness of PbS films was obtained by two means. The first one is the alpha-step apparatus which can measure a thickness-step up to 160 μm with a resolution of 200 A°. The second one is to measure the optical thickness, which is the geometric thickness multiplied by the refractive index, using interferometric method based on the spectral reflectivity curve recorded with two kinds of spectrometers: a Perkin Elmer grating spectrometer for the spectral range 400-1100 nm and the SYSTEM2000 FTIR Perkin Elmer spectrometer for the spectral range 1.5-20 μm. Measurement of the optical thickness provide also the refractive index of the deposited layer. The absorption coefficient for the thin films is calculated based on the experimentally recorded reflectivity $R(\lambda)$ and transmittance $T(\lambda)$ using the Beer-Lambert law (Linne, 2002):

$$T(\lambda) = (1 - R(\lambda)) \cdot \exp(-\alpha(\lambda) \cdot d) \tag{1}$$

$$\alpha(\lambda) = \frac{1}{d} \cdot \ln\left[\frac{1 - R(\lambda)}{T(\lambda)}\right] \tag{2}$$

where, α is the absorption coefficient (in cm^{-1}), d is the thickness of the layer (in cm), R and T are respectively the reflectivity and transmittance of the PbS thin films. Figures 6.1–6.3 display typical examples of R(λ), T(λ), and $\alpha(\lambda)$ respectively.

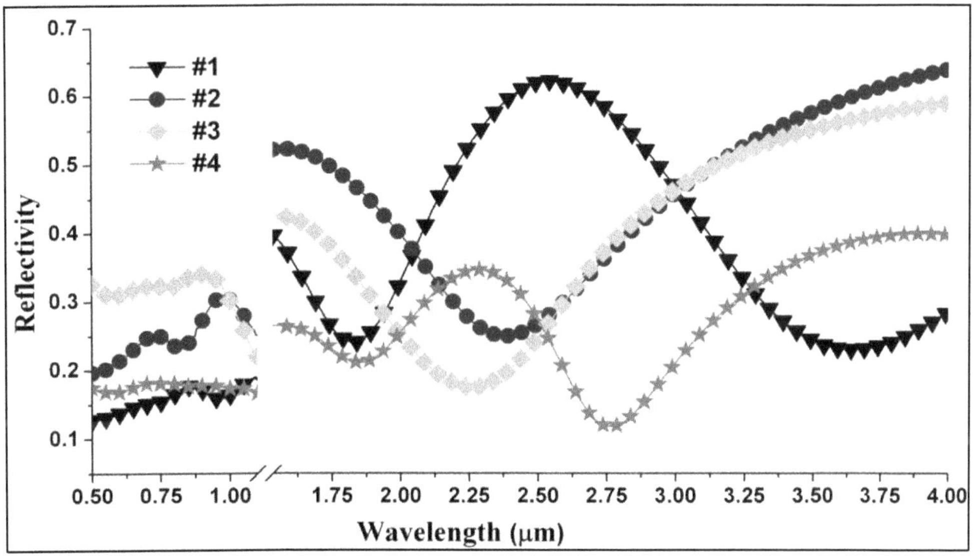

Figure 6.1: The Spectral Reflectivity of the PbS Thin Films

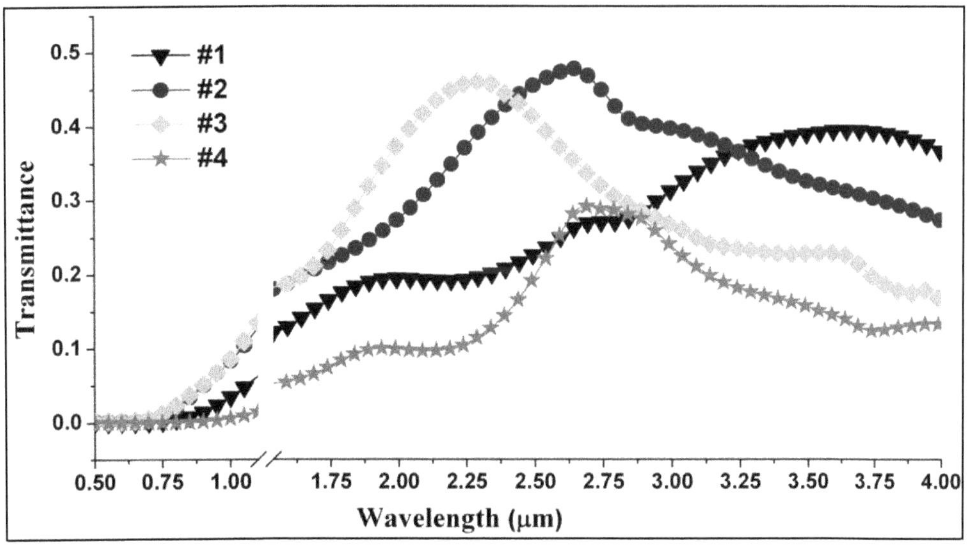

Figure 6.2: The Spectral Transmittance of the PbS Thin Films

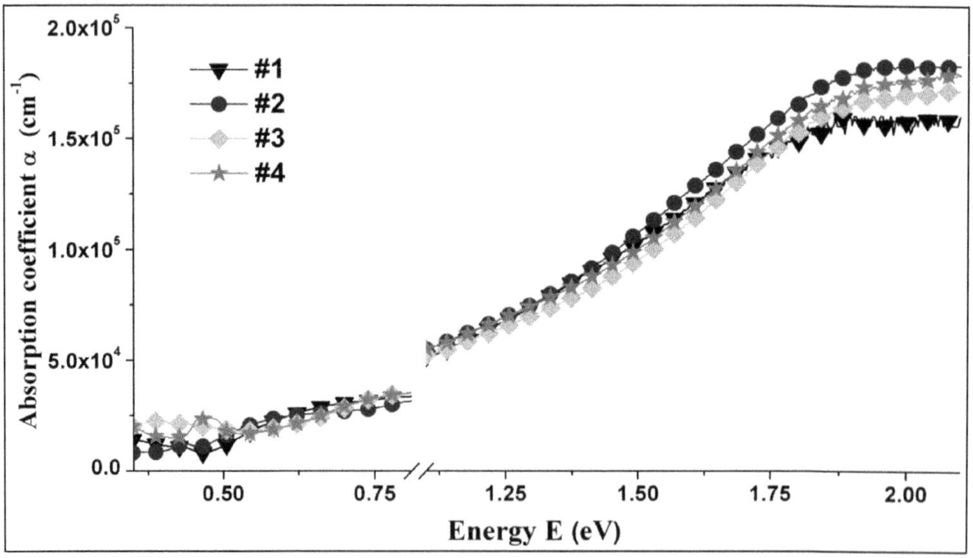

Figure 6.3: The Absorption Coefficient of the PbS Thin Films

The $(\alpha h\nu)^2$ versus $h\nu$ show linear behavior indicating presence of direct optical band gap in the deposited nanocrystaline thin films [Joshi, 2004]. Figure 6.4 shows typical examples of the $(\alpha h\nu)^2 = f(h\nu)$ curve.

For the electrical and photoelectrical measurements gold electrodes (E, F) in a coplanar configuration were evaporated in vacuum on the surface of PbS. The distance between electrodes was 5 mm. For current–voltage (I–V) characteristics and

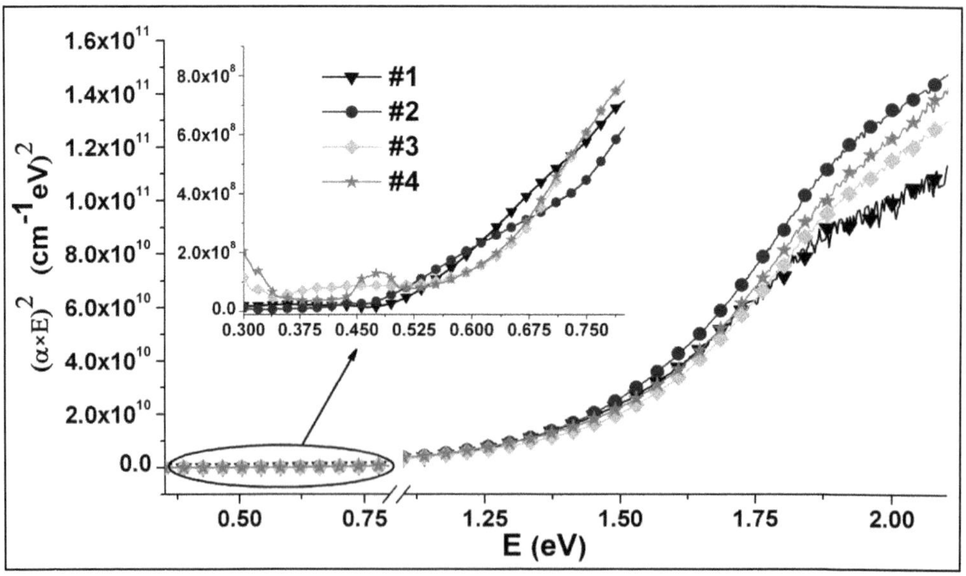

Figure 6.4: $(\alpha\times E)^2 = f(E)$ Spectra Used for Absorption Threshold Determination

electrometer (model PC500) and a digital oscilloscope (Tektronix, model TPS2014) were used. The photoconductive measurements were performed in modulated light, using a standard setup composed by a lock-in amplifier (STANFORD Research Systems, model SR830), a dc voltage source (model GPC-3030D), a black body as an IR source (GRASEBY IR Systems, model 830), and a mechanical chopper (InfraRed Industries Inc. model 830). A load resistance, with a value close to the dark resistance of the PbS film, usually a few hundreds kΩ, is series connected with PbS and is used to collect the photoconductive signal.

Results and Discussion

Optical and Geometrical Thickness

The average values of thickness for each set are summarized in Table 6.2. Note that those values are around 500 nm. Taking into account the preparation parameters, our values are compatible with previous studies in the literature (Pentia, 2001a; Mahmoud, 2001). Comparison between values in Table 6.2 and optical thickness given by the interferometric method enables us to find an average value over the studied spectral range for the refractive index of PbS films:

$$\overline{n} = \frac{\text{Optical thickness}}{\text{geometric thickness}} = 3.9 \pm 0.3$$

This index is somewhat lower than that of the bulk PbS (n = 4.1). This result is expected, the prepared thin films have a lower density than the bulk PbS due to the lower stoichiometry and poorer crystallinity.

Table 6.2: Thickness, Band Gap, Cutoff Wavelength, and Crystallite Size of PbS Films

Sample	*Thickness (nm)*	*Eg_1 (eV)*	*$\lambda_{cut\text{-}off1}$ (μm)*	*D_1 (nm)*	*Eg_2 (eV)*	*$\lambda_{cut\text{-}off2}$ (μm)*	*D_2 (nm)*
#1	478	1.39	0.89	24	0.51	2.43	64
#2	310	1.46	0.85	24	0.46	2.7	82
#3	325	1.48	0.84	23	0.6	2.07	50
#4	650	1.21	1.02	27	0.46	2.7	82

Thickness (nm): Total thickness of all layers in the concerned sample.

E_{g1} (eV), E_{g2} (eV): Optical band gap energy of the first and second layer respectively

$\lambda_{cut\text{-}off1}$ (μm), $\lambda_{cut\text{-}off2}$ (μm): The cutoff wavelength due to the first and second layer respectively

D_1, D_2: Crystallite diameter in the first and second layer respectively

Band-gap and Nanocrystallite Size Measurements

In order to determine the optical band gap energy and in the assumption that this gap is a direct one, the transitions can be characterized by the Tauc relation (Tauc, 1972):

$$\alpha.E = C.(E - E_g)^{1/2} \quad (3)$$

where, α is the absorption coefficient and C is a constant related to the details of conduction and valence bands. Based on this expression, the linear adjustment of $(\alpha h\nu)^2$ versus $h\nu$ give us the value of E_g. In fact, the close examination of the precedent curve shows that there are two absorption thresholds for each sample, the first one is in the visible and near infrared (at about 0.9 µm) and the second one is in the mid-infrared (at about 2.5 µm). This can be clearly seen with the linear adjustment for each region in Figure 6.5. The values of correspondent band gaps (E_{g1} and E_{g2}) are given in Table 6.2.

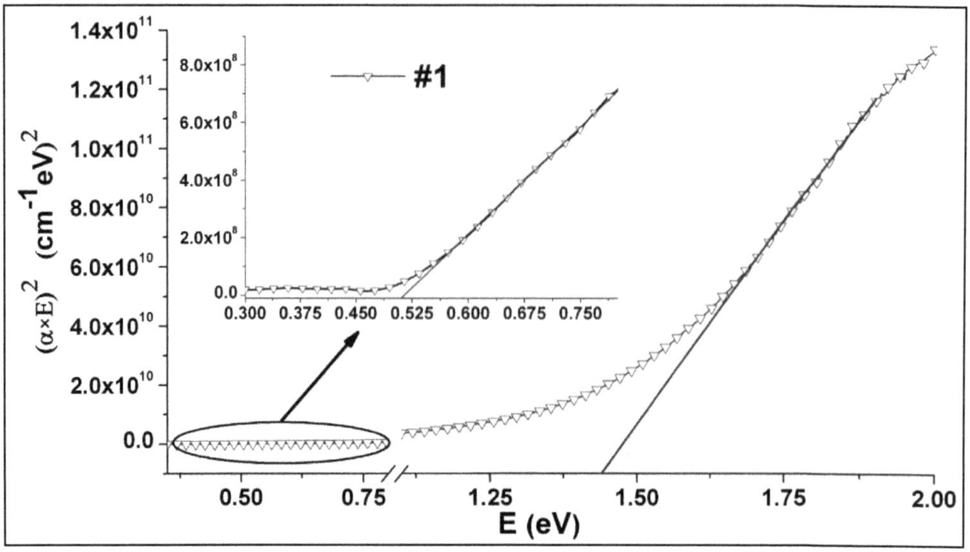

Figure 6.5: Example of the $(\alpha\times E)^2 = f(E)$ Spectra. The two optical band gaps with their linear adjustment are shown.

The optical band gap of bulk PbS is 0.41 eV. Using the first approximation parabolic bands model, and for a nanocrystalline radius R, the increase in band gap of the nanocrystalline thin film compared to bulk PbS is given by (Popa, 2006):

$$\Delta E_g = \frac{\hbar^2.\pi^2}{2.R^2}.\left(\frac{1}{m_e^*}+\frac{1}{m_h^*}\right) - \frac{1.786\times e^2}{\varepsilon.R} \quad (4)$$

$$\Delta E_g = 2.77\cdot10^{-35}\cdot X^2 - 3.47\cdot10^{-28}\cdot X \quad (X = \frac{1}{R},\ n = 3.9,\ \Delta E_g\ in\ Joule,\ R\ in\ m)$$

where, m_e^*, m_h^* are the effective masses of the electron in the conduction band and of the hole in the valence band respectively, and ε is the dielectric constant of PbS. Equation (4) includes the basic physical principles underlying the increase in band gap when the nanocrystalline size decreases but is quantitatively correct only for dimensions greater than 10 nm, which is the case for the thin films prepared in this study.

As it could be seen in Table 6.2, the two different absorption thresholds are correlated with the size of nanocrystallites in the thin film. The absorption threshold in the near infrared region (1.2-1.5 eV, $\lambda_{cut\text{-}off} \cong 0.9$ µm) corresponds to nanocrystallites of small size (at about 25 nm). The other one in the mid infrared (0.45-0.6 eV, $\lambda_{cut\text{-}off} \cong$ 2.5 µm) is due to nanocrystallites of larger size (at about 70 nm). Note that this observation is compatible with the absorption threshold of bulk PbS. This latter is composed of grains with size greater than 160 nm and it has a band gap of 0.41 eV ($\lambda_{cut\text{-}off} = 3$ µm). This confirms that, for all this cases, the band gap increases with deceasing grain size.

Electrical and Photoelectrical Properties

A typical example of the (*I–V*) curve is shown in Figure 6.6. As it could be clearly seen, there are two different slopes in this curve. This means that we have two different regions or layers in the PbS thin film. For relatively low voltages the current passes through the upper layer (resistance R_2 only) and for higher voltages the current passes through both layers (resistance R_2 in parallel with resistance R_1).

To confirm the hypothesis of presence of two different layers, we have deposited PbS on a glass substrate for 20 minutes and we have obtained a thin film 55 nm thick. Investigations show that there is only one optical band gap of 1.8 eV corresponding to a cutoff wavelength is in the visible spectral range (0.7 µm) and grain diameter of

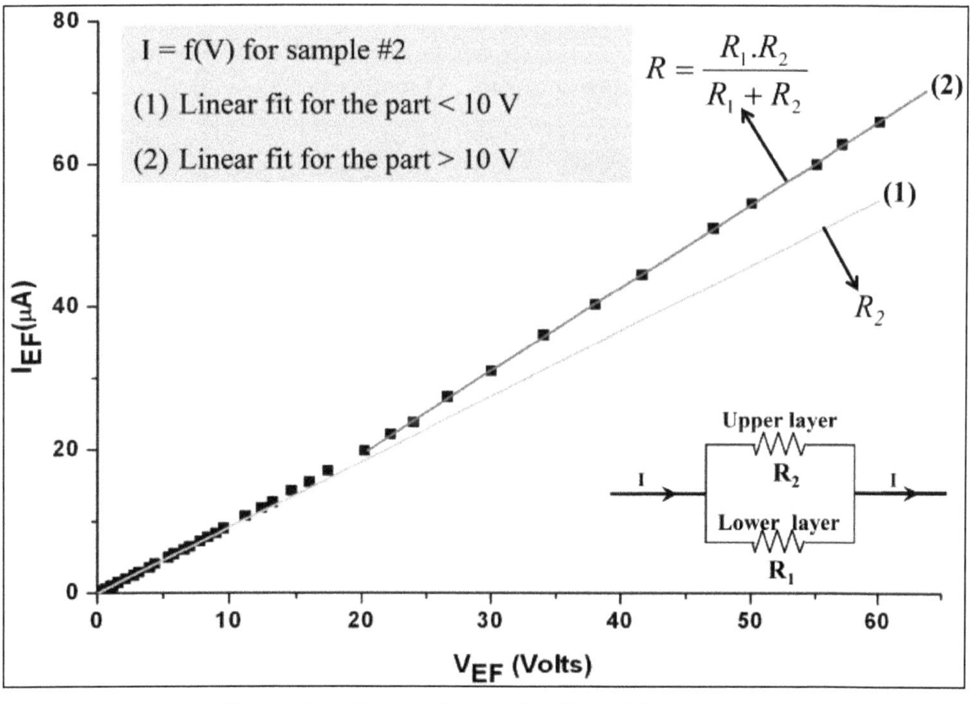

Figure 6.6: Typical Example of the (I-V) Curves

21 nm. Using the same sample, two new depositions each of 20 minutes were done to obtain a new thin film of total thickness 350 nm.

Characterization shows that there are two band gaps, the first one is identical to the precedent case and the second one, resulting from the new deposition (PbS on PbS), has a band gap of 0.49 eV ($\lambda_{cut\text{-}off}$ = 2.5 µm) with grain diameter of 70 nm. This experiment confirms that the prepared thin films are composed of two layers (PbS-on-glass and PbS-on-PbS) with different grain size. A scheme of the prepared samples can be found in Figure 6.7.

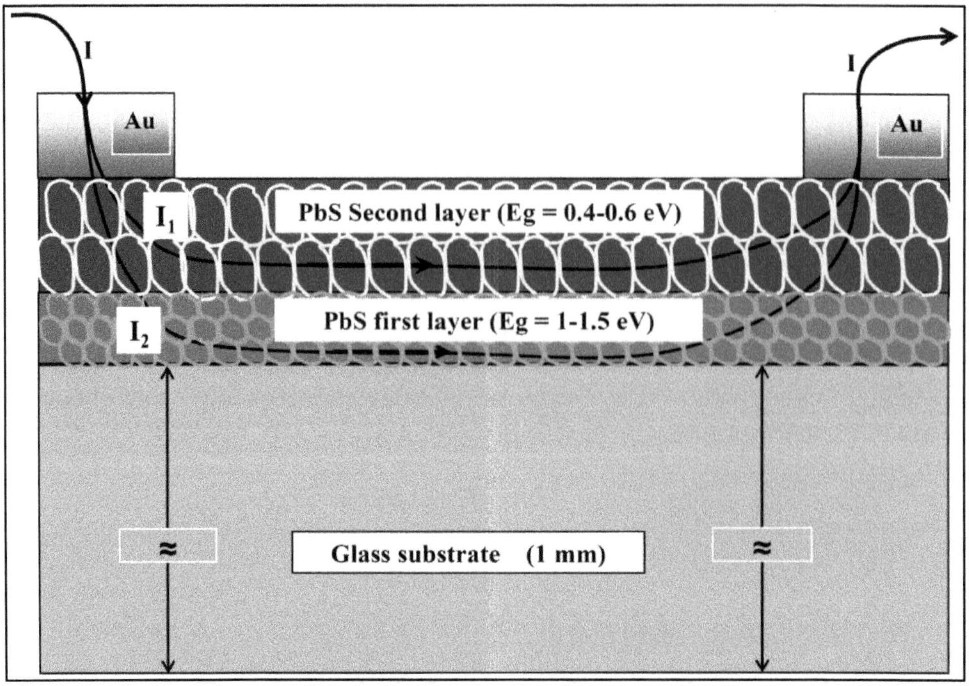

Figure 6.7: Scheme of the Two Layers of PbS Thin Film on Glass Substrate

The results of the photoconductive measurements are given in Table 6.3. The time constant was calculated using the relation:

$$\tau = \frac{1}{2\pi \cdot F_C} = \frac{1}{2\pi \cdot \Delta \nu} \qquad (5)$$

where, F_C is the cutoff frequency corresponding to a decrease of 3 dB in the photoconductive signal, and the specific detectivity was calculated using the expression (Fraden, 2004):

$$D^* = \frac{1}{NEP} \cdot \sqrt{A_d} \cdot \sqrt{\Delta \nu} \qquad (6)$$

where, A_d is the detector surface, Δv is the bandwidth, and NEP (Noise-Equivalent Power) is the light level required to obtain a signal-to-noise ratio equal to unity.

Table 6.3: Specific Resistance, Time Constant and Specific Detectivity of PbS Films

Sample	*Specific Resistance* ρ *(Ω cm)*	*Time Constant* τ *(μsec)*	*Specific Detectivity* D^* *(cm Hz$^{1/2}$ W^{-1})*
#1	33	—-	Non
#2	31	44	1.00×10^7 (Good)
#3	27	47	0.13×10^7 (Not good)
#4	11	42	1.46×10^7 (Good)

Parameters found in this study (ρ, τ, D^*) are compatible with those reported in catalogues and in the literature (Mahmoud, 2001; Bass, 1994; Hamamatsu, 2004). The specific detectivity is somewhat lower than the reported values and can be improved by coupling the sample with an adequate preamplifier.

Note that the first set of samples does not exhibit any photoconductivity; this is due to the absence of the reducing agent and the Bi salt in the bath. When the amount of Bi salt was doubled (set #3) the quality of the deposited film was degraded resulting in a worth specific detectivity compared with previous set (#2). We reduced the reaction temperature from 30 to 24 °C while keeping the same proportion between the reducing agent and Bi, the specific detectivity improved by a factor of 10 (set #4). This confirms that optimum selection of the reaction parameters (reducing agent, Bi ions as doping element, and temperature) is of crucial importance for obtaining films with a good photoconductivity that can be used in infrared detection. We know now that a reduced reaction temperature close to room temperature with good proportion between the reducing agent and the Bi salt (close to 300:1) enable us to obtain semi-conductor thin films of PbS with good photoconductivity.

Conclusions

In this article we have reported the preparation and characterization of PbS nanocrystalline thin films using Chemical Bath Deposition (CBD) technique. Optical band gap was measured and nanocrystallites size was calculated. The prepared thin films show good photoconductive properties.

Electrical and photoelectrical measurements indicate that the prepared thin films are composed of two layers with different grain size. Our results suggest that the shift of the maximum of photoconductivity to shorter wavelengths and the larger value of band gap can be ascribed to nanocrystallite size effect. We have attributed the band gap found in the near infrared (at about 0.9 μm) to the presence of nanocrystallites of 20 nm in size in the lower layer (PbS-on-glass) and that in the mid infrared (at about 2.5 μm) to the presence of nanocrystallites of 70 nm in size in the upper layer (PbS-on-PbS). The possibility to vary the band gap of PbS in the range 0.5–2.7 eV is an important property, which opens the door for the creation of new classes of versatile optoelectronic devices.

Acknowledgements

The authors wish to thank J. Saeed, Kh. Maya, and A. Farra for their help and for fruitful discussions.

References

Bass, M., Van Stryland, E.W., Williams, D.R. and Wolfe, W.L., 1994. *Handbook of Optics: Fundamentals, Techniques and Design*, 2nd Edition. McGraw-Hill Professional.

Blount, G.H., Schreiber, P.J., Smith, D.K. and Yamada, R.T., 1973. *J. Appl. Phys.*, 44: 978–981.

Chakravorty, D. and Giri, A.K., 1992. *Chemistry of Advanced Materials*, (Ed.) C.N.R. Rao. Blackwell Sci. Publications, Oxford, pp. 217.

Davis, L. and Norr, M.K., 1966. *J. Appl. Phys.*, 37(4): 1670.

Espevik, S., Wu, C. and Bube, R.H., 1971. *J. Appl. Phys.*, 42: 3513–3529.

Fraden, J., 2004. *Handbook of Modern Sensors: Physics, Designs, and Applications*. Springer-Verlag.

Hamamatsu Photonics, 2004. *Characteristics and Use of Infrared Detectors*. Technical Information SD–12.

Johnson, T.H., 1984. Lead salt detectors and arrays: PbS and PbSe. *Proc. SPIE*, 443: 60–94.

Joshi, R.K., Kanjilal, A. and Sehgal, H.K., 2004. *Applied Surface Science*, 221: 43–47.

Linne, M.A., 2002. *Spectroscopic–Measurement: An Introduction to the Fundamentals*. Academic Press.

Mahmoud, S. and Hamid, O., 2001. *FIZIKA A (Zagreb)*, 10: 21–30.

Moss, T.S., Burrell, G.J. and Elis, B., 1973. *Semiconductor Optoelectronics*. Butterworth, London.

Nair, P.K. and Nair, M.T.S., 1989. *Semicond. Sci. Technol.*, 4: 807.

Nascu, C., Vomir, V., Ionescu, V. and Pop, I., 1995. *Rev. Roum. Chem.*, 40: 127.

Pentia, E., Pintilie, L., Matei, I., Botila, T. and Ozbay, E., 2001. *Journal of Optoelectronics and Advanced Materials*, 3(2): 525–530.

Pentia, E., Pintilie, L., Tivarus, C., Pintilie, I. and Botila, T., 2001. *Mater. Sci. Eng. B*, 80: 23.

Pintilie, I., Pentia, E., Pintilie, L., Petre, D., Constantin, C. and Botila, T., 1995. *J. Appl. Phys.*, 78(3): 1713.

Popa, A., Lisca, M., Stancu, V., Buda, M., Pentia, E. and Botila, T., 2006. *Journal of Optoelectronics and Advanced Materials*, 8(1): 43–45.

Qadri, S.B., Yang, J., Ratna, B.R., Skelton, F.E. and Hu, J.H., 1996. *Appl. Phys. Lett.*, 69: 2205.

Steigerwald, M.L. and Brus, L.E., 1989. *Annu. Rev. Mater. Sci.*, 19: 471–495.

Tauc, J. and Menth, A., 1972. *J. of Non-Crystalline Solids*, 8–11: 569.

Thielsch, R., Bohme, T., Reiche, R., Schlafer, D., Bauer, H.D. and Bottcher, H., 1998. *Nanostruct. Mater.*, 10: 131.

Träger, F. (Editor), 2007. *Springer Handbook of Lasers and Optics*. Springer, New York.

Wang, Y., Harron, N., Mahler, W. and Suna, A., 1989. *J. Opt. Soc. Am. B*, 6: 808.

– *Section II* –

APPLICATIONS OF NANOTECHNOLOGY

Chapter 7

Global Scenario of Nanotechnology

Y.K. *Agrawal*

Institute of Research and Development, Gujarat Forensic Sciences University, Sector 18A, Near Police Bhavan, Gandhinagar – 382 007, Gujarat, India

ABSTRACT

Nanotechnology is multidisciplinary science which encompasses the creation and utilization of materials, devices and systems at the level of atoms and molecules. It concerns with very small particles which deal with the fundamental aspects of understanding the properties of nanoparticles and developing technological applications. Nanoproducts aim at creating new types of materials, devices and systems at the nanoscale for wide variety of applications from electronics to sensors to biomedical research. This article describes global scenario in the recent developments of revolutionary nanotechnology.

Introduction

Over the past few years, a little word with big potential has been rapidly insinuating itself into the world's consciousness. That word is "nano." The prefix nano is derived from the Greek word 'nanos' meaning dwarf. The concept of nanotechnology is attributed to Nobel laureate Richard Feynman in a lecture that he gave in 1959 and which was published in 1960 'There's Plenty of Room at the Bottom'.

In 1999 report of the Inter Agency Working Group on Nanoscience, Engineering and Technology formally defined nanotechnology as: concerned with materials and systems whose structure and components exhibit novel and significantly improved physical, chemical, and biological properties, phenomena, and processes because of their small nanosize.

Nanotechnology is likely to have a profound impact on our economy and society in the early twenty-first century, comparable to that of semiconductor technology, information technology, or cellular and molecular biology. Science and technology research in nanotechnology promises breakthroughs in such areas as materials and manufacturing, nanoelectronics, medicine and healthcare, energy, biotechnology, information technology, and national security. It is widely felt that nanotechnology will be the next industrial revolution. This field is expected to open new venues in science and technology.

Dozens of major universities across the world from Northwestern University in the United States to Delft University of Technology in the Netherlands and the National Nanoscience Center in Beijing, China, India are building new faculties, facilities, and research groups for nano. Nano research also crosses scientific disciplines. Chemists, biologists, doctors, physicists, engineers, and computer scientists are all intimately involved in nano development.

Nano is a big business. The National Science Foundation predicts that nano-related goods and services could be a $1 trillion market by 2015, making it not only one of the fastest-growing industries in history but also larger than the combined telecommunications and information technology industries at the beginning of the technology boom in 1998. Nano is already a priority for technology companies like HP, NEC, and IBM, all of whom have developed massive research capabilities for studying and developing nano devices.

In the media, nano has captured headlines at CNN, CNBC, and almost every online technical, scientific, and medical journal. The Nobel Prize has been awarded several times for nano research, and the Feynman Prize was created to recognize the accomplishments of nanoscientists. *Science* magazine named a nano development as Breakthrough of the Year in 2001, and nano made the cover of *Forbes* the same year, subtitled "The Next Big Idea." Nano has hit the pages of such futurist publications as *Wired Magazine*, found its way into science fiction, and been the theme of episodes of *Star Trek: The Next Generation* and *The X-Files* as well as a one-liner in the movie *Spiderman*.

In the midst of all this buzz and activity, nano has moved from the world of the future to the world of the present. Innovations in nano related fields have already sparked a plenty of commercial inventions from faster burning rocket fuel additives to new cancer treatments and remarkably accurate and simple to use sensors for glucose. Nano skin creams and suntan lotions are already on the market, and nano-enhanced tennis balls that bounce longer appeared at the 2002 Davis Cup. To date, most companies that claim to be nano companies are engaging in research or trying to cash in on hype rather than working toward delivering a true nano product, but there certainly are exceptions. There is no shortage of opinions on where nano can go and what it can mean, but both pundits and critics agree on one point–no matter who you are and what your business and interests may be, this science and its spin-off technologies have the potential to affect you greatly.

Approaches

Synthesis of nanomaterials is commonly thought of in terms of "top-down" or "bottom-up" processes. Top down approaches begin with larger starting materials and, in a more or less controlled fashion (depending on the technique), remove material until the desired structure is achieved. Most microfabrication techniques for inorganic materials (lithography and milling techniques, etc.) fit this description. In contrast, bottom-up approaches begin with smaller subunits that are assembled, again with varying levels of control, depending on technique, into the final product.

Now imagine taking one of the eight 40-centimeter cubes and cutting it the same way. Each of the eight resulting cubes will now 20 centimeters on a side and will have the same properties as the parent cube. If the gold is continue cutted in this way and proceed down in size from feet to inches, from inches to centimeters, from centimeters to millimeters, and from millimeters to microns, we will still notice no change in the properties of the gold. Each time, the gold cubes will get smaller. Eventually we will not be able to see them with the naked eye and we'll start to need some designed tools to keep cutting. Still, all the gold bricks' physical and chemical properties will be unchanged. This much is obvious from our real world experience at the macroscale chemical and physical properties of materials are not size dependent. It doesn't matter whether the cubes are gold, iron, lead, plastic, ice, or brass (Figure 7.1).

When we reach the nanoscale, though, everything will change, including the gold's color, melting point, and chemical properties. The reason for this change has to do with the nature of the interactions among the atoms that make up the gold, interactions that are averaged out of existence in the bulk material. Nano gold doesn't act like bulk gold.

Different Nanomaterials

Nanofibers, nanorods, dendrimers, nanospheres, nanobubbles, nano diamonds, polymeric nanoparticles, lipid nanoparticles, peptide nanotube, polymer nanotube, carbon nanotube, nanoclays, nanowires, nanopowders, nanosuspension, nanoemulsions, metal nanoparticles, halloysite nanotube, magnetic nanoparticles, inorganic nanoparticles, fullerenes, nanogel.

Research and Achievement in Nanotechnology

Nanobomb

Single-wall carbon nanotubes (SWCNT) as potent therapeutic nanobomb agents for killing breast cancer cells. Researchers showed that by adsorbing water molecules in SWCNT sheets or loosely adsorbed on top of cells, potent nanobombs were created that heated the water molecules inside them to more than 100°C upon exposure to laser light of 800 nm at light intensities of approx 50–200 mW/cm2. Conversion of optical energy into thermal energy, and the subsequent confinement of thermal energy in SWCNT, caused the water molecules to evaporate and develop extreme pressures in SWCNT causing them to explode in solutions. Co-localized nanobombs killed human BT474 breast cancer cells in physiological phosphate-buffered saline (PBS) solution. Cells that were treated with nanobombs exploded into fragments, while the surrounding cells not treated with nanobombs were viable. SWCNT-based nanobomb agents can potentially outperform most nanotechnological approaches in killing cancer cells without toxicity.

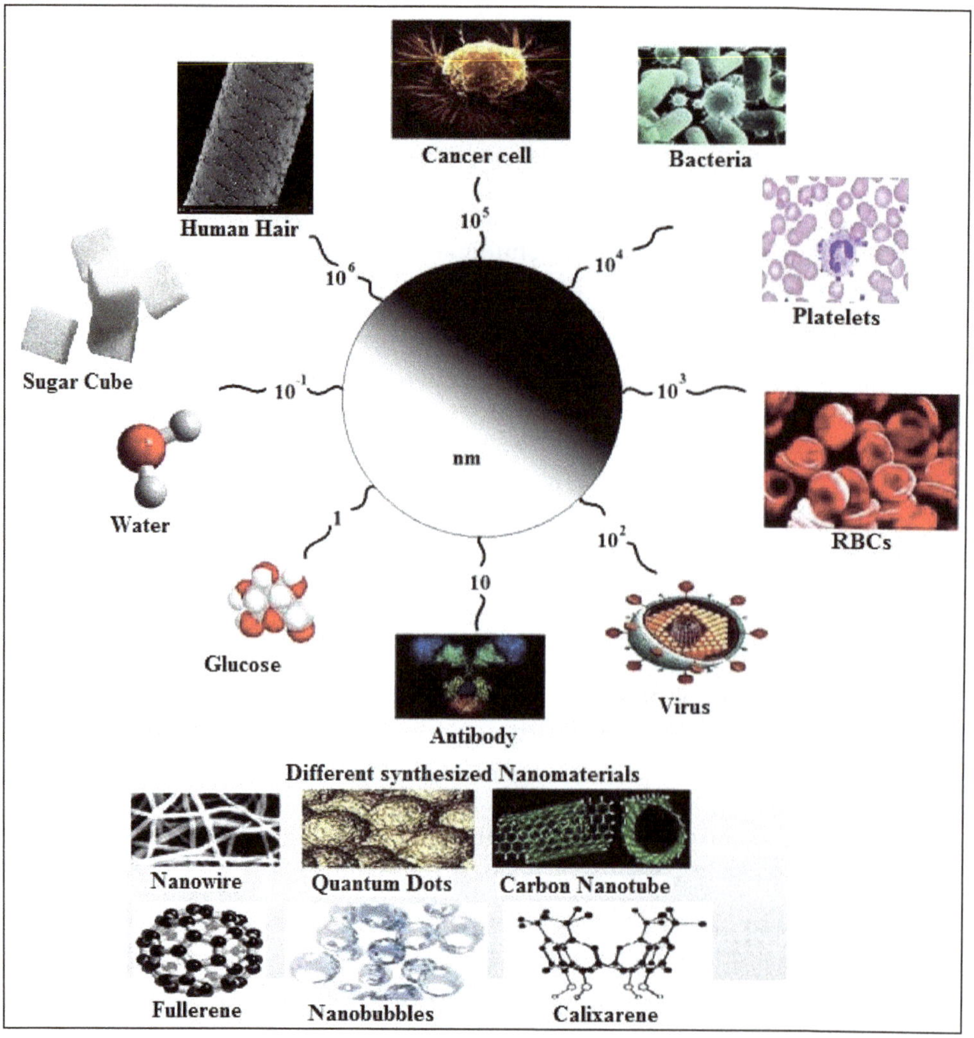

Figure 7.1: Comparison of Man Made Nanomaterials with Naturally Occurring Substances

Nanodiagnostics

The ability to analyze simultaneously multiple nucleic acid sequences for variation in a rapid and accurate fashion is pivotal in all areas of diagnostics. Several studies suggest that nanoparticle-based technology could enable sensitive detection of sequence variation in DNA.

CNT Based Tape

Wall climbing robot requires strong adhesive forces in their feet and also need to mimic the qualities of gecko foot. Such requirements also needed in the manufacture of special adhesives gloves for astronauts. These needs have led to the development of a special adhesive tape which unlike ordinary tape would stick to surfaces like permanent glue and also could be removed and reused at the same time. Not only these tapes stay glued to the walls but they could also be stuck on surfaces such as glass, Teflon and many others. The tape has been named Gecko Tape. There are no viscoelastic glues used in the tape, so it is reusable and doesn't dry up quickly or fall off walls and ceilings. The basic working principle is that the CNT within tape make use of Vander Waals forces, which help in holding microscale molecules even within a short range.

Nanodiamond

Nanoscale diamond particles have become an interesting material. Due to their inertness, small size and surface structure, they are well-suited for biological applications, such as labelling and drug delivery.

Electrophoresis

The utility and versatility of carboxylic single-walled carbon nanotubes (c-SWNT) in capillary electrophoresis (CE) is demonstrated, using as model solutes homologues and structural isomers. In the case of homologues of caffeine and theobromine, distinct changes in the electrophoretic parameters occur at a critical concentration of c-SWNT in the run buffer. It is suggested that the c-SWNT of a definite concentration could form a network in the run buffer as a pseudostationary phase on the basis of the unique tubule structure, providing a different separation from sodium dodecyl sulfate (SDS) micelles. In the case of structural isomers of catechol and hydroquinone, differing from the homologues, it is mainly attributable to the functional groups on the c-SWNT that have an effect on the electrophoretic behaviors by forming intermolecular hydrogen bonding with analytes. Furthermore, aggregated c-SWNT serve as anticonvective media and minimize solute diffusion contributing to zone broadening. The presence of charged c-SWNT suppressed the electrodiffusion and decreased the adsorption between capillary wall and solutes, which led to better peak shapes of isomers.

Nanoinjector

Technologies for introducing molecules into living cells are vital for probing the physical properties and biochemical interactions that govern the cell's behavior. Researchers have developed Nanoscale cell injection system (termed the nanoinjector) that uses carbon nanotubes to deliver cargo into cells. A single multiwalled carbon nanotube attached to an atomic force microscope (AFM) tip was functionalized with cargo via a disulfide-based linker. Penetration of cell membranes with this "nanoneedle" was controlled by the AFM. The following reductive cleavage of the disulfide bonds within the cell's interior resulted in the release of cargo inside the cells, after which the nanoneedle was retracted by AFM control. The capability of the nanoinjector was demonstrated by injection of protein-coated quantum dots into live human cells.

UK-based researchers from the School of Pharmacy, University of London, chemically modified carbon nanotubes by adding a range of different functional groups and found that they were able to enter a variety of cell types, including human cancer cells.

Pharmaceutical Application

Traditionally, almost half of new molecular entities identified by pharmaceutical industry screening programmes have failed to be developed because of poor water-solubility, which makes their formulation difficult or even impossible. However, important advances are being made in improving the bioavailability of hydrophobic compounds, so that promising candidate drugs need no longer be abandoned or have their development stalled by sub-optimal formulation. In addition to more conventional methods, tackling this problem has become one of the main applications of nanotechnology in drug delivery. This is also the most advanced approach commercially, as formulation scientists increasingly turn to a range of nanotechnology-based solutions to improve drug solubility and bioavailability.

For poorly water soluble compounds, Elan's proprietary NanoCrystal technology can enable formulation and improve compound activity and final product characteristics. The NanoCrystal technology can be incorporated into all dosage forms both parenteral and oral, including solid, liquid, fast-melt, pulsed release and controlled release dosage forms.

Nano Test Tubes

Carbon Nanotubes make Guinness Book of Records as Smallest Ever Test Tubes. The nanotube served as a confining environment for the researchers to initiate reactions between the molecules inside it and then directly observe the resulting material using an electron microscope. It also allowed them to improve the way the molecules connected to each other. Without the tiny test tube, the molecules made up a twisted, branched polymer but the tube seemed to have an ordering effect, forcing them into a linear chain. The nanotube has an inner diameter of approximately 1.2 nanometres, and a length of about 2 micrometers. Its volume is one yoctolitre, which is 10-24 litres, and around 2,000 molecules react in that space.

Electronics

Japan Science and Technology Corporation and the Japanese Institute of Research and Innovation have developed a fuel cell for mobile use with platinum-coated carbon nanotubes as electrodes. The fuel cell performance is reported to be much improved compared with the conventional material of activated carbon.

This was accomplished by using a type of carbon nanotube called the "carbon nanohorn". The fuel cell thus produced has about 10-times the energy capacity compared with a lithium battery, and if used for personal computers, in the future, a continued usage time of several days can be expected.

Samsung research group has shown a 4.5" device displaying full-color images and later a 9" full color display with 576 x 242 pixels.

Nano Radio

A team of researchers with the U.S. Department of Energy's Lawrence Berkeley National Laboratory (Berkeley Lab) and the University of California at Berkeley have created the first fully functional radio from a single carbon nanotube, which makes it by several orders of magnitude the smallest radio ever made. The nanotube vibrates at radio frequencies to receive the signal, and then acts as both amplifier and demodulator. With only a battery and sensitive earphones, it can pick up AM or FM.

Artificial Muscle

Scientists have designed the muscles to run on fuels rather than batteries or electricity from power cords. While the artificial muscles don't look like bulging biceps or well-defined quads, they are, nonetheless, like natural muscles in important ways: they run on a combination of oxygen and energy-rich fuels and are capable of doing work–such as lifting a metal weight up and down.

Nanoclays

An overview of nanoclays or organically modified layered silicates (organoclays) is presented with emphasis placed on the use of nanoclays as the reinforcement phase in polymer matrices for preparation of polymer/layered silicates nanocomposites, rheological modifier for paints, inks and greases, drug delivery vehicle for controlled release of therapeutic agents, and nanoclays for industrial waste water as well as potable water treatment to make further step into green environment. A little amount of nanoclay can alter the entire properties of polymers, paints, inks and greases to a great extent by dispersing 1nm thick layered silicate throughout the matrices. The flexibility of interlayer spacing of layered silicates accommodates therapeutic agents which can later on be released to damaged cell. Because the release of drugs in drug-intercalated layered materials is controllable, these new materials have a great potential as a delivery host in the pharmaceutical field. The problem of clean water can be solved by treating industrial and municipal waste water with organoclays in combination with other sorbents like activated carbon and alum. Organoclays have proven to be superior to any other water treatment technology in applications where the water to be treated contains substantial amounts of oil and grease or humic acid.

Nano Chip

In constructing a computer, or any other electronic device, it is important to store information on a temporary or a long-term basis. Information storage is done in memory devices, and different sorts of memories have been used since the introduction of the early magnetic core memories. Indeed, the increase in the capability of memory that is the amount of information that can be stored in a given volume of space has improved even faster than.

Using nanostructures, it is possible to reduce the size of memory bits substantially more and thereby to increase the density of magnetic memory, increase its efficiency, and lower its cost. Chris Murray's work at IBM is typical of the best research being done in this area.

Using a soft lithography method such as dip-pen nanolithography, nanocontact printing, or controlled feedback lithography, it is possible to reduce individual features down to the size of a few nanometers. If such dots each contained one bit of information (either a zero or a one), and if they were spaced by distances equal to ten times their own sizes, then on a piece of paper as large as the one you are reading, you could easily store 100,000 sets of the *Encyclopedia Britannica.*

Nanotubes can, however, exhibit even more interesting behavior. Scientists such as Cees Dekker in Delft, Paul McEuen at Cornell, Phaedon Avouris at IBM, and Charles Lieber at Harvard have demonstrated that single nanotubes can actually act as transistors. Pairs of nanotubes, or crossed nanotubes, have been shown to work as logic structures. These experiments constitute a proof of principle that nanotube logic, at an unprecedentedly small scale, can actually provide a modality for computation.

Artificial Nose

We looked at how biological noses work using molecular recognition to send a neural signal to the brain. In the artificial nose, the most common replacement for the nasal membrane is an electrically conducting polymer. When the polymer is exposed to any given molecule in the vapor phase, its conductivity will change a little bit. In the electronic nose, a random polymer, or mix of polymers, is spread between electrodes. When the molecules to be smelled land on the polymer(s), the conductivity properties in particular regions will change in a particular way that is specific to any given analyte. The nature of the detection is interesting, since it is based on so-called neural nets.

Nano Fibers

A company called Nano-Tex, which uses nanoscale structures to change the physical properties of clothing. Its Nano-Dry product purports to combine the strength of synthetic material such as nylon with the comfort of natural open-weaves by carrying away perspiration and moisture. It does this by wrapping strong, durable, synthetic core fibers with specially made absorbent fibers to produce what could be called a smart heterogeneous composite nanomaterial.

Gore-Tex fabric is a smart material with two functions: it allows water vapor to escape (so that the wearer does not feel clammy), and it keeps the liquid water out (so that the wearer does not get wet). In this material, a polymeric sheet is pierced by tiny holes. The sheet is a fluorinated carbon, like Teflon. The holes are nearly nanoscale in size and will allow vapor molecules or small clusters of molecules to pass, while preventing liquid from passing.

Nano Filters

Applications of nanostructures in separations can be as simple as the cellophane that is used in packaging, which permits small molecules to pass through its nanoscopic pores while blocking other larger species. Precisely the same principle applies in dialysis, an important but extremely taxing procedure by which sufferers from kidney disease literally has impurities washed out of their blood. In current dialysis technology, blood must be pumped out of the body, filtered through a dialysis machine, and then pumped back into the body. There have been advances in dialysis membrane technology, but this technology remains a major challenge to the chemical industry.

The Air Products Corporation derives much of its income from using chemical ways to separate oxygen and nitrogen from air and then selling those pure gasses for industrial, chemical, and medical uses. Nanofiltration has other applications ranging from water purification to removal of toxins from waste streams. All these applications can be facilitated by the construction of nanoscale porous structures, and the development of such porous structures.

Nano Protectant

The National Institute of Advanced Industrial Science and Technology (AIST) have succeeded in the suppression of light-degradation of a photo-functional organic molecule, β-carotene, by encapsulating it in single-wall carbon nanotubes (SWCNT).

Nano Therapy

Cancer cells treated with carbon nanotubes can be destroyed by non-invasive radio waves that heat up the nanotubes while sparing untreated tissue, a research team led by scientists at The University of Texas M. D. Anderson Cancer Center and Rice University has shown in preclinical experiments. Researchers show that the technique completely destroyed liver cancer tumors in rabbits. There were no side effects noted. However, some healthy liver tissue within 2-5 millimeters of the tumors sustained heat damage due to nanotube leakage from the tumor.

Nanotubes, tiny hollow carbon filaments about one ten-thousandth the diameter of a human hair, are already famed as one of the most versatile materials ever discovered. A hundred times as strong as steel and one-sixth as dense, able to conduct electricity better than copper or to substitute for silicon in semiconductor chips, carbon nanotubes have been proposed as the basis for everything from elevator cables that could lift payloads into Earth orbit to computers smaller than human cells.

Nanotubes in Brain Repair

Thin films of carbon nanotubes deposited on transparent plastic can also serve as a surface on which cells can grow.

Flat Displays

One of the many promising types of applications for carbon nanotubes is flat electronic display devices. In 1997 a joint research team from the National Institute of Materials and Chemical Research and Mie University discovered that carbon nanotubes emit ring-shaped electron discharges when they are exposed to a weak electrical field. The ring-shaped discharges are nanometer sized and are emitted from the ends of the tubes. In 1999 Ulvac Japan Ltd developed a process for growing carbon nanotubes which are both aligned vertically and located at the desired locations on base plates of field emission displays (FED). (FEDs are a new type of flat panel display device which is currently being developed. They combine the high image quality of CRT with the thinness and low power consumption of LCD.)

Miniature X-Ray Machine

The electron field-emission properties of carbon nanotubes enable the fabrication of cold cathodes for a variety of vacuum device applications. The utilization of these cathodes is an attractive alternative for the replacement of thermionic or hot cathodes for generating X-rays. Miniature X-ray tubes have been fabricated using triode-style carbon nanotube based cathodes. In this article we report the results of characterization studies, such as beam current dependence on the control gate voltage. Also, results on focal spot measurements and electron-beam modeling allow the possibility of reducing focused spot sizes. Driving gate voltages below 1000 volts for easy pulsing has been achieved, and the extended lifetime data suggests that a regulated power supply would be ideal for a constant AC operation mode. The 1mm focal spot size achieved so far is suitable for most XRF applications.

Conclusion

Although there is a general sense that the technological base in nanotechnology is developed sufficiently to enable researcher to make ready use of these tools and materials, there are still fundamental questions about these materials that must be answered if nanotechnology is ultimately to have a significant impact extending beyond the laboratory.

Chapter 8

Detection of Explosives Using Nanotechnology

Vinod Kumar Jain
Amity Institute of Advance Research and Studies (M&D)
Amity University, Sector-125, Noida, U.P.
E-mail: vkjain@amity.edu

ABSTRACT

There is immediate requirement to detect hidden explosive materials in luggage, vehicles, and aircrafts or at crowded places like railway stations and airports. The present detection system of explosive materials involves sophisticated technological procedures and is of high cost and also more time consuming. Many methods are based on fluorescence measurements, Raman spectroscopy, electrochemistry, ion mobility, spectroscopy, and IR spectroscopy, etc. Each method has inherent advantages and disadvantages, but there is common drawback of having high degree of complexity and less possibility of miniaturization of these devices. In recent years, several alternative detection techniques have been proposed to develop early warning systems. The urgent need is to develop a new type of hand-held detector with the performances of high sensitivity, good recognition ability, fast response and low cost. It should be capable to continuous real time detection, even when the vapor pressure and concentration in air is very low at ambient temperature. Several groups are actively engaged in developing these detectors based on Micro Electro Mechanical System (MEMS) technology. In last few years, work on silicon microcantilevers has shown the possibility of achieving high degree of detection. The detection is based on either the measure of the change in the piezoresistance on the cantilever or deflection produced in the micro cantilever due to adsorption of the explosive vapors on the cantilever surface. It can also change the resonant frequency of the cantilever on loading with vapors. A new sensing technique, which is based on heat absorption property of explosive at melting temperature, has also been tried. This can be achieved by making a micro heater on the cantilever. Secondly, a step heating method has also been proposed. Thirdly, nano explosion can also be produced on the surface of the micro cantilever to have more deflection and then make optical measurements. Our studies and theoretical calculations on the mass sensitive factor of the microcantilever shows that 7.5 pgm mass can be detected using the particular dimensions of the microcantilever. These micro-cantilevers can be made using bulk as well as surface micromachining. The paper also presents a new technique developed here using corona based electrostatic charge for depositing thin films of MWCNTs on the surface of the dielectric fiber or on any other substrate of any shape. This sensor measures the electrical conductivity and allows scientists to detect the presence of explosives

Keywords: *Explosion detection, Nanodevices, Security.*

Introduction

Manifold increase of international terrorism and due to the use of most advanced explosive materials by terrorists has raised the challenge to the scientists to find an immediate solution to it. The urgent requirement is to detect hidden explosive materials in luggage, vehicles, and aircraft or at crowded places like railway stations and airports before they reach their target. The present detection systems of explosive materials involve using specially trained dogs or costly and sophisticated technological procedures, which are also more time consuming.

Many of the methods available are based on fluorescence measurements, Raman spectroscopy, electrochemistry, ion mobility, spectroscopy, and IR spectroscopy, etc. (Yinon, 2003). There are also some recent technologies to detect explosives using highly penetrating radiations. Each method has inherent advantages and disadvantages, but there is common drawback of having high degree of complexity and less possibility of known miniaturize these systems. In recent years, several alternative detection techniques have been proposed to develop early warning systems. The need is to develop a new type of handheld detector with the performances of high sensitivity, good recognition ability, fast response and low cost. It should be capable of continuous real-time detection, even when the vapor pressure and concentration in air are very low at ambient temperature. The analysis of the sample is highly dependent on sample collection, analysis, and comparison with the standards.

Among the major detection techniques, trace detection methods suffer from the fact that available vapor plumes are normally too dilute for detection at a distance. These issues set the parameters for the design and development of mechanical micro electro system (MEMS) and nanotechnology-based (Rasmussen *et al.*, 2003; Thaysen, 2001; Nickolay *et al.*, 2004; Pinnaduwage *et al.*, 2004, Senesac *et al.*, 2008) next generation sensors for explosive detection. At present several groups are actively engaged in developing these sensors based on micro- and nanotechnology to achieve high reliability, highly sensitivity and low cost detection systems. In last few years, work on silicon micro cantilevers has shown the possibility of achieving high degree of detection. The detection is based on either measuring the change in the piezoresistance on the cantilever or deflection produced in the micro cantilever due to adsorption of the explosive vapors on the cantilever surface. Our studies and theoretical calculations on the mass sensitive factor of the microcantilever show that up to 7.5 pgm mass can be detected using the particular dimensions of the microcantilevers. This micro-cantilever can be made using bulk as well as surface micromachining.

A new class of sensors is also coming up very fast, which can be made using carbon nanotubes for sensing the chemical and biological applications. Research in the last one decade has shown that carbon nanotubes (CNTs) are promising nanomaterials for the development of sensors especially for gas detection, pressure, flow, thermal, optical, mass, position, and nerve gas and also for sensing bio-molecular recognition. Nanomaterials based sensors are comparable to or even better than thin film and MEMS based sensors. The technology is simple, low cost and also requires low power. This paper presents a new technique developed using corona based electrostatic charge for depositing thin films of MWCNTs on the surface of the dielectric fiber substrate and functionalize it to covert it in to a sensor of very high sensitivity.

Detection Systems

Based on Dogs

Today, the cheapest, very reliable detection of explosive is done by using the specially trained dogs. These dogs are used to identify the scents of several common explosive materials and can also find the explosive handler. The dog brain is largely dominated by an olfactory cortex. The olfactory in dogs is roughly forty times bigger than the olfactory in humans, with 125 to 220 million smell-sensitive receptors.The olfactory ability of dogs is sensitive enough to detect trace amounts of many compounds, which makes them very effective in screening objects. Dogs can discriminate odors at concentrations nearly 100 million times lower than humans can. A dog can search an entire airport in a couple of hours. These trained dogs detect explosives very fast, but their usefulness becomes less as a dog becomes tired or bored.

Based on Honey Nees

A new approach is to train the honey bees and use them for explosive detection. Honey bees may be more reliable than dogs. According to Dr. Grant "Dogs treat their training as a game while "Bees don't treat it as game; it is a response. They are like little micro machines." In case of dogs, they get bored and may tell you that there are drugs present even when they are not. Basically honey bees can be trained to detect explosives by giving food associated with a sugar reward," said Dr David Grant. When they give them sugar and a smell at the same time they learn the two things go together so the next time they get the smell they expect sugar and they stick out their tongues (ref). Researchers say the bees are much more effective than sniffer dogs. This proven system is not yet commercially available.

Spectrometry

Several systems have been developed to detect trace signatures for various explosive materials. The most common technology is Ion Mobility Spectrometry (IMS). This method is similar to mass spectrometry (MS), where molecules are ionized and then moved in an electric field in a vacuum, except that IMS operates at atmospheric pressure. The time that it takes for an ion, in IMS, to move a specified distance in an electric field is indicative of that ion's size-to- charge ratio: ions with a larger cross section will collide with more gas at atmospheric pressure and will therefore be slower. Several attempts are being made to miniaturize, ruggedize and make MS affordable for field application.

Another technique is based on the comparison in reflected ultraviolet, infrared and visible light measurements on multiple areas of the suspect material. In another detection technique using gas chromatography (GC), molecules have to be separated before detection. This not only improves the performance of the detector, but also adds another dimension of data, as the time it takes for a molecule to pass through the GC may be used as an indicator of its identity.

MEMS Based Microcantilever

Micro Electro Mechanical Systems (MEMS) are intelligent micro machines having sensors, actuators and signal processing electronics at one place. This technology has brought together silicon based microelectronics and micromachining technologies. These systems have the capability of performing a variety of functions like physical and chemical sensing, analyzing the information and communication. The most important factors are that these devices are very small in size and also of low cost as these devices can be batch fabricated in thousands at a time. MEMS devices have applications in every field, but major areas are coming like in aerospace, defense, automobiles, biomedical and communications and in many other sectors. MEMS are made using the same standard process steps used in IC fabrication facilities like diffusion, oxidation, metallisation and photolithography. In addition to these other unit processes, such as micromachining, plating, molding and substrate bonding, are also required to make MEMS.

Micromachining can be classified in two major categories, *i.e.*, bulk micromachining and surface micromachining. Bulk micromachining, which is isotropic and anisotropic types, can be used to make beams, thin diaphragms, cantilevers, etc. Surface micromachining is generally used to make freely suspended structures microcantilever, microturmine, etc. on the silicon chip. Some times both the technologies have to be used to make many complicated structures. Deep reactive ion etching system (DRIE) is also used in case high aspect ratio is required.

Micro cantilever (MC) sensor technology is a next generation electrochemical technique with a large number of applications in chemical, physical and biological detection. These micro cantilevers (MCs) can be made using bulk or surface micromachining of silicon, silicon nitride or silicon dioxide. In addition to this MCs have an advantage of making of multi-element sensor arrays and also can also be integrated with electronic circuitry on the same chip. The basic principle of operation of a MC is quit simple. When a molecule adsorb at its surface, cantilever beam bends due to the induced stress shown in Figure 8.1. The stress is produced due to change in temperature or mass. The bending of the cantilever will be in nanometer scale and can be measured either as the shift in the resonance frequency of the cantilever or by optical deflection method. In the optical detection method, the light beam from a low

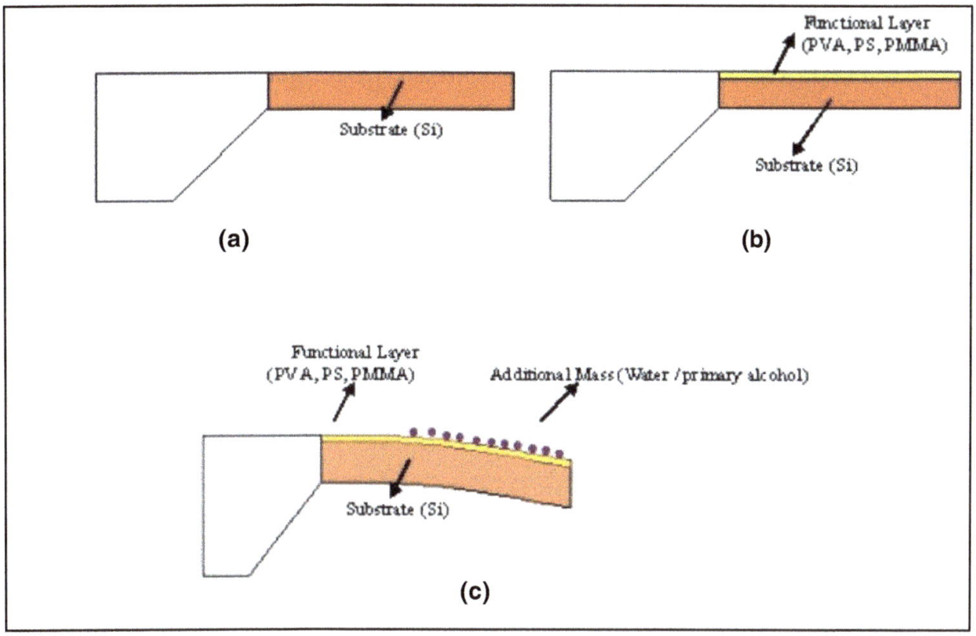

Figure 8.1: Schematic of Microcantilever Based Chemical Sensor (a) Unexposed beam without functional layer; (b) Unexposed beam with functional layer; (c) Beam exposed to chemical

power laser diode is focused on the free end of the cantilever and its displacement can be known by measuring the reflected light spot with a position sensitive detector. It has certain limitations, *e.g.*, poor reflectivity of light, long term drift of the cantilever bending and reproducibility in the binding of the analyte to the cantilever surface. Although some of the difficulties have been overcome.

Rasmussen *et al.* made the MC with piezoresistive based measurement system.The resistors and electrical wiring were encapsulated using silicon nitride and wiring is done through titanium silicide. Piezoelectric thin films can be easily applied to produce the resonant biosensor in which the added mass of the target bio molecules changes the resonant frequency of the cantilever. The piezoelectric film on the cantilever converts the mechanical energy into electrical signal. The sensor can be produced on the micrometer scale by using lithographic techniques.

MCs can also be operated in vibration mode. By loading the cantilevers with extra mass of vapors also decreases the frequency. It consists of nanometer-sized cantilevers fully integrated with CMOS circuitry. Electro statically exited MC resonance frequency can be monitored by measuring the capacitance between the vibrating cantilever and a fixed parallel electrode using integrated CMOS. By measuring the change in resonance frequency, the mass change can be measured. One can say that electrostatic excitation and capacitive detection is the most simple and easiest possible combination, among all possible techniques used for excitation and detection on MEMS and NEMS. The typical dimensions of commercially available micro machined cantilevers are 50-200 μm long, 10-40 μm width and 0.5 μm-5μm thick for the mass in the range of a few nanograms. The resonant frequencies of these cantilevers are in the range of a few kilo hertz to a few hundred kilo hertz. Depending upon the dimensions the cantilever spring constant varies from 0.008 N/m to 30 N/m. Our work (Ganesan *et al.*, 2006) is an effort to optimize mass sensitivity factor for polymer coated silicon cantilevers for chemical sensing applications. Simulations are carried for mass sensitivity factor and spring constant change before and after chemical reaction in these cantilevers. The results give the guidelines for designing high sensitive micro cantilevers for chemical sensing applications.

Analysis

The basic mechanical parameters of a cantilever are the spring constant and the resonance frequency. These parameters are very important for the use in AFM. A soft cantilever will simply bend due to its own weight and mechanical noise from the surroundings. From the Hooke's Law;

$$F = -kZ \tag{1}$$

where, F is the applied force, k is the spring constant, Z is the bending of a cantilever. Spring constant for a rectangular cantilever of length L is expressed as (Nickolay *et al.*, 2004)

$$k = \frac{3EI}{L^3} \tag{2}$$

where,

$$I = \frac{WT^3}{12} \tag{3}$$

where, W is the width, of the cantilever, T is the thickness, I is the moment of inertia and E is the Young's modulus. These expressions can be used in Eq. (1) to find the spring constant of a simple cantilever, *i.e.*

$$k = \frac{EWT^3}{4L^3} \tag{4}$$

The resonance frequency f_{res} *of a long and thin rectangular cantilever beam is*

$$f_{res} = \frac{k}{2\pi}\left[\frac{EI}{\lambda}\right]^{1/2} \tag{5}$$

where, λ is the linear mass density.

Consider a microcantilever of length L_c, width W_c, thickness T_{cs}. The spring constant of the cantilever can be calculated as follows:

$$k = \frac{E_S T_{CS}{}^3 W_C}{4L_C{}^3}\left(L_c \rangle\rangle\rangle T_{cs}\right) \tag{6}$$

(Ensure to have the subscripts as corrected in the preceding line). Ensure you also define E_s and ρ_s.

The resonance frequency of the cantilever is

$$f_{res} = \frac{k}{4\pi}\left[\frac{T_{cs}{}^{3}E_{s}}{3\rho_{s}}\right]^{1/2} \tag{7}$$

Chemical Detection Based on Resonance Frequency Shift

In this section we present results from theoretical analysis where we utilized the resonance frequency shift of a cantilever beam for chemical sensing. In this analysis, we have taken silicon as a substrate material and functional layer as polyvinyl alcohol (PVA), polystyrene (PS), polymethylmethacrylate (PMMA) whose mechanical properties are known. The functional layer is coated on top of the substrate and whose thickness should be less than the thickness of substrate.

After surface modifications, the mechanical characteristics of microcantilever beam changes due to the formation of functional layer. The linear mass density of the modified cantilever now will become

$$\lambda_f = \rho_s A_s + \rho_f A_f \tag{8}$$

And its flexural rigidity will be redefined as $EI = E_s I_s + E_f I_f$, where s and f represent substrate and functional layers respectively. Now the moment of inertia of the modified beam will also change to (Ganesan *et al.*, 2006)

$$I_s = \frac{W_c T_{cs}{}^{3}}{12} + W_c T_{cs}\left(Y_c - \frac{T_{cs}}{2}\right)^2 \tag{9}$$

$$I_f = \frac{W_c T_{cf}{}^{3}}{12} + W_c T_{cf}\left(\frac{T_{cs}}{2} - T_{cs} + Y_c\right)^2 \tag{10}$$

where, Y_c is the centre of mass. The new resonance frequency after surface modification will be expressed as

$$f_{new} = \frac{k}{2\pi}\left[\frac{E_s I_s + E_f I_f}{\lambda_f}\right]^{1/2} \tag{11}$$

This expression shows that both the stiffness and mass loading affect the resonance frequency. If the functionalized cantilever is exposed to vapor this results in swelling of the polymer layer coated on the top of cantilever surface. The increase in mass implies decreasein resonance frequency since it is inversely proportional to applied mass. This changes the surface stress which results in the bending of the cantilever. The amount of mass increase after reaction is directly related to the number of molecules deposited or absorbed by the functional layer. Now the final resonance frequency due to the increase of mass can be expressed as

$$f_{final} = \frac{k}{2\pi}\left[\frac{3\left(E_s I_s + E_f I_f\right)}{\left(0.2235\left(\rho_s A_s + \rho_f A_f\right)L + M\right)L^3}\right]^{1/2} \qquad (12)$$

where, ρ_s, and ρ_f are the densities of the substrate and functional layers respectively. A_s, and A_f are the cross sectional areas of substrate and functional layer respectively, M is the additional mass added at the end of a cantilever beam due to chemical reaction. If we know the order of applied mass during reaction then we can estimate the shift in resonance frequency from Eq. (11) and (12)

$$\Delta f = f_{new} - f_{final} \qquad (13)$$

Cantilever Fabrication

Materials used to fabricate micro machined cantilevers are silicon, silicon nitride, silicon carbide and silicon dioxide. Commercially available cantilevers are made from silicon-based materials coated with thin layer of gold or silver or aluminum for better optical readout. Composite cantilevers are very much prone to thermal noise due to biomaterial effects. This can be eliminated by increasing the ratio of substrate to functional layer thickness.

The fabrication procedure of simple silicon nitride cantilever is shown in Figure 8.2. Wet chemical etching has been used. In the first step, a small thin layer of silicon nitride was deposited by low-pressure chemical vapor deposition (LPCVD) onto a

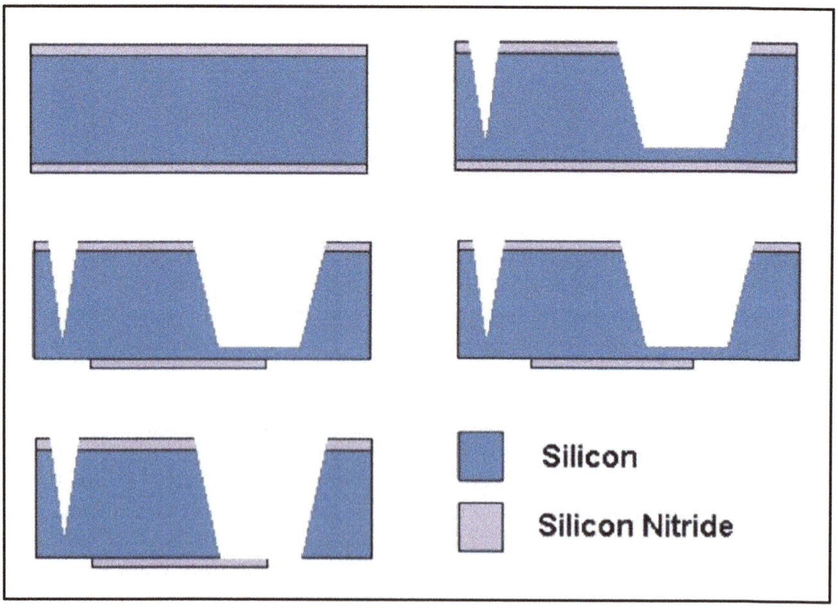

Figure 8.2: Schematic of Silicon Nitride Cantilever

double side polished silicon crystal wafer. In the second step, a deep etch pattern was prepared on the silicon nitride layer on the non-cantilever side of the wafer using photolithography and reactive ion etching. Next step was etching of anisotropic surface with KOH until the silicon wafer attained the required thickness. The required shape was achieved on the silicon nitride surface by reactive ion etching and followed by evaporating few nanometers of gold onto the cantilever side of the wafer. Finally the cantilever was realized by etching the remaining portion of silicon in KOH and by removing the plasma enhanced chemical vapor deposited (PECVD) silicon nitride with buffer solution.

To optimize the microcantilever for water vapor detection, the cantilever should be coated with water sensitive polymer like polyvinyl alcohol (PVA), carboxy methyl cellulose (CMC) or polyvinyl pyridine (PVP). PVA coated sensor shows larger response to water vapor (frequency shift up to 120Hz). Typical cantilever resonance frequency was calculated using equations (10–13) and the same were used for calculating the mass sensitivity factor. Mass sensitivity factor is a ratio between shifts in the resonance frequency of the cantilevers loading per nanogram. Figure 8.3 shows a plot of mass sensitivity factor versus applied sensitivity factor. The mass on the cantilever of different thickness (Vague sentence). A PVA layer of 2.5 μm thickness on the cantilever has been used as the functional layer to adsorb the water vapors. The results in Figure 8.4 show that sensitivity is almost stable for a particular range of applied mass from 0.41–1.00 nanogram for a cantilever of dimensions of 500 μm length, 100 μm width and 1 μm thickness. For all other values, the sensitivity increases for applied mass of less than 0.41 nanograms or increases if the applied mass is more than 1 nanogram. Calculations also show the minimum mass which this cantilever can sense is 7.5 pictogram. It also shows that maximum shift in frequency per nanogram is 55.92 Hz. The stable range of sensitivity increases with increase in micro cantilever thickness up to 7 μm but thereafter it starts to decrease. Further, sensitivity is insensitive to low mass changes. Hence if the change in mass due to the absorption of water molecules on the top of micro cantilever is in the order of 0.41–1.00 ngm then the 500 μm x 100 μm x 1 μm will be the most suitable dimensions for designing the sensor. Similarly a large number of calculations have been done to optimize the thickness and keeping the length and width as constant parameters. The details results have been given in our previous publications (give details). Our studies and theoretical calculations on the mass sensitive factor of the microcantilever shows that 7.5 pgm (pictogram) mass can be detected using the particular dimensions of the microcantilever. This micro-cantilever can be made using bulk as well as surface micromachining.

The detection of an explosive is based on either measuring the change in the piezoresistance of the cantilever or deflection produced in the micro cantilever due to adsorption of the explosive vapors on the cantilever surface or the change of the resonant frequency of the cantilever on loading with vapors. Apart from these other techniques can also be used to detect explosive or other chemicals using micro cantilevers. For example, every explosive has its unique melting point; TNT and RDX are two main explosives having their melting points as 80 °C and 204 °C respectively. In this technique the property utilized is the absorption of heat by explosive particles

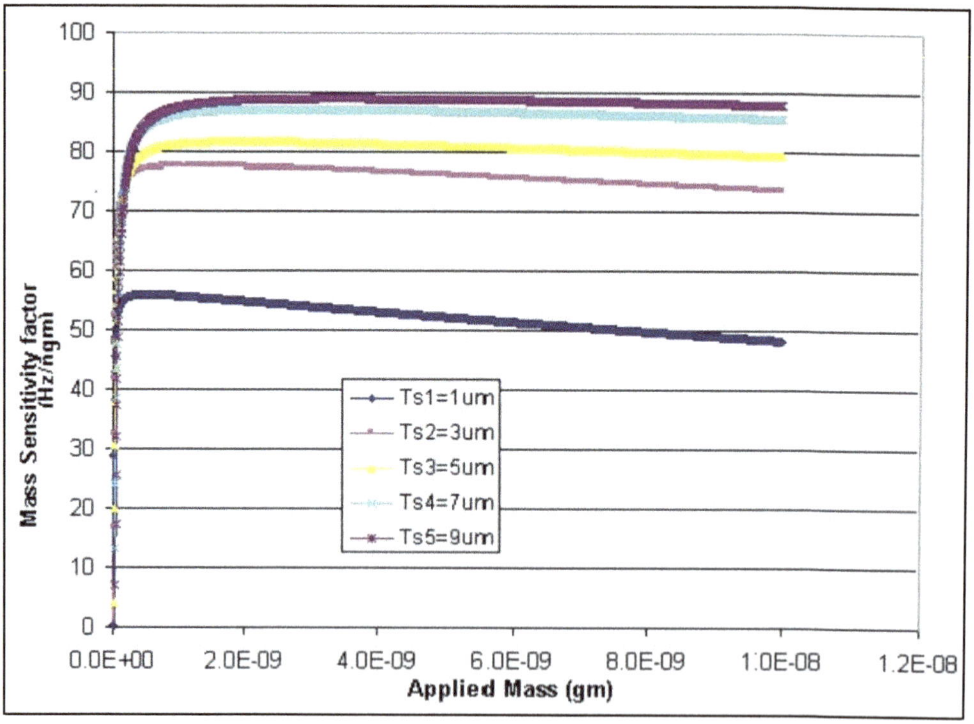

Figure 8.3: Variation in Sensitivity Factor with Change in Applied Sensitivity Factor

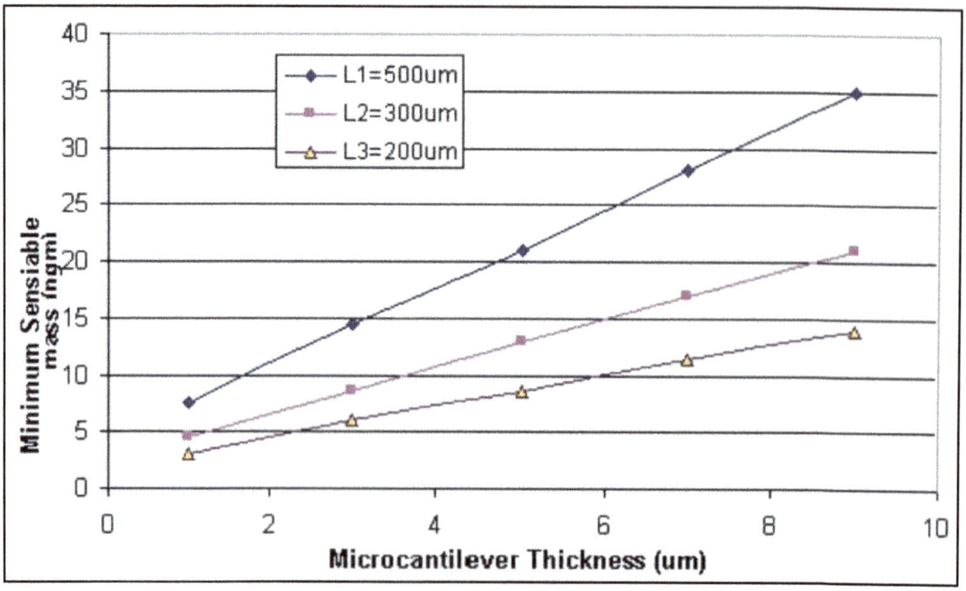

Figure 8.4: Variation of Minimum Sensible Mass with Change in Substrate Layer Thickness for Different Values of Cantilever Lengths (PVA width 100 μm and functional layer thickness 2.5 μm)

during melting process on the surface of the cantilever. The sensor is heated up to 80°C (melting point of TNT) with the help of a micro heater on the microcantilever. The TNT particles on the beam surface will absorb heat, melt and evaporate, then take heat away from the sensor, so that the temperature distribution of the beam will change and make the corresponding electrical signal change.

Besides the absorption of heat as described above, a step heating method has also been proposed (Figure 8.5). Here the sensors made on the freely suspended membranes with micro heaters have been made. The membrane is 50 x 50 microns and has a air gap of two micron between the membrane and the silicon substrate to provide a better heat isolation (Jain, 2007; Nagapriya, 2003). By increasing the applied voltage in steps to the micro heater of MC, the temperature of the sensor will also increase up to the melting point and produce a response graph called base graph respectively. The same step voltages are applied to the sensor before and after putting the explosive on the sensor. The explosive will absorb heat from the beam during melting causing its temperature to decrease transitorily and it will shift the response graph from the previous results. By using this concept, other types of explosives can be detected, because different explosive will produce different response graphs as demonstrated in Figure 8.6.

Figure 8.7 shows the sensitivity detection technique undergoing nano explosion on the surface of the micro cantilever, which produces more deflection and to make optical measurements (Pinnaduwage *et al.*, 2004).

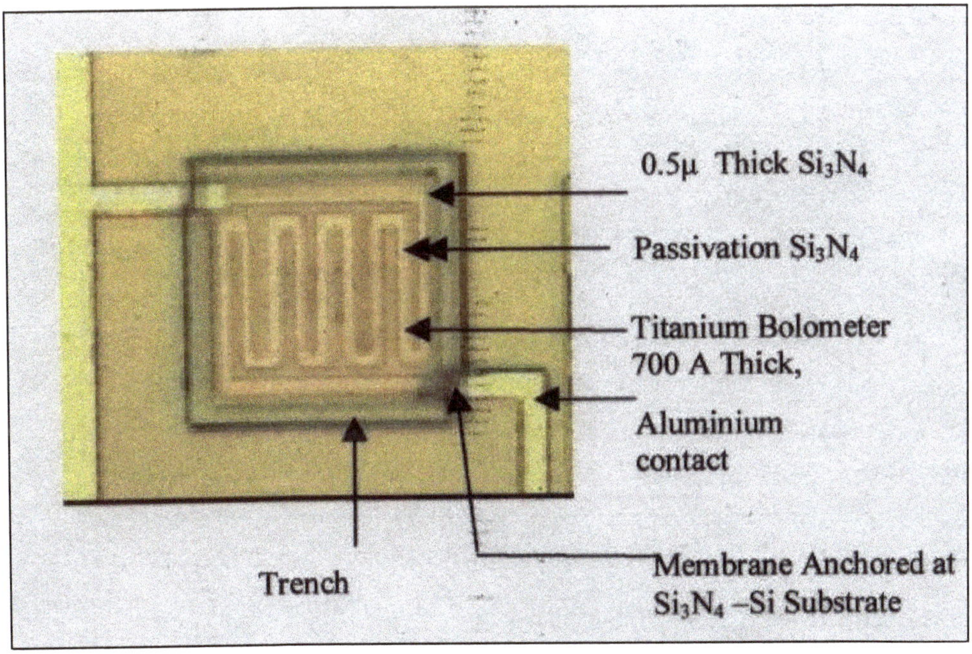

Figure 8.5: Micro Heater on a Silicon Nitride of the Dimension of 50 x50 x 0.1 μm

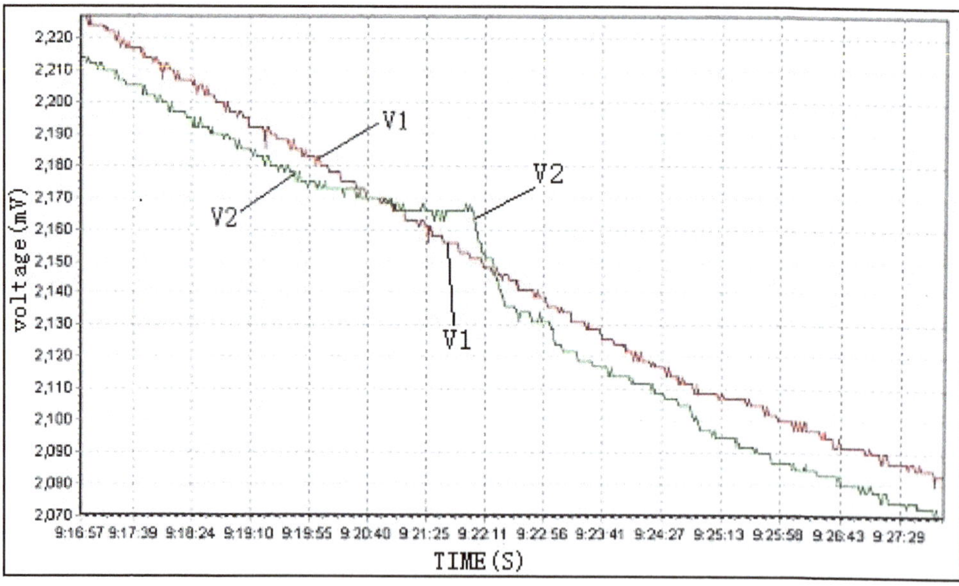

Figure 8.6: Response Graph of a Cantilever

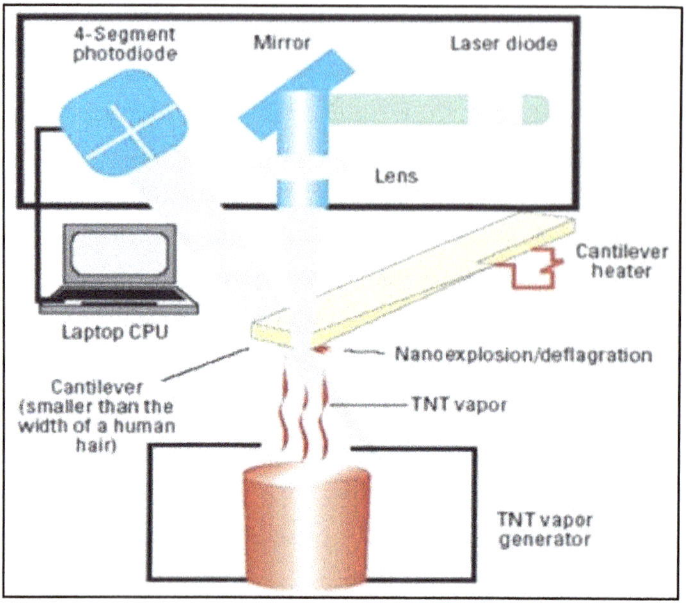

Figure 8.7: Cantilever Detection System (Pinnaduwage *et al.*, 2004)

Nanotechnology Based Sensors

Carbon nanotubes (CNT) based sensors have shown very high functional properties and provide high sensitivity and selectivity. They can detect up to the level of sub-ppb levels, require very low power and work at room temperature as compared to metal oxide and conductive polymer based sensors. The sensing mechanism of these sensors is based on the changes in the electrical conductivity of the nanotubes caused by the charge transfer. The sensitivity can be greatly enhanced by functionalization of the CNTs with different functional groups. It was found that the response times of nanotube sensors are at least an order of magnitude faster than those based on solid state sensors. These next-generation nanotechnology based sensors detect in two ways, either by measuring the change in resistivity of the sensor during the exposure to the explosive vapors or by fluorescence response. A tiny sensor that monitors electrical conductivity allows for specific molecule detection. All these characteristics make the corona based electrostatic self-assembly (CESA)-CNTs based devices very appealing for commercial applications.

CNT based sensors also offer many technical advantages such as the capability to deposit on almost any substrate, including plastics, irregularly shaped or on very large substrates. Further

Commercially available multi-walled CNTs with 60 per cent purity were used in fabricating thin films without further purification. Thin films of CNTs were coated on the surface of the Teflon fibers using corona electrostatic discharge technique. Variable dimensions of the Teflon fibers were selected for the growth of thin films to detect the presence of explosives. The nanocomposite film has been developed, and it shows very fast fluorescence response to trace of vapors of explosives such as TNT, DNT, etc.

Vasuda *et al.* (2009) developed a new technique for growing thin films of CNTs as well a method to functionalize these films using electrostatic charging techniques on to flexible substrates and thin dielectric wires. The self-assembled films have been further converted into sensors for the detection of alcohol vapors and for many other chemicals with high sensitivity and high reproducibility. The detections are based on the changes in conductance of the sensors. The developed sensors have shown recovery of less than a minute. The corona based electrostatic self-assembly (CESA) pattering technique for batch fabrication of the sensors is fast, simple, low-cost, and does not require much specialized equipment. These films can be deposited at room temperature, and can be self-functionalized CNTs. In these studies the length of the fibers were varied between 8-10 mm and the diameter of the fiber selected was 500 mm. The fibers were coated with a dielectric paste as binder to provide a strong bond between CNT and the surface of the fiber. Maintaining a gap of 10-12mm between the fibers and CNT powder, electrostatic field of few 1000's of volts was applied to self-assemble CNT on to the fiber under atmospheric conditions. These films were then dried under ambient environmental conditions for 10-12 hours. Commercial grade silver paste was used to fabricate electrical contacts. Figure 8.8 shows the photographs of fabricated device.

Functionalization of the Thin Films of Carbon Nanotubes

Generally functionalisation of these nanotubes has been done using strong acid treatment for a longer period of time. Here we have developed a fast and solution-free electrostatic technique in presence of ozone to functionalise these sensor films in 10-15 seconds. The active oxygen was adsorbed on the surface of CNTs and provided

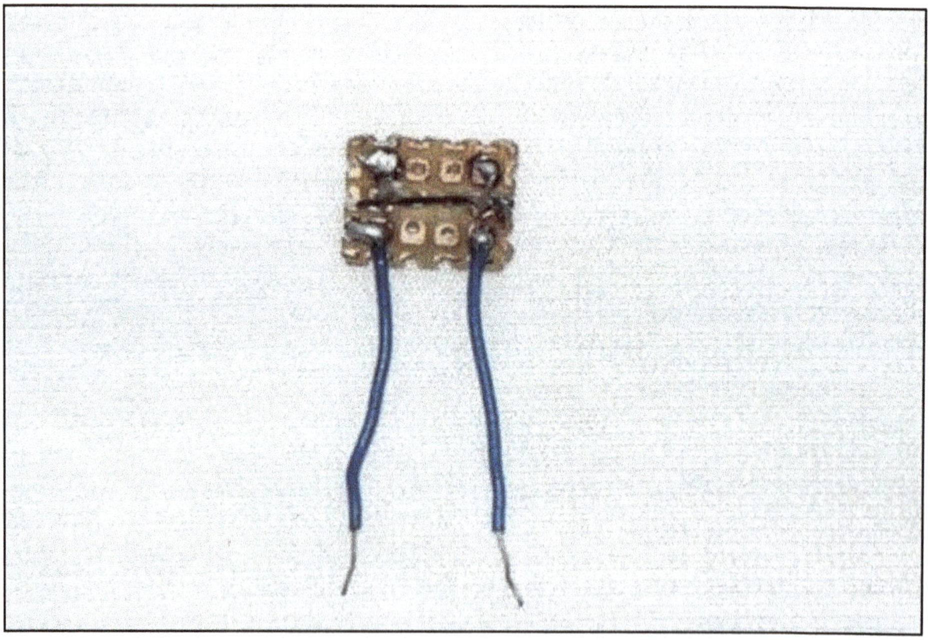

Figure 8.8: Photograph of a CESA-CNT Film Fabricated for Device Applications

active oxygenated sites on the surface of CNTs for further bond formations. Functionalization of the corona-based electrostatic self-assembly (CESA)-CNT films was confirmed with infrared spectroscopy (IR) and Raman spectroscopy and an analysis of the obtained data has been published in our previous publication (specify this publication). Figure 8.9 shows the result of alcohol sensing using this new sensor. By changing the functionalisation, the sensor senses the NO^-_2 and NH_3 and the results can be seen in Figure 8.10. The resistivity changes in the opposite directions.

Due to the interaction between nanotube devices and gas molecules, the electron configuration is changed in the nanostructured sensing device, therefore, the changes in the electronic signal such as current or voltage can be observed. By measuring the conductivity change of the CNT device, the concentration of the chemical species, such as a certain type of molecule, can be measured. By combining with MEMS technology, lightweight and compact size sensors can be made in wafer scale with low cost. This Nano sensor technology can extend its application in civilian areas such as explosives detection, monitoring filter bed breakthroughs, personnel badge detectors, embedded suit hermit city sensors, and other applications. Additionally, a wireless capability with the sensor chip can be used for networked mobile and fixed-site detection and warning systems for military bases, facilities and battlefield areas.

Hydrogen Peroxide Based Explosive Sensor

Many terrorists are using the the hydrogen peroxide based explosives. There is no easy way to detect these chemicals in the field, but now this new technology (Bohrer, *et al.*, 2008) has made it possible to detect it in very short time. The present

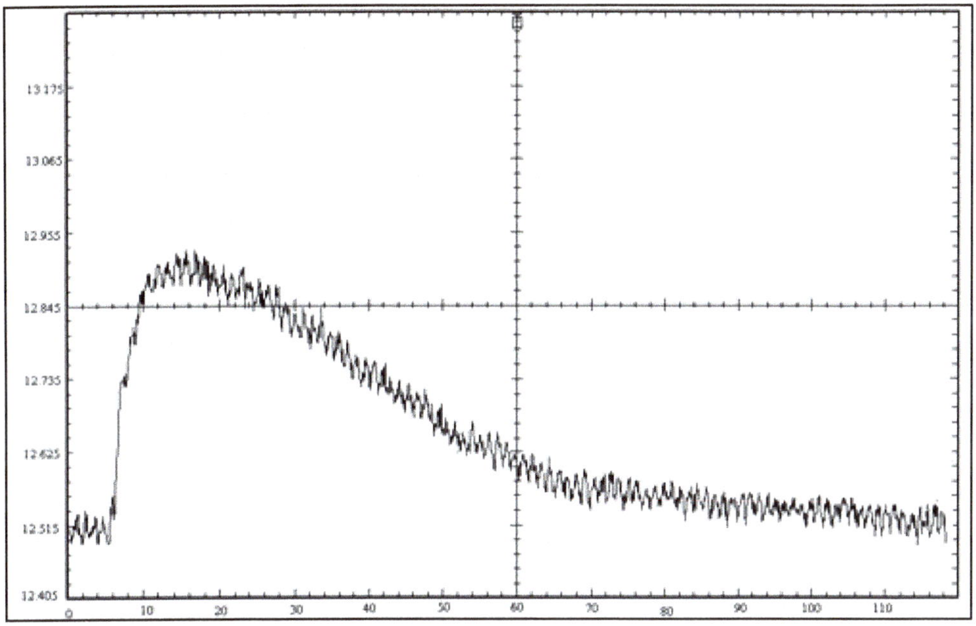

Figure 8.9: Change in Resistivity in CNT Film on Exposure to Alcohol Vapors

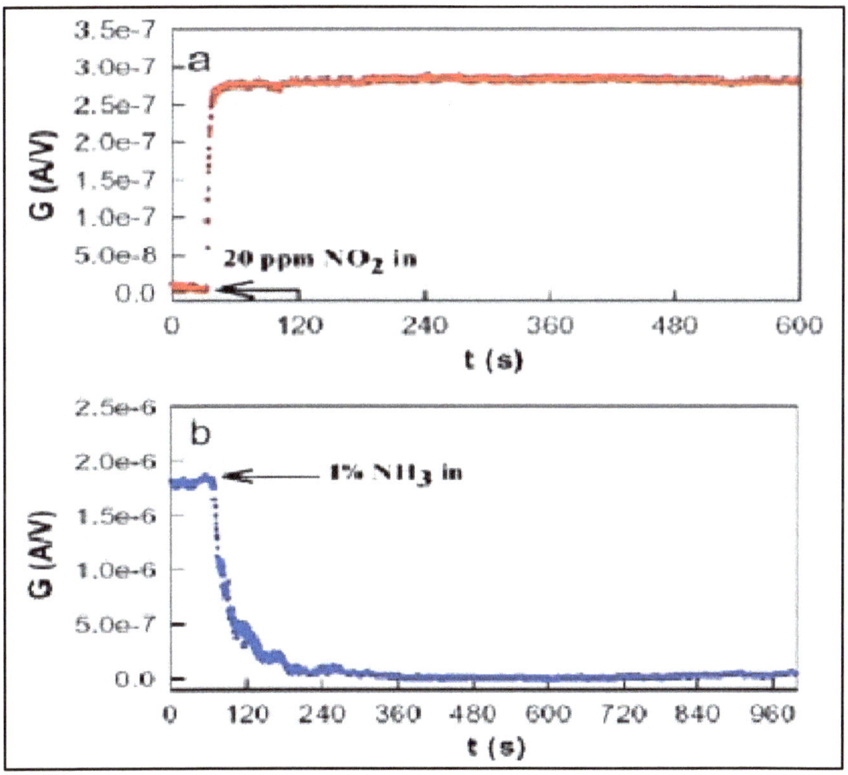

Figure 8.10: Change of Resistivity in CNT Film on Exposure to NO_2 and Ammonia

sensor consists of two thin films, one of copper phthalocyanine and other is of cobalt phthalocyanine. Hydrogen peroxide is an oxidizing agent and cobalt films will respond to it by showing decreasing current, and copper or nickel films will show increased current. It is this unusual reaction that made the sensor possible. The tiny sensor can detect very small amounts of hydrogen peroxide. We have made a new type of sensors based on carbon nanotubes for the detection of hydrogen peroxide. The detectivity is much higher in comparison to thin films sensors.

The present investigations relates to the development of a process for functionalization of multi-walled carbon nanotubes by metal-phthalocyanines and their use as highly selective and sensitive sensor for detection of hydrogen peroxide in the vapour phase. The new investigations provide a process in which multiwall carbon nanotubes (CNT) are first functionalized using corona electrostatic discharge technique and then functionalized with metal-phthalocyanines to detect H_2O_2 vapors with very high selectivity and sensitivity. A combination of cobalt-Pc and copper-Pc thin film sensors has been found to show opposite conductivities to H_2O_2 vapors making this paired sensor as a highly reliable method to selectively identify the presence of H_2O_2 vapors unambiguously. The technique for fabrication of sensors is simple, fast, and low cost and does not require any expensive or specialized equipment. The sensors work at room temperature.

Acknowledgements

I am thankful to Dr Ashok Chauhan, Founder President Amity University for his continuous guidance and support. I am also thankful to Dr. Vasuda Bhatia, Dr. A.L.Verma, Dr. Kanchan Saxena and Mr. Vikesh for their guidance and support.

References

Bhatia, Vasuda, Gaur, Vikesh and Jain, Vinod K., 2009. New technique to grow thin films of carbon nanotubes based on electrostatic charge deposition and their application for gas detection. Accepted in *Nano Science.*

Bohrer, Forest I. *et al.*, 2008. *J. Am. Chem. Soc.*, 130(12): 3712–3713

Ganesan, M., Sharma, N.K. and Jain, V.K., 2006. *Indian Journal of Pure and Applied Physics*, 44: 407–413.

Jain, V.K., 2007. *Indian Journal of Pure and Applied Physics*, 45: 304–310.

Lavrik, N.V., Sepaniak, M.J. and Datskos, P.G., 2004. *Review of Scientific Instruments*, 75(7): 1–25.

Nagapriya, K.S., Raychaudhuri, A.K., Jain, V.K., Jalwania, C.R. and Kumar, V., 2003. *Applied Physics Letters*, 82(16): 2721–2723.

Pinnaduwage, L.A., Hedden, D.L., Gehi, A., Yi, d., Thundat, T. and Lareau, R.T., 2004. *Physical Review*, 95(10): 5871–5875.

Pinnaduwage, L.A., Hedden, D.L., Gehi, A. *et al.*, 2004. *Review of Scientific Instruments*, 75(11): 4554–4557.

Rasmussen, P.A., Thaysen, J., Hansen, O., Eriksen, S.C. and Boisen, A., 2003. *Ultramicroscopy*, 7: 371–376.

Sanchez, Jason C. and Trogler, Wlliam C., 2008. *Journal of Materials Chemistry*, 18: 3143–3156.

Senesac, L. and Thundat, T.G., *Materials Today*, 11(3): 28–36.

Thaysen, 2001. Cantilever for bio-chemical sensing integrated in a microliquid handling system. *Ph.D. Thesis*, Mikroelektronik centre, Denmark Technical University.

Yinon, Jehuda, 2003. *Analytical Chemistry*, 99A–105A.

Chapter 9

Drug Delivery and Drug Targeting Using Therapeutic Nanoparticles

Ranjit Jagadale, Niraj Sakhrani, Harish Padh*

Department of Biotechnology,

B.V. Patel Pharmaceutical Education and Research Development (PERD) Centre,

Thaltej, Ahmedabad – 380 054, Gujarat, India

**E-mail: hpadh@yahoo.com*

ABSTRACT

The rapid development in the current pharmaceutical drug discovery resulted in emergence of increasing number of novel therapeutic drugs for the treatment of variety of diseases. However, at present the main problem associated with systemic drug administration mainly includes distribution of pharmaceuticals throughout the body, the lack of drug-specific affinity towards a pathological site, nonspecific toxicity, and other side effects resulting from high doses. An attractive strategy to enhance the therapeutic index of drugs is to specifically deliver these agents to the defined target cells, thereby, keeping them away from healthy cells. This would allow for more effective treatment achieved with a better tolerance. Cancer chemotherapy strives to cause a lethal cytotoxic lesion that can arrest a tumor's progression. The attack is generally directed against metabolic sites essential to cell replication. Ideally these drugs should interfere only with cellular processes unique to the tumor cells. However, currently available anticancer drugs do not specifically recognize neoplastic cells, but rather affect all proliferating cells both normal and abnormal. Therapeutic nanoparticles are used for passive tumor targeting based on enhanced permeability of tumor vasculature, where the increased permeability of the vasculature allows passage of nanosized carriers loaded with drug to accumulate in the tumor tissue. Apart from this, tumor cells overexpress membrane receptors such as folate, transferrin, lipoprotein, asialoglycoprotein and others. Ligands or monoclonal antibodies to these receptors can be conjugated with the therapeutic nanoparticles to specifically target tumor cells and to increase their intracellular uptake based on receptor mediated endocytosis (RME). This chapter specifically discusses various therapeutic nanoparticles used for drug delivery and designing drug delivery system for targeting cancer cells.

Keywords: *Therapeutic nanoparticles, Receptor mediated endocytosis (RME), Drug delivery, Tumor targeting.*

Introduction

Drug delivery is a challenge to a researcher. A drug if not rendered to its target site in its active form, it is considered to be useless, or possibly a curse in some of the cases such as in Adverse Drug Reactions (ADR). Non-specificity to the desired site in the cell may lead to development of resistance against that drug by affected cells, thus making target cells in-effective to that drug (Shashindran and Gitanjali, 2006).

Taking a case of cancerous tissue, most of the chemotherapy in market lacks specificity and hence limiting the efficacy of the chemotherapeutic drug. Chemotherapeutic drug show an effect in tumorous tissue and an undesired effect on normal tissue as well, thus there is a need for targeted delivery of anti-cancer drugs. Targeted delivery would help in better compliance by avoiding toxicity and reducing dosage requirement of drugs. Nanoparticles plays an important role, to act as a carrier to deliver therapeutic/imaging agent to its desired tissue, desired cell or if possible, desired organelle. Nanotechnology refers to development of a material or device having one of its dimensions as 100 nanometer or smaller. It has various diverse applications as describe in Figure 9.1. They are being used for drug screening, gene delivery, drug delivery, imaging and diagnosis. Iron oxide nanoparticles for example have been used for enhancing the tissue contrast in Magnetic Resonance

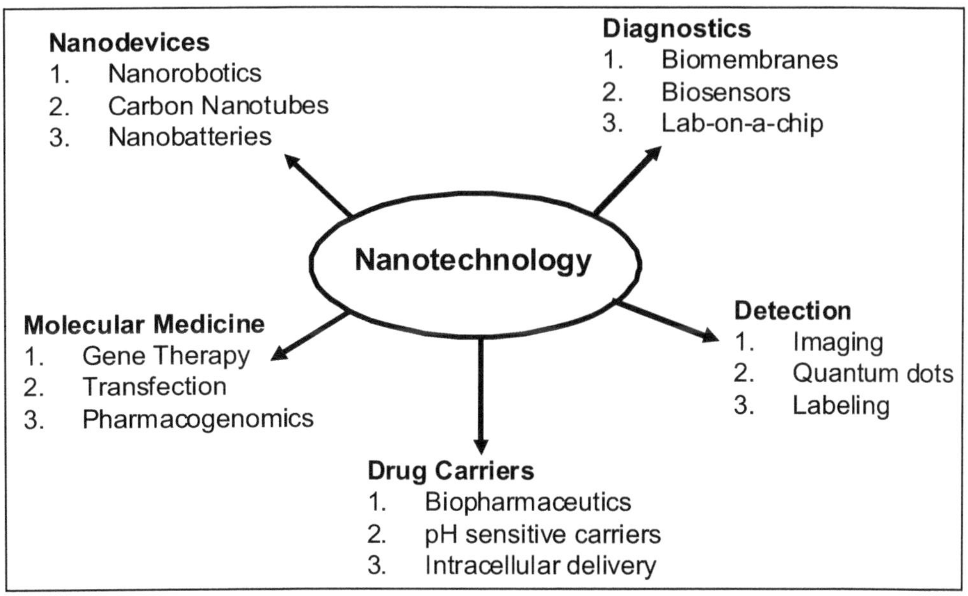

Figure 9.1: Applications of Nanotechnology

Imaging (MRI), hyperthermia for killing tumor cells, drug delivery and drug targeting. The beauty of the iron oxide nanoparticles is 'paramagnetism' shown by the nanosized particles *i.e.* they demonstrate the magnetic properties only in presence of external magnetic field. These particles are known as Superparamagnetic Iron Oxide Nanoparticles or SPIONs. SPIONs are also used for improving the cancer drug delivery and magnetic fluid hyperthermia in cancer. They can be conjugated with the drugs and targeted to tumor tissue by applying external magnetic field. Combining drug targeting approach with hyperthermia induced by high magnetic field gives additive effect in tumor eradication. As far as drug delivery is concerned, they are in their initial developmental stage but the results in experimental animals are promising.

Types of Nanoparticles

Proteins and few other molecules in the human body are of nanosize range for example Lipoprotein (≈ 20 nm), γ-globulin (≈ 4.3x26 nm), hemoglobin (≈ 4.5x7 nm), and fibrinogen (≈ 4x76 nm). Due to their sizes they are excluded from glomerular filtration (glomerular filtration cut off: 5 nm or 45 kD) and their life in plasma is more than other micro molecules. This property can be extrapolated to certain nanoscale particles which are biodegradable or biocompatible. The size of these particles avoids their renal clearance ultimately leading to increase in plasma half life. If such particles are able to carry drug along with them then the plasma half life of drug will increase. 'Abraxane®', a Paclitaxel conjugated Albumin is nice example of the nanoparticulate drug delivery system where plasma protein is used as drug carrier. Nanoscaled particles used for drug delivery and drug targeting can be classified according to material used for their preparation. Table 9.1 represents few examples of nanoscaled drug delivery systems for systemic cancer therapy and their status of approval.

Table 9.1: Nano-vehicles in Market (Davis *et al.*, 2008)

Nanoparticle	*Latest Stage of Development*	*Examples*
Liposomes	Approved	DaunoXome, Doxil, Myocet
Albumin nanoparticle	Approved	Abraxane
PEGylated proteins	Approved	Oncospar, PEG-intron, PEGASYS, Neulasta
Polymeric micelles	Clinical trials	Genexol-PM, SP1049C, NK911, NK012, NK105, NC-6004
Polymer drug conjugate based particles	Clinical trials	XYOTAX (CT-2103), CT-2106, IT-101, AP5280, AP5346, FCE28068 (PK1), FCE28069 (PK2), PNU166148, PNU166945, MAG-CPT, DE-310, Pegamotecan, NKTR-102, EZN-2208
Gold nanoparticles	Clinical trial	CYT-6091

Polymeric Nanoparticles (Duncan, 2003)

Biocompatible and biodegradable polymers conjugated with drug are used for improving bioavailability, decreasing toxicity, and sometimes for improving therapeutic activity. Different types of polymers, such as homopolymer, block copolymers, graft copolymers, and multivalent copolymer are used for preparation of polymeric nanoparticles. Homopolymer or straight chain polymer is a simple polymer with single repeating monomer. Polyethylene glycol (PEG) is a good example of homopolymer. PEG can be activated at its either ends by replacing the hydroxyl group with more reactive amine group and conjugation with drug/ligand molecules can be done at either ends. Block copolymers contains two or more monomeric units with hydrophilic or hydrophobic properties. They are used for the preparation of micelles and polymerosomes (polymer vesicles). Graft copolymers are basically the polymers with branching, which provides more than two functional groups for attaching multiple numbers of drug molecules per molecule of a polymer and forms nanoaggregates when dispersed in appropriate solvent. Multivalent copolymers have more than one functional group that can be attached with the drug therefore the number of drug molecules per polymer is increased which boost the efficiency of conjugation and delivery. The schematic representation of all these polymers is shown in Figure 9.2.

Polymeric nanoparticles are further classified into polymeric conjugates, polymeric micelles or polymerosomes.

Polymeric Conjugates

Polymer is conjugated to drug or a targeting ligand with a covalent bond with or without spacer molecule. PEG-intron® is a conjugate of PEG and interferon approved by US FDA for antiviral therapy.

Polymeric Micelles

Polymeric micelles are produced from hydrophilic and hydrophobic block copolymers which contains hydrophobic core and hydrophilic corona in aqueous

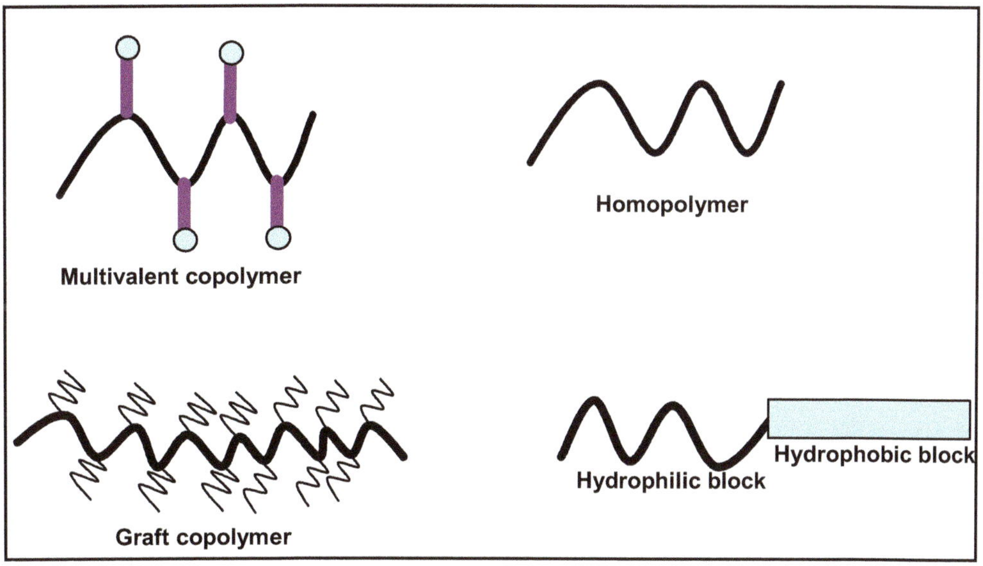

Figure 9.2: Schematic Diagram of Different Types of Polymers

solution. Drug can be incorporated in the core or conjugated to hydrophilic end of the block copolymer.

Polymeric Vesicles

Polymer vesicles are also formed from block copolymer containing hydrophilic and hydrophobic monomers. Basically the formation of micelles or polymer vesicles depends on the ratio of the hydrophilic and hydrophobic chain length of block copolymer.

Metallic Nanoparticles

Metallic nanoparticles are obtained from the inorganic materials. SPIONs are obtained from co-precipitation of ferric and ferrous chloride and are in the preclinical developmental stage for drug delivery and drug targeting. Gold nanoparticle CYT-6091 is in clinical trial.

Lipid Based Particles

Liposomes

Phospholipids in aqueous system can form bilayer structure known as liposome. Both lipophilic and hydrophilic drug can be encapsulated in liposomes. Alec Bengham first discovered these particles approximately 40 years back which are extensively studied thereafter for the drug delivery and drug targeting by different scientist groups worldwide. Stealth® liposomes are coated with polyethylene glycol to avoid uptake by reticuloendothelial system and thus increasing their plasma residence time. Doxil® is the first liposomal formulation of PEG coated liposomes approved by the US FDA for the treatment of Kaposi's sarcoma. Myocet™ and DaunoXome® are two other examples of liposomal formulations which are approved for systemic drug delivery of doxorubicin and daunorubicin respectively.

Solid Lipid Nanoparticles

These are the nanoparticles formed from various lipids which are in solid state at physiological temperature. Lipophilic drugs can be efficiently encapsulated in solid lipid nanoparticles. The choice of lipid depends on nature of drug and its compatibility with the selected lipid.

Dendrimers

Dendrimers are star shaped polymeric particles with large number of functional group available for modification. Dendritic polymers are emerging as potentially ideal drug delivery vehicles because they are easily manipulated and they provide a large density of functional groups. They possess unique properties and are considered to be most promising candidates for delivery of drug and bioactives. The following characteristics of dendrimers make them ideal candidates for conjugation with biomolecules. They are monodisperse (~1.0), nanosize (~20 nm), have multiple functional groups at the terminal, and possess end group 'tailorability' for conjugation.

Miscellaneous

Fullerene are a family of carbon allotropes, molecules composed entirely of carbon, in the form of a hollow sphere, ellipsoid, tube, or plane. Spherical fullerenes are also called buckyballs, and cylindrical ones are called carbon nanotubes or buckytubes.

Quantum dots are small devices that contain a tiny droplet of free electrons. They are fabricated in semiconductor materials and have typical dimensions between nanometres to a few microns. Quantum dots have proven themselves as a fluorescent imaging probes for detection of tumor. Polymer encapsulated quantum dots are applicable for drug delivery and drug targeting in cancer chemotherapy (Qi and Gao, 2008).

Fullerenes and quantum dots may prove to be potential agents for drug delivery and drug targeting however their toxicity remains to be evaluated. Figure 9.3 represents the schematic illustration of all the above mentioned nanoparticles.

Characteristics of Common Routes of Drug Administration for Systemic Absorption

Intravenous Route

The drugs which are not absorbed orally can be given by intravenous route. With IV administration, the drug enters directly into the systemic circulation and avoids the GI tract and, therefore, first-pass metabolism by the liver. This route permits a rapid effect and a maximal degree of control over the circulating levels of the drug. However, unlike drugs present in the GI tract, those that are injected cannot be recalled by strategies such as emesis or binding to activated charcoal. Intravenous injection of some drugs may introduce bacteria through contamination, induce hemolysis, or cause other adverse reactions by the too rapid delivery of high concentrations of drug to the plasma and tissues. Therefore, the rate of infusion must be carefully controlled.

Subcutaneous Route

Drug is injected in the subcutaneous layer from where it is absorbed through blood capillaries and enters into the systemic circulation. Thus the direct entry of the drug into systemic circulation is avoided. Also the rate of absorption is decreased and duration of action is prolonged. The onset of action is slightly increased as compared to the IV administration. Lidocain, levonorgestrel are the few examples of drugs administered through subcutaneous route.

Intramuscular Route

Drugs administrated through intramuscular route can be aqueous solutions or specialized depot preparations. Drug is rapidly absorbed from the aqueous solution whereas the depot preparation which is basically the suspension of drug in non-aqueous vehicle like ethylene glycol or peanut oil is absorbed slowly over a longer period of time. An example is sustained-release haloperidol decanoate, whose slow diffusion from the muscle produces an extended neuroleptic effect.

Oral route: Despite very poor absorption the oral route is the most preferred route of drug administration due to its compliance and safety. The rate of absorption is slow, the onset of action is increased and the duration of action is also increased.

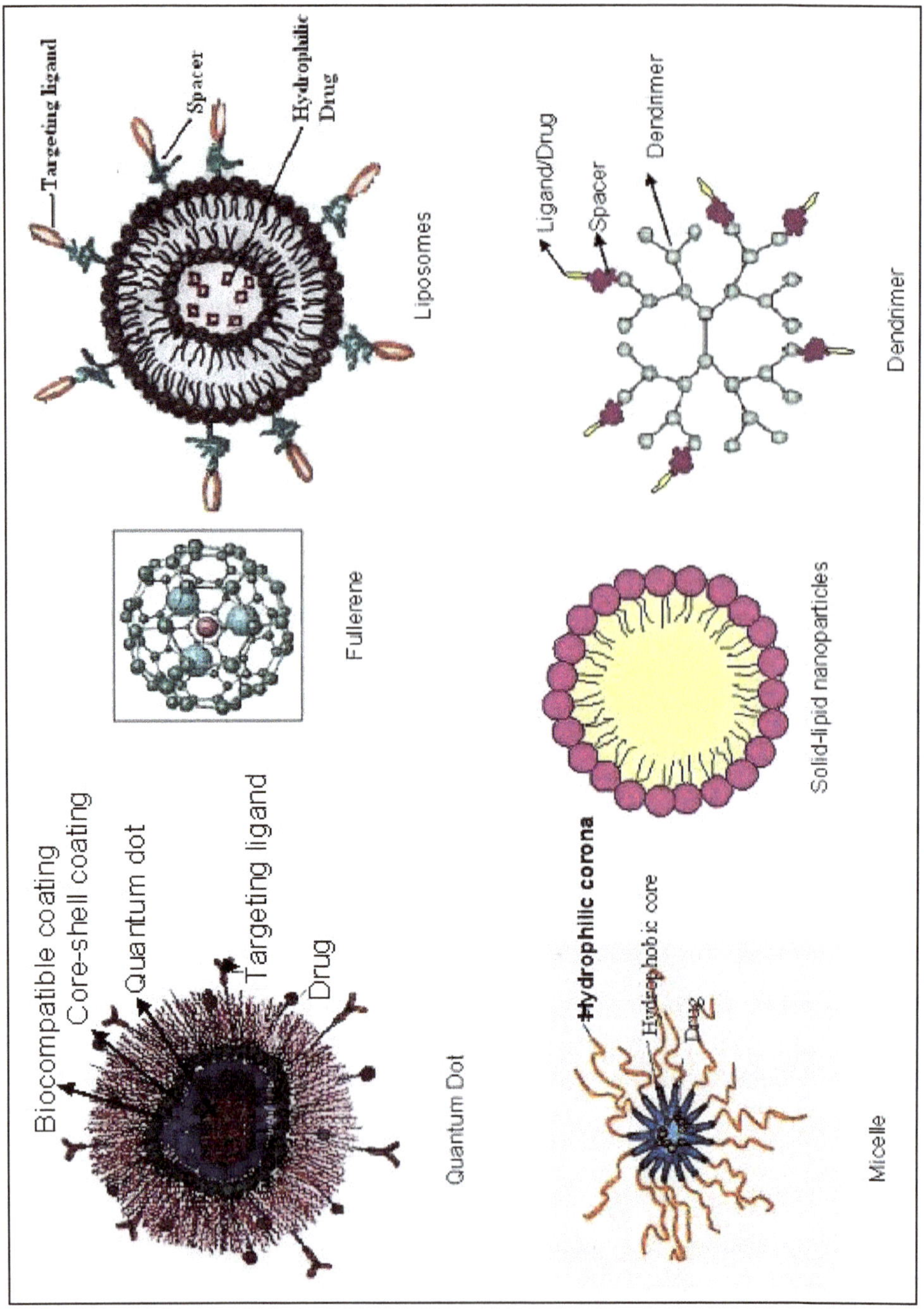

Figure 9.3: Schematic Illustration of Various Nanoparticles

First pass metabolism due to liver and gut microsomal enzymes is involved. Also drug has to cross the gut barrier for entry into the systemic circulation.

Other routes of drug administration like nasal route, intrathecal, intraspinal, trasdermal, intracerebrospinal, vaginal, rectal posses their unique advantages and applications but they are not the common routes of drug administration for achieving systemic absorption. Table 9.2 shows the comparison of the characteristics of the common routes of drug administration.

Table 9.2: Characteristics of Common Routes of Drug Administration

Route	*Absorption Pattern*	*Special Utility*	*Limitations and Precautions*
Intravenous	Absorption circumvented potentially immediate effect	Valuable for emergency use permits titration of dosage, Suitable for large volumes and irritating substances if diluted	Increased risk of adverse effects, must be injected slowly, not suitable for oily solutions or insoluble substances, Chance of infection
Subcutaneous	Prompt from aqueous solutions, slow and sustained from repository preparations	Suitable for some insoluble suspensions and for implantation of pellets	Not suitable for large volumes, possible slough from irritating substances
Intramuscular	Prompt from aqueous solutions, slow and sustained from repository preparations	Suitable for some oily vehicles, moderate volumes and some irritating substances	Precluded during anticoagulant medication, may interfere with interpretation of some diagnostic tests *e.g.* (creatinin posphokinase)
Oral ingestion	Variable, Depends upon many factors like gut barrier, first pass metabolism	Most convenient, safe and economical	Requires patient cooperation, absorption potentially erratic and incomplete for drugs that are poorly soluble and absorbed slowly

Oral Drug Delivery

Oral route is the most preferred route of drug administration. It is most economical, safe and patient compliant route, non invasive as needle is not required for its administration, self administered as the patient himself can take the dose without visiting the physician whereas other routes are supposed to be used under strict medical supervision. Besides its advantages the oral route possess several disadvantages. Poor aqueous solubility, chemical instability in stomach or gastric pH, metabolic instability in the presence of proteolytic enzymes and poor permeability through gastrointestinal membrane limits the absorption of the drug through oral route.

Some drugs demonstrate the local absorption phenomenon where they are absorbed from specific locations of the gastrointestinal tract. It is a challenging task to target the drugs at particular sites of the GI tract for optimum absorption. Several other challenges associated with the oral delivery are the dilution of the administered drug and it is difficult to maintain optimum concentration in body by using oral route of drug delivery. So in many cases there is no other choice for the parentral route. Targeted drug delivery to the specific tissue or organ in the body seems to be difficult through oral route but targeting intestinal sites for increasing absorption has been studied extensively. Lymphatic targeting where the drug is absorbed into the lymph nodes and further enters into thoracic lymph duct. Thoracic lymph duct empties into the blood venous system at the juncture of internal jugular vein and subclavian vein. The rate of absorption is very slow as the flow rate of lymph is not more than 100 ml/min but the first pass effect is avoided and hence more amount of the drug enters into the circulation. Colon targeting using specific polymers which are metabolized by the bacterial flora is another example of targeted drug delivery through oral route. Floating tablets, osmotic pumps, self emulsified drug delivery systems are few novel drug delivery techniques used to overcome some of the limitations of oral drug delivery.

Targeted Drug Delivery Using Nanoparticles

The problems associated with systemic drug delivery are drug distribution throughout the body, lack of drug specific affinity, non-specific toxicity and harmful side effects. Drug targeting can be an effective solution to eliminate these problems associated with the conventional drug delivery systems. With targeted drug delivery system several objectives can be achieved such as therapeutic index of the drug increases a number of folds, nondiscriminate uptake of the toxic agents is decreased, distribution of the drug is restricted to the target site, the dose required is reduced and finally toxicity of the drug to other non pathological sites is minimized.

Drug targeting can be classified according to the mechanism of targeting such as:

1. Physical
2. Biological, or
3. Chemical systems

Physical targeting was basically started with the prime objective of modifying the pharmacokinetics of the drug without achieving any specificity. Biodegradable polymers and osmotic pumps are few examples of physical targeting. Direct application of the drug at target site such as pilocarpine drug delivery to the eye using polymer device (Occusert®) and contraceptive sustained release of progesterone from vaginal polymeric inserts is a form of physical targeting.

The biological targeting system is based on antibodies, in principle, could result in a highly desirable delivery profile. Drug conjugated antibodies form a good targeting system with high specificity; however, drug conjugation is prone to affect the specificity and affinity. Other issues like problems of stoichiometry, that is, how many molecules can be conjugated and the enzymatic or immunological cleavage of the antibody.

Chemical system of drug targeting consists of wide variety of targeting systems possible both theoretically and practically. They can be divided into three classes *viz.*

1. Enzymatic physical-chemical-based targeting,
2. Site-specific enzyme-activated targeting, and
3. Receptor-based chemical targeting.

With the targeted drug delivery the various objectives to overcome the limitations of conventional drug delivery system can be achieved. There is a very high demand for targeted drug delivery systems due to their safety and efficacy. The current drug delivery market size is given in Table 9.3.

Table 9.3: The Current Drug Delivery Market Size

Year	*Market Size*
2003	$50 billions
2006	$67 billions
2008	$90 billion
Total pharma market	$500 billion in 2008

Targeted Drug Delivery in Cancer

Cancer chemotherapy strives to cause a lethal cytotoxic lesion that can arrest a tumor's progression. The attack is generally directed against metabolic sites essential to cell replication, for example, the availability of purine and pyrimidine precursors for DNA or RNA synthesis. Ideally these drugs should interfere only with cellular processes unique to the malignant cells. However, currently available anticancer drugs do not specifically recognize neoplastic cells, but rather affect all proliferating cells both normal and abnormal. Almost all antitumor agents have a steep dose-response curve for both toxic and therapeutic effects, thus it is important to tailor the drug doses to the physical state of the patient. Tumor targeted drug delivery is the best solution for avoiding the cytotoxic effects of the chemotherapeutic agents to normally proliferating cells.

Passive Tumor Targeting

Passive targeting relies on the properties of tumor tissue with a consideration of disease pathology for preferential accumulation of drug in tumor tissue and hence decreases non-specificity. It is based on difference in endothelial structures called fenestration between normal vasculature and tumor associated vessels. Tumor results due to uncontrolled proliferation of the cells. To maintain increased growth rate and the metabolic needs it requires lots of nutrients and oxygen. Tumor vasculature is anatomically different than the normal vasculature in various aspects like, disrupted basement membrane, increased gap junction between the endothelial cells, and irregular diameter of the blood capillaries. Various vascular mediators such as Bradykinin, prostaglandins, nitric oxide, matrix metalloproteinase and vascular endothelial growth factor found to be increased in the plasma of the tumor patients. The intracellular space between the endothelial cells of normal blood vessels is on an average 6 nm whereas that in the tumor vasculature it increases up to 1.7 μm (ranging from 0.3 to 4.7 μm).

Increase in the permeability of the tumor vasculature is very useful phenomenon as far as passive drug targeting is concerned (Figure 9.4). The intracellular space between the endothelial cells is increased up to 1.7 μm allowing the particles in the

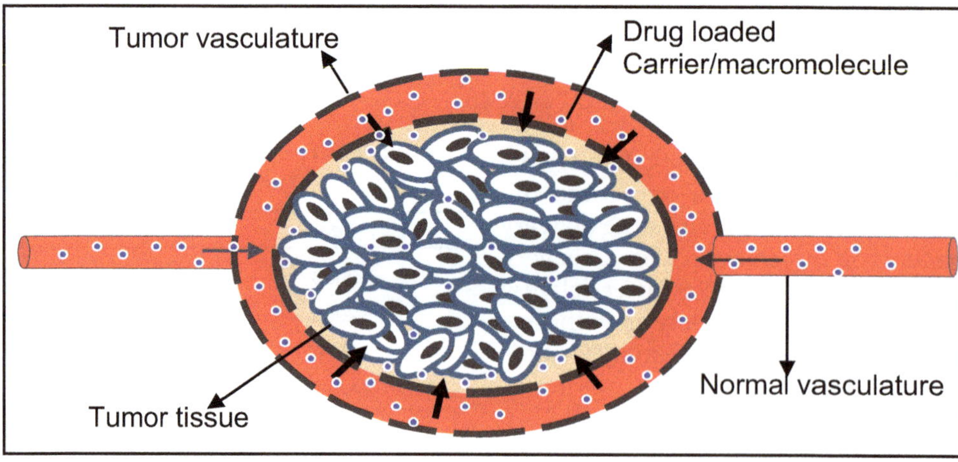

Figure 9.4: Passive Targeting by Enhanced Permeability and Retention: A schematic diagram showing increased in the porosity of endothelial cells lining the tumor vasculature leading to increase in the permeability of tumor vasculature.

nanosize range to easily extravasate through capillaries and accumulate in tumor tissue. Vascular permeability in tumor is quite high compared to glomerular filtration. Particles above 5 nm size are not filtered though glomeruli, however they can still enter into the tumor tissue. There is no question of extravasation through normal blood vessels as the endothelial cells are tightly bound. Hence it is possible to target the cancer cells by using drug loaded nanoparticles or polymer conjugated macromolecules. For passive targeting the drug carrier must be larger than 5 nm (or >45 kD) otherwise they would be rapidly cleared by the kidney. The upper limit of the nanoparticle size is yet not clear but 100 nm is fairly acceptable as the larger particles are rapidly taken up by the macrophages. Polymer coated nanoparticles escapes from the macrophages as there is no adsorption of serum opsonins on their surface (Greish, 2007; Maeda *et al.*, 2004).

Active Tumor Targeting

Receptor Mediated Endocytosis Pathway for Targeted Drug Delivery

Endocytosis is defined as a process of energy dependant uptake of extracellular material(s) into cells and their multiphasic vesicle enclosed transport into lysosomes for complex enzymatic processing. It include pinocytosis, phagocytosis and receptor mediated endocytosis (RME). RME occurs in pitted membrane regions lined either by the assistor proteins caveolin-1 or clathrin. Characterization of proteins involved in the distinct paths for a number of RME receptors has been studied using biochemical modifiers along with fluorescence imaging and confocal microscopy. It has been reported that, drug loaded nanoparticles when conjugated to the ligands or monoclonal antibodies against the receptors involved in RME, their uptake inside the cell increases. Numerous receptors, including transferrin (Yang *et al.*, 2008a), asialoglycoprotein (Biessen *et al.*, 1999), epidermal growth factor (EGF) (Mamot *et al.*, 2003), folate (Yang *et al.*, 2008b) and chemokine (Biragyn *et al.*, 2004), serve as high affinity binding sites and have been investigated for their use in targeting to different cell types. A schematic illustration of RME based drug targeting is shown in Figure 9.5.

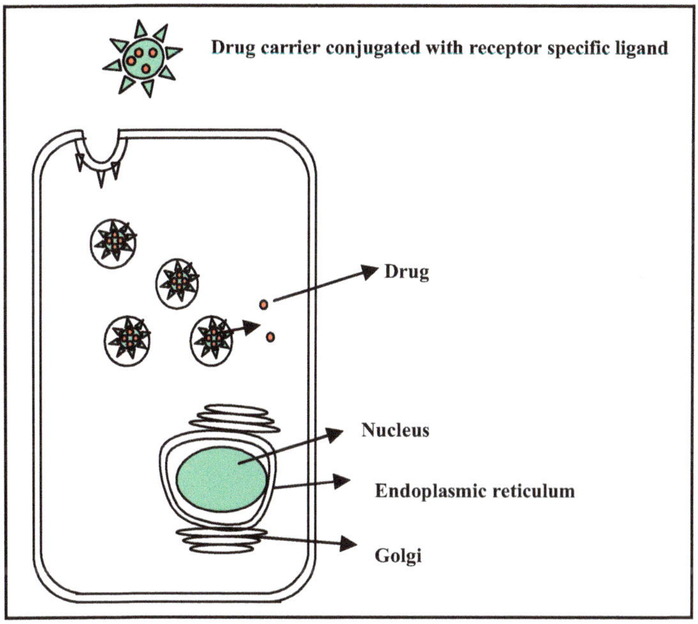

Figure 9.5: Receptor Mediated Endocytosis for Intracellular Drug Delivery: A schematic diagram showing the endocytosis of drug loaded carrier conjugated with the receptor specific ligand. Drug loaded carrier is taken up by the RME mechanism and the drug release from the carrier system takes place due to acidic pH of late endosomes.

Hyperthermia Therapy

Tumor cells are very vulnerable to small increase in temperature. Based on this characteristic property tumor hyperthermia therapy is developed. Whole body hyperthermia is used for cancer therapy. SPIONs when injected in the body they accumulate into tumor tissue due to enhanced permeability of the tumor vasculature. The external magnetic field can also be used for guiding the SPIONs towards the tumor. Local hyperthermia can be achieved by generating therapeutic heat by external magnetic field, using SPIONs. Also it has been proved that hyperthermia along with chemotherapy gives an additional effect in eradication of tumor cells (J. van der Zee, 2002). Figure 9.6 schematically represents the hyperthermia therapy using external magnetic field.

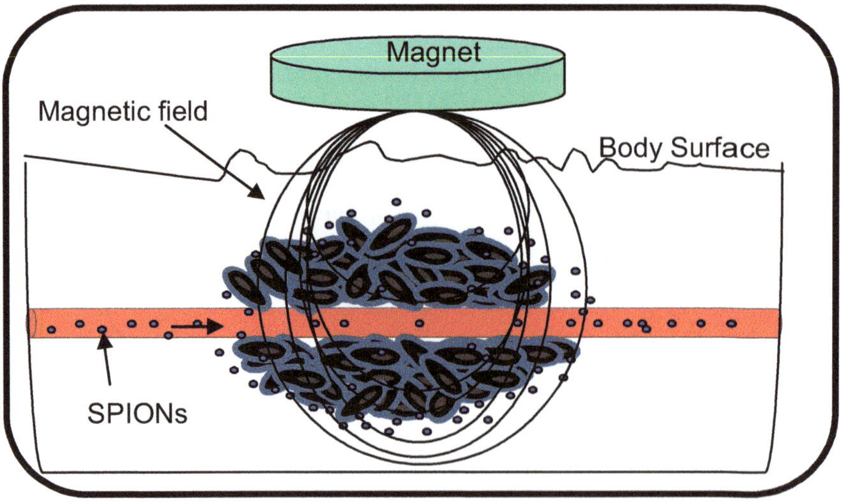

Figure 9.6: Hyperthermia Therapy Using External Magnetic Field

Earlier Work

Recently in our laboratory, using dextran coated superparamagnetic iron oxide nanoparticles. Pinocytic organelles have been isolated and characterized from rat peritoneal macrophages. To achieve the purification of pinosomes in adequate quantity and purity, 1 mg/ml colloidal iron-dextran particles along with 1 mg/ml FITC-dextran was fed to rat peritoneal macrophages for varied time and the endocytic organelles of different time-points were isolated by retention on column in presence of electromagnetic field. The pinosomes isolated were highly electrodense in nature.

Another off-shoot of the project on endocytosis in our laboratory was of synthesis of complex receptor-specific fluorescent labeled nano-sized superparamagnetic iron oxide nanoparticle which is shown as in Figure 9.7. The first step of the multi-step ligand preparation was the synthesis of magnetite nanoparticles (Fe_3O_4), which were characterized by their paramagnetic nature *i.e.* they show magnetic properties only in presence of magnetic field. The next step was coating of the magnetite particles to stabilize them as ferrofluid by preventing the agglomeration of the particles. To achieve the same the magnetite particles were coated with fluorescence labeled protein *i.e.* RBITC-BSA which would act as an antigen in the ligand and would be recognized by its specific antibody to be further recognized by Fc receptors present on the macrophage cell surface. Finally, the coated iron oxide particles were incubated with antibody against BSA to achieve synthesis of a receptor-specific ligand.

Transmission electron microscopic examination of the indigenously prepared iron oxide nanoparticle ligand stained with 1 per cent uranyl acetate revealed presence of evenly spaced 17-20 nm particles (Figure 9.8). Receptor-specific nature of the ligand was established using low temperature (10° C). Rat peritoneal macrophages were incubated with the indigenously prepared receptor-specific superparamagnetic iron oxide ligand for 30 minutes at low temperature, following which the cells were washed

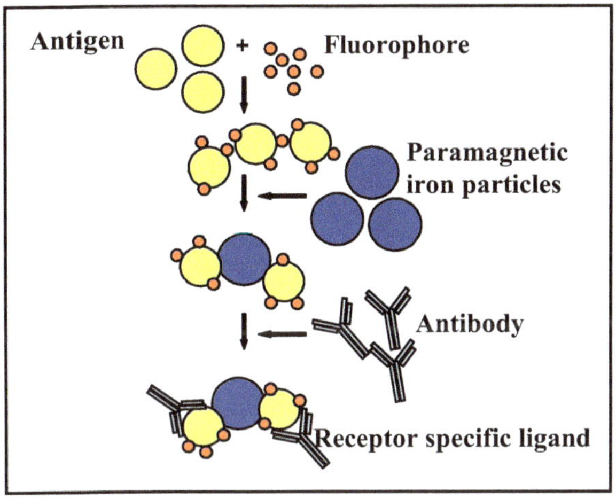

Figure 9.7: Preparation of Receptor-specific Ligand

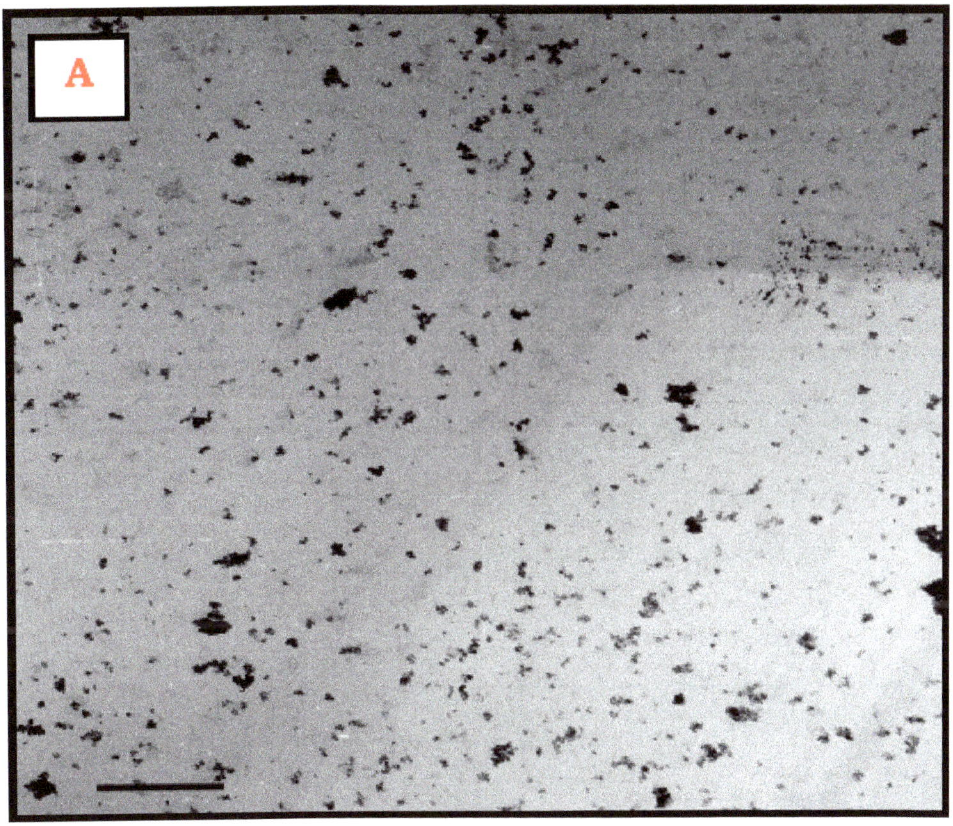

Figure 9.8: Transmission Electron Micrograph of Receptor-Specific Superparamagnetic Iron Oxide Ligand

free of ligand and shifted to 37° C for 20 minutes to allow internalization of the receptor bound ligand. Subsequently cells were homogenized and the post nuclear supernatant was passed through glass column (containing steel wire matrix) in presence of electromagnetic field. The endocytic vesicles thus isolated were characterized microscopically by fluorescence as well as transmission electron microscope (Figures 9.9 and 9.10). The magnetically retained fraction was also subjected to various biochemical and immunological characterization, which revealed presence of pure endosomal preparation devoid of lysosomal contamination in the magnetically retained fraction. Thus using the indigenously synthesized superparamagnetic iron oxide nanoparticle ligand and by employing electromagnetic purification technique, a pure endosomal preparation with ~ 10-22-fold purification, up to 65 per cent yield and <2 per cent contamination of other sub-cellular organelles could be obtained (Padh *et al.*, 1989a; Padh *et al.*, 1989b; Padh *et al.*, 1993; Neekhara and Padh, 2004; Nolta e.al, 1991; Nolta *et al.*, 1993; Padh 1995; Padh and Tanjore, 1995; Vats, 2007).

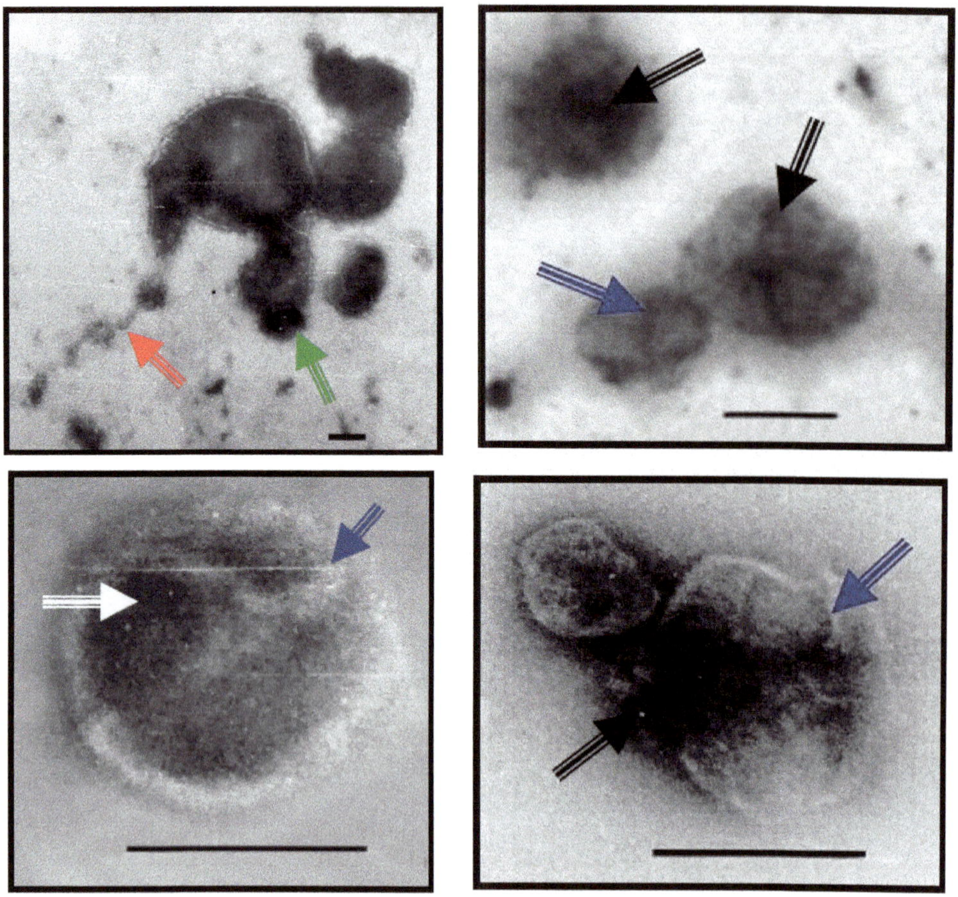

Figure 9.9: Transmission Electron Micrograph of Magnetically Retained Fraction

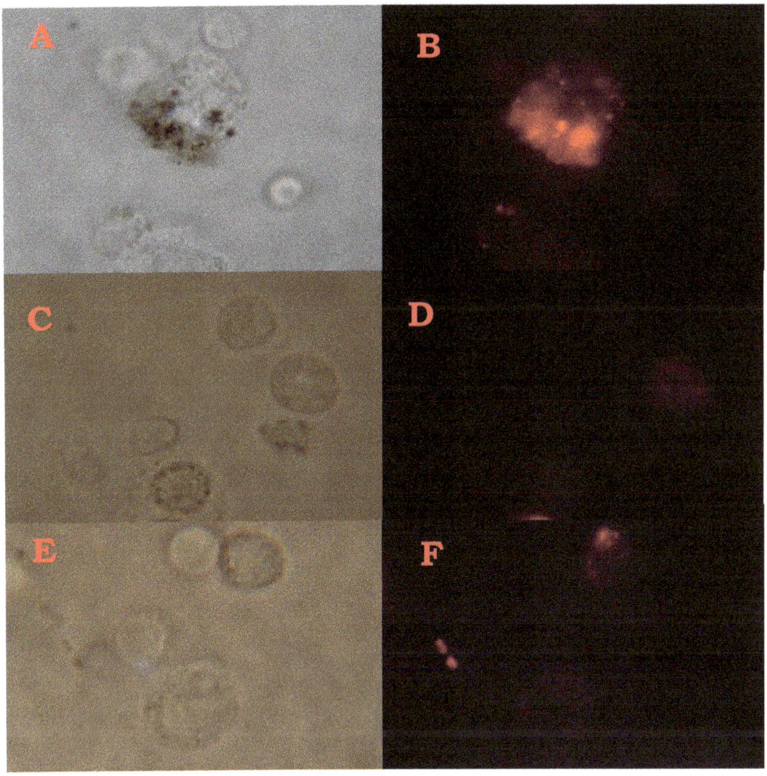

Figure 9.10: Cell Uptake of Receptor Specific SPIONs

SPIONs conjugated to folate increases the intracellular uptake in folate receptor (FR) overexpressing tumor cells as the rate of receptor mediated endocytosis is higher than other endocytosis processes. Das *et al.* (2008) have reported a strategy for combined cancer targeted magnetic resonance and therapy, using 3-aminopropylphosphonic acid (APPA) coated SPIONs as the vector. APPA, as a surface coupling agent has excellent biocompatibility and strong affinity towards transitional metal containing supports. They have synthesized a complex multifunctional nano-conjugate, based on the APPA functionalized magnetite core. Rhodamine isothiocynate (RITC, an imaging agent), folic acid, and methotrexate (MTX, a chemotherapeutic agent) were subsequently coupled to the hybrid nanomaterial to endow simultaneous targeting, imaging and intracellular drug delivery capability. Further to demonstrate the cancer targeting ability of these particles they performed *in vitro* studies using human cervical carcinoma HeLa cells, known to significantly over-express the folate receptor and MG-63 cells which do not express folate receptors. A significant increase in the uptake of SPIONs with folate conjugation was observed in FR^{+ve} HeLa cells whereas non-folate targeted SPIONs have almost similar uptake in FR^{+ve} HeLa cells and FR^{-ve} MG-63 cells (Figure 9.11a). Confocal microscopy (Figure 9.11b) demonstrates the cellular internalization and apoptotic effect of the nano-conjugate against FR^{+ve} human cervical HeLa cancer cell line. The study clearly demonstrates that receptor mediate

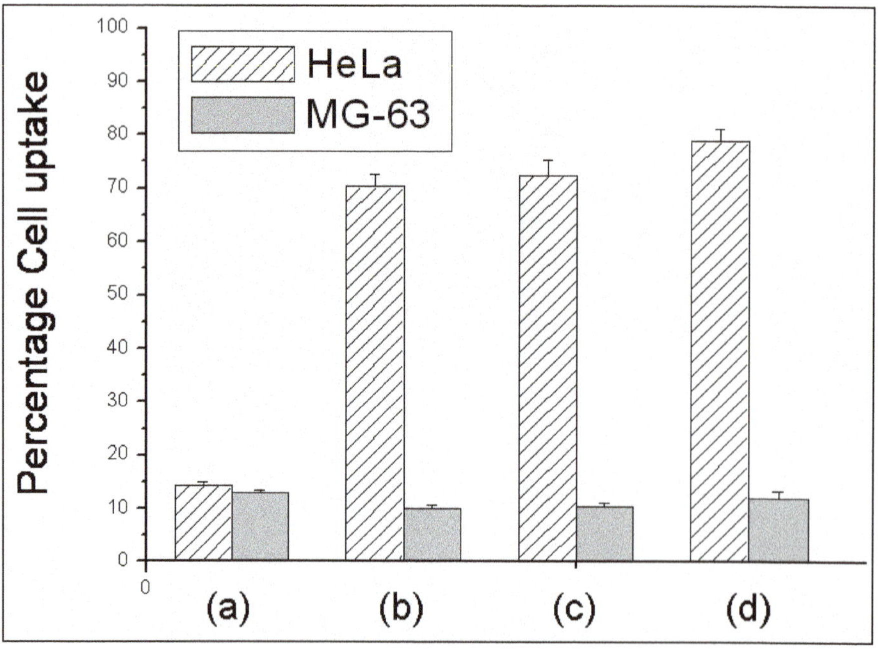

Figure 9.11(a): Uptake of (a) Fe_3O_4-APPA-RITC-OH (b) Fe_3O_4-APPA-RITC-FA (c) Fe_3O_4-APPA-RITC-FA-OH-MTX and (d) Fe_3O_4-APPA-RITC-FA-OH-MTX under an external magnetic field by folate receptor (FR) positive HeLa cells vs. FR-negative MG-63 cell lines quantified by flow cytometry (Das and Pramanik, 2009).

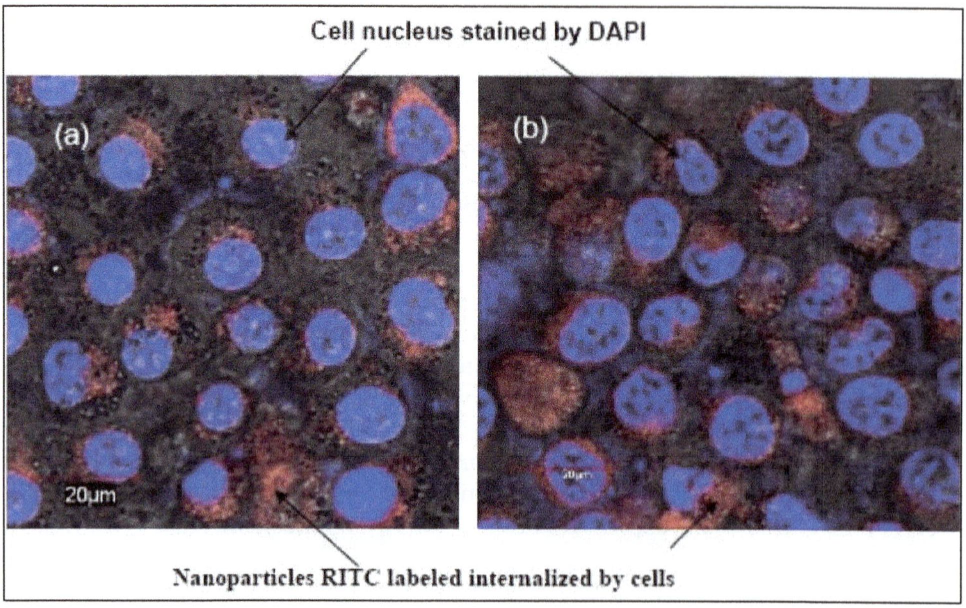

Figure 9.11(b): Phase Contrast Confocal Images of Folate-Receptor (FR) Positive Human Cervical HeLa Cancer Cells Incubated with (a) Fe_3O_4-APPA-RITC-FA (taken as control) and (b) Fe_3O_4-APPA-RITC-FA-OH-MTX for 4 h (Das *et al.,* 2008).

endocytosis significantly increases the uptake of nanoparticles thereby increasing its availability at intracellular target site and increasing the efficacy as well.

Hypothesis

Today hyperthermia is considered as "main stream" as surgery, chemotherapy, and radiation and is now recognized as the "fourth modality" in approved cancer treatment. However, due to its relatively recent acceptance in major medical circles, it is not well known. Although potential synergism and benefit of combined hyperthermia and radiation for cancer treatment is well established, but is yet to be optimized clinically *i.e.* the heat delivery via external applicators or interstitial antennas has not demonstrated appropriate positive therapeutic ratio. Even magnetic fluid hyperthermia involves local injections of iron oxide nanoparticles. To achieve a relatively non-invasive treatment of cancer and a positive therapeutic ratio we intend to prepare an iron oxide nanoparticle based targetable anticancer therapeutic which would target the folate receptors over-expressed on cancer cells. It would be a dual approach where the chemotherapeutic as well as folate/transferrin molecules will be conjugated to iron oxide nanoparticles. Owing to the presence of folate/transferrin molecules on the tri-conjugate anticancer therapeutic, we thereby assume that its uptake would be via receptor-mediated endocytosis. Targeted delivery of SPIONs and the chemotherapeutic would facilitate their accumulation in metastatic cancer cells. Thereby, use of hyperthermia in conjunction with the conjugate would be employed to ensure complete killing of cancer cells. Secondly, we are also preparing the liposomal particles conjugated with folate/transferrin on the outer surface and encapsulating the drug into its aqueous core or organic bilayer, depending on the solubility of the drug. Some synthetic phospholipids such as Distearoyl phosphatidyl ethanolamine can be conjugated with the receptor specific ligands. The conjugated phospholipid is easily incorporable into the liposomes during its synthesis.

Specific Objectives

The basic aim was to synthesis a target specific tri-conjugate biosafe anticancer therapeutic which cures cancer by dual approach using SPIONs as nanocarrier. First approach being Chemotherapy, as we incorporated anticancer therapeutic Paclitaxel as well as a targeting moiety which is over expressed in various cancer cells *i.e.* Folate along with stabilizing/solubilizing polymer Polyethylene Glycol (PEG) on to the surface of SPIONs. Whereas, the second approach would be Hyperthermia, in which we can achieve localization of magnetic particles to tumor tissue using an external magnet and also can generate heat with the help of external magnetic field by making the particles to vibrate. The steps involved in the preparation of tri-conjugate biosafe superparamagnetic iron oxide nanotherapeutic would be in four steps, first would be standardization of synthesis of core which is superparamagnetic iron oxide nanoparticle (SPIONs), second would be activation of polymer Polyethylene Glycol (PEG) to PEG bis-amine, third would be activation of folic acid and paclitaxel for reaction with polyethylene glycol bis-amine, and the last step would be adsorption of conjugated polyethylene glycol onto the synthesized SPIONs. The complete assembly of tri-conjugate biosafe anticancer nanotherapeutic will be as shown in Figure 9.12B.

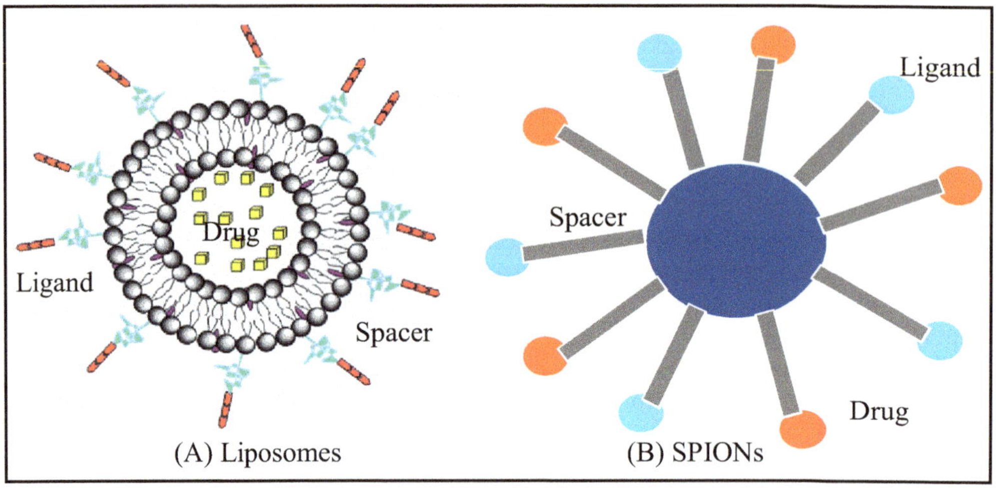

Figure 9.12: Design of the RME Based Targeted Drug Delivery System with (A) Liposomes and (B) SPIONs as Carriers

Liposomes, which are prepared from the phospholipids, are relatively non toxic. They can carry more amount of drug into its aqueous core. Two drugs can be given simultaneously using single liposomal formulation. The design of targeted liposomes is shown in Figure 9.12A.

Our Preliminary Studies

Synthesis of SPIONs Conjugated with Folic Acid and Paclitaxel

SPIONs were synthesized by co-precipitation of ferrous and ferric chloride using liquid ammonia. Further they were coated with PEG-bis-folate and PEG-bis-paclitaxel for stability and folate specific delivery. PEG-bis-amine was selected for in house synthesis of PEG-bis-folate and PEG-bis-paclitaxel, as the two amino groups of PEG-bis-amine can be easily attached with folate and drug. Since the hydroxyl groups of polyethylene glycol are hindered and are less reactive, they need to be converted to amine as a terminal functional group for PEG which is comparatively more reactive. Firstly, hydroxyl group of polyethylene glycol was modified by a leaving group 4-Florobenzene sulphonyl Chloride (FBSCl) resulting in bis-florobenzene sulphonyl PEG, which in presence of ammonium hydroxide gets converted to PEG bis-amine (Figure 9.13). The formation of PEG bis-amine was confirmed using Thin Layer Chromatography (TLC). PEG-bis-folate was synthesized by treating activated folate with PEG-bis-amine according to previously reported method (Lee and Lows, 1994). For conjugation of Paclitaxel with polyethylene glycol, 2'OH group (Figure 9.13) needs to be activated. Paclitaxel was reacted with Succinic anhydride which gives 2'OH succinyl Paclitaxel. The activated Succinyl Paclitaxel was reacted with PEG-bis-amine to obtain PEG-bis-paclitaxel.

PEG-bis-paclitaxel and PEG-bis-folate were attached to the surface of SPIONs and further characterization will be done by transmission electron microscopy and

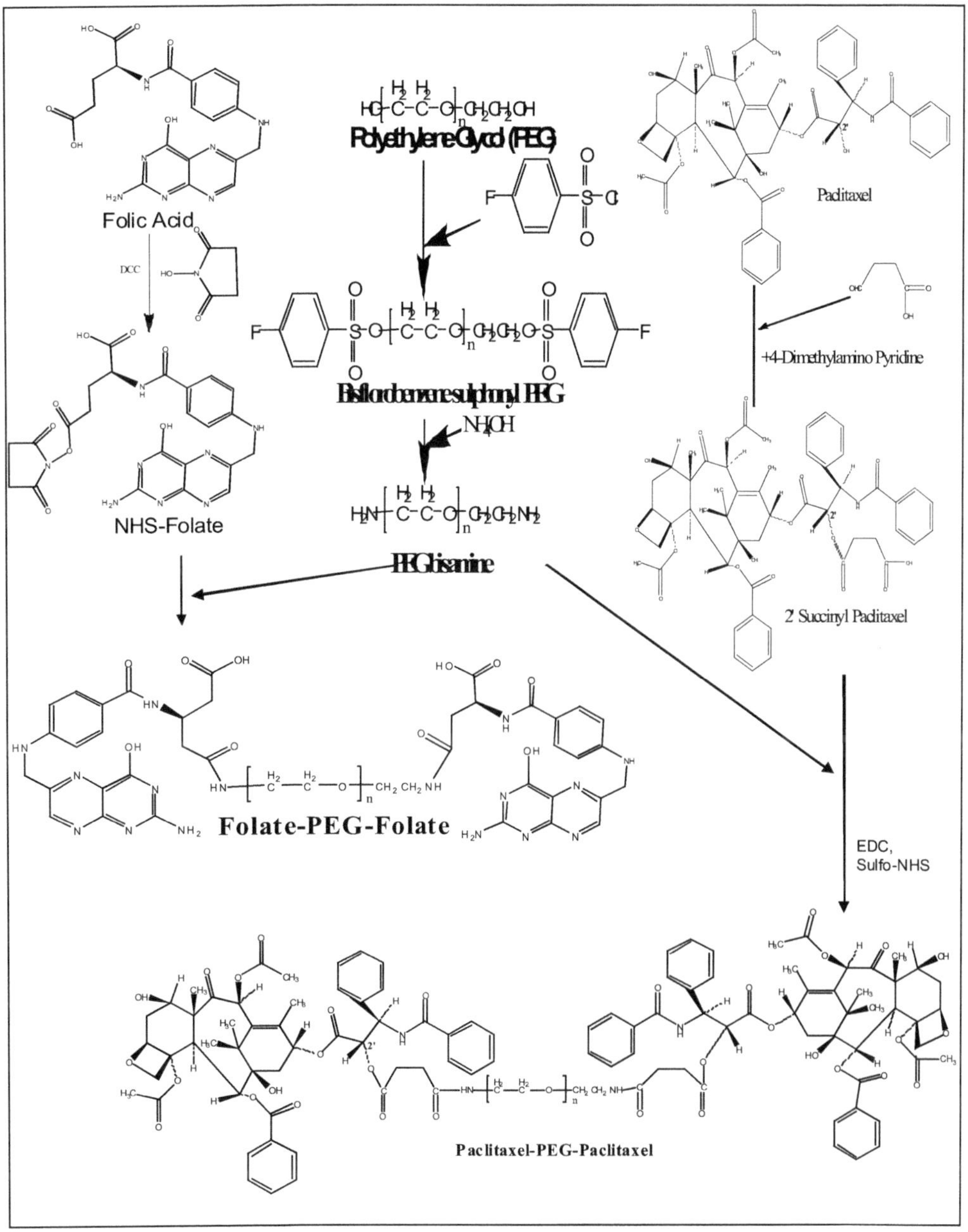

Figure 9.13: Chemical Reactions for Synthesis of PEG bis-amine, Folate-PEG-Folate and Paclitaxel-PEG-paclitaxel

dynamic light scattering. Cytotoxicity and cell uptake studies will be done in cancer cell lines. Figure 9.13 shows the reaction mechanisms of the above mentioned reactions.

Liposomes

Liposomes were synthesized by lipid dehydration rehydration method using phosphatidylcholine and cholesterol using 1:1 and 2:1 molar ratio of PC:CH and gemcitabine as a drug candidate. 30 per cent entrapment of gemcitabine was obtained at 1:1 molar ratio whereas at 2:1 molar ratio the entrapment was only 11.04 per cent indicating that cholesterol does increase the entrapment. This may be due to inhibition of leaking of drug as it is reported that cholesterol increases membrane rigidity. Size of the particle was determined by dynamic light scattering technique and was found to be 120 nm. Further we are going to synthesize folate conjugated liposomes and test them in human cancer cell lines.

Conclusion

Today targeted drug delivery in case of cancer chemotherapy is a challenge to researchers. In our study we are targeting cancerous tissue based on two approaches, first being ligand-receptor interaction and second one being magnetically directed. With that basis ~10 nm SPIONs and about 50-100 nm of PEG coated SPIONs were prepared. Our strategy was to couple ligand as well as chemotherapeutic drug to PEG and then coat it on the SPIONs. For the same, firstly PEG bis-amine was synthesized from Polyethylene Glycol. Simultaneously activation of Folic acid by N-hydroxysuccinimide and activation of Paclitaxel by converting it into 2′ succinyl paclitaxel was also achieved. This activated Folic acid and activated Paclitaxel were then reacted with PEG bis-amine to form Folate-PEG-Folate and Paclitaxel-PEG-Paclitaxel respectively.

Summary

Targeted Drug delivery will ensure better availability of drug to the target site increasing its efficacy, simultaneously decreasing the side reactions and also there would be decrease in drug dosage requirement for a desired effect. One of the important issues is the drug loading uniformity which ultimately decides the dose and their safe clearance from the body without any toxicity. In some cases extensive chemical modifications in the drug affects their therapeutic activity which is not acceptable. In spite of all this, most of the polymer therapeutics and liposomal formulations are currently in the different phases of clinical trial and few of them are approved for clinical use. Let us hope that in near future we will see more number of targeted drug delivery system in clinics and subsequent decrease in the horrible toxic effects of chemotherapeutic agents making more patients friendly. Our objective here provides promising approach for targeted drug delivery using magnetic nanoparticles. But, there are various issues which needed to be taken care of. Will the tri-conjugate anticancer therapeutic synthesized and stabilized? Will it be specifically taken up via ligand receptor interaction? Will hyperthermia in conjugation with chemotherapy prove beneficial? Will there be any toxicity using iron oxide nanoparticles?

Acknowledgement

The authors gratefully acknowledge DBT, New Delhi for financial support for this work.

References

Biessen, E.A., Vietsch, H., Rump, E.T., Fluiter, K., Kuiper, J., Bijsterbosch, M.K. and Berkel, T.J.V., 1999. Targeted delivery of oligodeoxynucleotides to parenchymal liver cells in vivo. *Biochem. J.*, 340: 783–792.

Biragyn, A., Ruffini, P.A., Coscia, M., Harvey, L.K., Neelapu, S.S., Baskar, S., Wang, J.M. and Kwak, L.W., 2004. Chemokine receptor–mediated delivery directs self–tumor antigen efficiently into the class II processing pathway in vitro and induces protective immunity in vivo. *Blood*, 104: 1961–1969.

Das, M. and Pramanik, P., 2009. Development of superparamagnetic nanoparticles for targeted cancer therapy. Abstract no. 46, *NIPER–Winter School*, February 24–28.

Das, M., Mishra, D., Maiti, T.K., Basak A. and Pramanik, P., 2008. Bio-functionalization of magnetite nanoparticles using an aminophosphonic acid coupling agent: new, ultradispersed, iron-oxide folate nanoconjugates for cancer-specific targeting. *Nanotechnology*, 19: 415101.

Davis, M.E., Chen, Z. and Shin, D.M., 2008. Nanoparticle therapeutics: an emerging treatment modality for cancer. *Nat. Rew. Drug Discov.*, 7: 771–782.

Duncan, R., 2003. The dawning era of polymer therapeutics. *Nat. Rew. Drug Discov.*, 2: 347–360.

Greish, K., 2007. Enhanced permeability and retention of macromolecular drugs in solid tumors: A royal gate for targeted anticancer nanomedicines. *J. Drug Target.*, 15: 457–464.

J. van der Zee, 2002. Heating the patient: A promising approach? *Annals of Oncology*, 13: 1173–1184.

Lee. R.J., and Low, P.S., 1994. Delivery of Liposomes into CulturedK B Cells via Folate Receptor-mediated Endocytosis. *J. Biol. Chem.*, 271: 8481–8487.

Maeda, H., Fang, J., Inutsuka, T. and Kitamoto, Y., 2003. Vascular permeability enhancement in solid tumor: various factors, mechanisms involved and its implications. *Int. Imunopharmacol.*, 3: 319–328.

Mamot, C., Drummond, D.C., Greiser, U., Hong, K., Kirpotin, D.B., Marks, J.D. and Park, J.W., 2003. Epidermal growth factor receptor (EGFR)–targeted immunoliposomes mediate specific and efficient drug delivery to EGFR- and EGFRvIII overexpressing tumor cells, *Cancer Res.*, 63: 3154–3161.

Neekhra, N. and Padh H., 2004. An insight into molecular mechanism of endocytosis. *Indian J. Biochem. Biophys.*, 41: 69–80.

Nolta, K.V., Padh, H. and Steck, T.L., 1991. Acidosomes from Dictyostelium: initial biochemical characterization. *J. Biol. Chem.*, 266: 18318–18323.

Nolta, K.V., Padh, H. and Steck, T.L., 1993. An immunocytochemical analysis of the vacuolar proton pump in Dictyostelium discoideum. *J. Cell Sci.*, 105: 849–859.

Padh, H., Ha, J., Lavasa, M. and Steck, T.L., 1993. Post–lysosomal compartment in Dictyostelium discoideum. *J. Biol. Chem.*, 268: 6742–6747.

Padh, H., Lavasa, M. and Steck, T.L., 1989b. Characterization of a vacuolar proton ATPase in Dictyostelium discoideum. *Biochimica et Biophysica Acta*, 982: 272.

Padh, H., Lavasa, M., and Steck, T.L., 1989a, Prelysosomal acidic vacuoles in Dictyostelium discoideum. *J. Cell Biol.*, 108: 865–874.

Qi, L. and Gao, X., 2008. Emerging applications of quantum dots for drug delivery and therapy. *Expert Opin. Drug Deliv.*, 5: 263–267.

Shashindran and Gitanjali, 2006. Adverse Drug Reaction Monitoring. *Health Administrator, Indian Society of Health Administrators*, 19: 20–21.

Vats, N., 2007. Electromagnetic Purification and Characterization of Endosomes from Mammalian Cells. *A Thesis*, The Bhavnagar University.

Yang, X., Chen, Y., Yuan, R., Chen, G., Blanco, E., Gao, J. and Shuai, X., 2008b. Folate-encoded and Fe_3O_4-loaded polymeric micelles for dual targeting of cancer cells. *Polymer*, 49: 3477–3485.

Yang, X., Koh, C.G., Liu, S., Pan, X., Santhanam, R., Yu, B., Peng, Y., Pang, J., Golan, S., Talmon, Y., Jin, Y., Muthusamy, N., Byrd, J.C., Chan, K.K., Lee, L.J., Marcucci, G. and Lee, R.J., 2008a. Transferrin receptor-targeted lipid nanoparticles for delivery of an antisense oligodeoxyribonucleotide against Bcl–2. *Mol. Pharmaceutics*, 6: 221–230.

Abbreviations

RME: Receptor mediated endocytosis

MTX: Methotrexate

SPION: Superparamagnetic iron oxide nanoparticle

FR: Folate receptor

APPA: 3-aminopropylphosphonic acid

PEG: polyethylene glycol

TLC: Thin layer chromatography

MRI: Magnetic resonance imaging

RITC/RBITC: Rhodamine isothyocynate

BSA: Bovine Serum Albumin

GI: Gastrointestinal

IV: Intravenous

EPR: Enhanced permeability and retention

4-FBSCl: 4-Florobenzenesulphonyl chloride

PC: Phosphotidyl Choline

CH: Cholesterol

Chapter 10
The ICPC Nanonet Project

Linsey Robertson
Technology Analyst, Institute of Nanotechnology, United Kingdom
Email: admin@icpc-nanonet.org

Introduction

Research at the nanoscale has the potential to solve many fundamental scientific problems and lead to new developments in different disciplines and application areas such as medicine, electronics, energy and the environment. Successful outcomes to such research will have a measurable impact on the future well being of our global society; however this can only be achieved through improving access to information and opportunities for international collaboration. The ICPC NanoNet is one such project working towards these goals.

European Commission 7th Framework Programme

The European Commission 7th Framework Programme for Research and Technological Development (FP7) is the EU's main instrument for funding research in Europe. It runs from 2007-2013 and has a budget of € 50.5 million. One of the main themes of FP7 is cooperation. In its Action Plan for Nanotechnologies (Nanosciences and nanotechnologies: An action plan for Europe 2005-2009, COM (2005) 243) the European Commission states the importance of "International Cooperation" and sets out recommendations to "secure their [developing countries'] access to knowledge and avoid any 'nano divide' or knowledge apartheid".

International Cooperation Partner Countries (ICPC)

There are 145 International Cooperation Partner Countries (ICPC) divided into 8 regions: Africa, Caribbean, Pacific, Asia, Eastern Europe and Central Asia (EECA), Latin America, Mediterranean Partner Countries (MPC), and Western Balkan Countries (WBC). (See Appendix A for the complete list).

The ICPC NanoNet project

ICPC NanoNet is a support action funded by the European Commission under FP7, lasting for four years from June 2008. It brings together partners from the EU, China, India and Russia (see Appendix B for details) and aims to provide wider access to published nanoscience research and opportunities for collaboration between scientists in the EU and ICPC regions. ICPC NanoNet aims to do this in a variety of ways including developing online tools, compiling regional reports and organising meetings and workshops.

Tools Provided by ICPC Nanonet

Electronic Archive

The Nano Archive (www.nanoarchive.org) is an online, open-access archive of nanoscience and nanotechnology publications for the benefit of the global scientific community, and in particular those scientists in developing countries who may not have access otherwise.

The Nano Archive provides a fully searchable database and includes full text articles where available. Where full text is not provided, the author's contact details are given so that they can be contacted directly for a copy.

Nano Archive is based on E Prints open-source software, and allows individual researchers to upload their own articles (in compliance with publishing restrictions). It provides links to other electronic archives, publishers and journals, and by doing this maximises information provision to researchers in ICPC regions.

Databases

The ICPC NanoNet website provides two databases to assist research development and new collaborations. Use of these databases will increase the visibility of organisations and researchers working in nanotechnology in all ICPC regions. Both databases build on similar initiatives within the Nanoforum project. (http://www.nanoforum.org/)

Organisation Database

The first of these is a database of organisations conducting N&N research within ICPC. This database has been populated and administrated by the project partners with entries providing a description of the organization, information on its facilities and capabilities, and contact details.

The database is designed to provide information about the locations and facilities of nano organisations across the globe, thereby encouraging researchers to visit other institutions and form new collaborations.

Researcher Database

The second database provides information about N&N researchers and their experience, application areas and expertise. This database is populated by individuals who can upload their own details after registering on the ICPC NanoNet website and are then contactable by this means. This facility is particularly useful if a researcher does not have a personal homepage as it allows promotion of interests and expertise, and ease of contact.

Once again, it is hoped that this database will promote collaboration and increased contact between individuals working in nanotechnology.

Reports on Nanoscience and Nanotechnology Developments in ICPC

Annual reports will be produced that describe nanoscience and nanotechnology (N&N) initiatives and activities in eight ICPC regions. Each of these reports provides an overview of current N&N research and development detailing major research centres, government programmes, industrial activity, networks, regional initiatives, funding initiatives and bibliometric data relating to research output. More detailed sections are included on countries that are pursuing a national nano initiative, including country specific information on research centres, programmes and leading national figures. The first of these will be published in July 2009 and will be freely available to registered users of the website.

Workshops and Networking

Networking Opportunities

There are a number of networking opportunities facilitated by the ICPC NanoNet. Encouraging networking at various levels will facilitate the formation of new collaborations. The different opportunities to network include:

- ☆ Contacting authors of publications in the archive;
- ☆ Sharing news, publications or research interests through the various sections of the website;
- ☆ Searching for and contacting institutions with specific facilities in the database of organisations;
- ☆ Searching for individuals with specific skills in the database of researchers;
- ☆ Making use of the online discussion boards;
- ☆ Participating in and booking online workshops (using AT&T Connect) for more deliberative discussions;
- ☆ Participating directly or online at the annual, physical workshops.

Researchers can share their research results with the international community through the Nano Archive, or by emailing news items (to be published on our partner site, www.nanoforum.org). Details of relevant FP7 funding calls will also be posted as news items to raise awareness of these and encourage the formation of consortia involving partners from ICPC regions.

From July 2009 online networking tools will be available, including a forum and access to AT&T Connect software for online workshops and meetings. This software allows the creation of virtual "rooms" where participants are invited to gather for online workshops. Using a headset and microphone, or telephone, participants can listen and contribute to the discussions while viewing Powerpoint presentations and documents on a shared space. Moderated workshops can be conducted with up to 25 participants, with the potential for arranging smaller meetings on request.

It is envisaged that these systems will be useful for researchers in different geographic regions to share ideas more effectively for the building of project consortia, and to share other useful information such as educational material, protocols, etc. Attending online meetings substantially reduces time and financial costs.

Annual Workshop

The first annual ICPC NanoNet Workshop will be held in Prague on Monday 1st June 2009. Project partners will be presenting the results and developments from the first year of the project and look ahead to subsequent years. Seven invited guest speakers will be highlighting specific nano-activities while stakeholders from across the globe can follow the proceedings live via a webcast. Videos of the workshop, PDFs of presentations, and the proceedings will be made available on the website for registered users to download. A small number of DVDs will also be pressed from the workshop to be sent to ICPC researchers on request. Three subsequent annual workshops will be held in China, India and Russia.

Future of the Project

Much of the first year of the project has been spent developing the tools described. The databases and archive are currently available online and the online networking tools will be available in the next month. The first annual workshop takes place on 1st June 2009 and the first annual reports will be published in July 2009.

Following this, efforts will be focussed on the development of the online workshops and the publicising of the project to augment the archive and databases.

Each year an updated report for each region will be published, highlighting current developments. These reports will be excellent sources of information on each of these regions. The annual workshops will disseminate the results and progress from the preceding year.

In conclusion, the ICPC NanoNet project has developed a number of tools to facilitate better contact, visibility and collaboration between researchers within the EU and ICPC regions. These are already proving invaluable in bringing together different researchers who may otherwise be unaware of each other. This can bring about new sharing of information or access to resources, ensuring that through cooperation nanotechnology research can thrive in all areas.

These tools are available to all registered users of the ICPC NanoNet website and registration is free.

References

Nanosciences and Nanotechnologies: An Action Plan for Europe 2005-2009, COM (2005) 243

http://www.nanoforum.org/

http://cordis.europa.eu/fp7/who_en.html#countries

http://www.jncasr.ac.in/

http://www.nano.org.uk

http://www.csmnt.org.cn/

http://www.malsch.demon.nl/

http://www.eltech.ru/english/

Appendix A: List of International Cooperation Partner Countries (ICPC)[1]

ACP *

AFRICA

- Angola
- Benin
- Botswana
- Burkina-Faso
- Burundi
- Cameroon
- Cape Verde
- Central African Republic
- Chad
- Comoros
- Congo (Republic)
- Congo (Democratic Rep. of)
- Côte d'Ivoire
- Djibouti
- Equatorial Guinea
- Eritrea
- Ethiopia
- Gabon
- Gambia
- Ghana
- Guinea
- Guinea-Bissau
- Kenya
- Lesotho
- Liberia
- Madagascar
- Malawi
- Mali
- Mauritania
- Mauritius
- Mozambique
- Namibia
- Niger
- Nigeria
- Rwanda
- Sao Tome and Principe
- Senegal
- Seychelles
- Sierra Leone
- Somalia
- South Africa
- Sudan
- Swaziland
- Tanzania
- Togo
- Uganda
- Zambia
- Zimbabwe

CARIBBEAN

- Barbados
- Belize
- Cuba
- Dominica
- Dominican Rep.
- Grenada
- Guyana
- Haiti
- Jamaica
- Saint Kitts and Nevis
- Saint Lucia
- Saint Vincent and Grenadines
- Suriname
- Trinidad and Tobago

PACIFIC

- Cook Islands
- Timor Leste
- Fiji
- Kiribati
- Marshall Islands
- Micronesia, Federal States
- Nauru
- Niue
- Palau
- Papua New Guinea
- Solomon Islands
- Tonga
- Tuvalu
- Vanuatu
- Samoa

ASIA

- Afghanistan
- Bangladesh
- Bhutan
- Burma/Myanmar
- Cambodia
- China[2**]
- India[2**]
- Indonesia
- Iran
- Iraq
- Lao People's Democratic Republic
- Malaysia
- Maldives
- Mongolia
- Nepal
- Oman
- Pakistan
- Philippines
- Sri Lanka
- Thailand
- Vietnam
- Yemen

EASTERN EUROPE AND CENTRAL ASIA (EECA)

- Armenia[3]
- Azerbaijan[3]
- Belarus[3]
- Georgia[3]

Contd...

Appendix A–Contd...

- Kazakhstan
- Kyrgyz Republic
- Moldova[3]
- Russia[2**]
- Tajikistan
- Turkmenistan
- Ukraine[2,3]
- Uzbekistan

LATIN AMERICA

- Argentina[2]
- Bolivia
- Brazil[2**]
- Chile[2]
- Colombia
- Costa Rica
- Ecuador
- El Salvador
- Guatemala
- Honduras
- Mexico[2]
- Nicaragua
- Panama
- Paraguay
- Peru
- Uruguay
- Venezuela

MEDITERRANEAN PARTNER COUNTRIES (MPC)

- Algeria[3]
- Egypt[2,3]
- Jordan[3]
- Lebanon[3]
- Libya[3]
- Morocco[2,3]
- Palestinian administered areas[3]
- Syrian Arab Rep.[3]
- Tunisia[2,3]
- Turkey[4]

WESTERN BALKAN COUNTRIES (WBC)

- Albania[4]
- Bosnia- Herzegovina[4]
- Croatia[4]
- Former Yugoslav Republic of Macedonia (FYROM) [4]
- Kosovo[5]
- Montenegro[4]
- Serbia[4]

*: In the 'Specific international cooperation actions', Africa can also be considered as a region on its own, while the Caribbean countries can also participate with Latin American and the Pacific countries with Asia.

**: For participation in the 'Specific international cooperation actions' each of Brazil, China, India and Russia may be considered individually as a region on its own. Thus, the required two or more partners can be located in these countries. However, in this case, at least two different partners from different provinces, oblasts, republics or states within Brazil, China, India or Russia are necessary.

1: Up-to-date information on the status of countries is available at: http://cordis.europa.eu/fp7/who_en.html#countries

2: Signed an agreement with the EC covering Science and Technology

3: These countries are also part of the European Neighbourhood Policy (ENP).

4: Until the country become associated to FP7.

5: As defined by UNSC resolution 1244 of 10 June 1999.

Appendix B: ICPC NanoNet Partners

The following organisations make up the ICPC NanoNet project consortium:

- ☆ Institute of Nanotechnology, UK–project coordinator
  http://www.nano.org.uk
- ☆ Chinese Society of Micro-Nano Technology, China
  http://www.csmnt.org.cn/
- ☆ Jawaharlal Nehru Centre for Advanced Scientific Research, India
  http://www.jncasr.ac.in/
- ☆ Malsch TechnoValuation, Netherlands
  http://www.malsch.demon.nl/
- ☆ MERIT, Universiteit Maastricht, Netherlands
  http://www.merit.unu.edu/
- ☆ Sociedade Portuguesa de Inovaç o (SPI), Portugal
  http://www.spi.pt/
- ☆ St Petersburg Electro technical University, Russia

For information about ICPC NanoNet please contact the project coordinator, Ms. Lesley Tobin lesley.tobin@nano.org.uk

Chapter 11

Application of Nano-Science and Nanotechnology in Uganda's Economy

Henry M. Bazira
Executive Director,
Water Governance Institute (WGI),
Uganda; P. O. Box 23704 Kampala, Uganda
E-mail: bazirah2@gmail.com

ABSTRACT

Nanotechnologies commercial application has only been recent (21st Century). Although there are products of nanotechnology in public use in several countries of the world, including Uganda, the popularity is still very low. There are several concerns about the lack of adequate policy, legal and institutional frameworks to effectively govern nanotechnology and its products; concerns about the social, economic and environmental viability, efficacy and safety; the lack of sufficient information to justify its application; and the fact that, in many cases, it is not introducing new ways of doing business, but rather improving the existing technologies and approaches, sometimes at greater cost. There is, therefore, need for further research to be undertaken to justify its application in various processes; and put in place adequate and effective governance mechanisms, especially in developing countries where related information on its application and governance structures are often lacking.

Keywords: *Nanoscience, Nanotechnology, Uganda, Economy, Risk, Opportunity.*

Introduction

Uganda has a population over 30 million (CIA Fact book, 2008) on a land area of 241,038 km^2. It is land locked by Kenya to the east, Tanzania to the south, Rwanda to the southwest and Democratic Republic of Congo to the west. The natural forest cover used to be 44 per cent of the land in colonial times, but has gradually been lost through encroachment for urban and rural settlements, industry, and agriculture and harvesting of forest products (timber, firewood, charcoal, medicinal plants and wild game) and now stands at less than 18 per cent of the land cover. Therefore, the growing population is exerting enormous increasing pressure on the natural forest for settlements, food, energy and other products.

The country is essentially an agriculture-based economy with a significant proportion of its national budget supported by external donor assistance. Over 34 per cent of the population is known to live below the US$1.0 poverty line[2]. This is not an enviable situation. Agriculture employs over 80 per cent of the population. The remaining 20 per cent are involved various sectors of the economy. Every nation wishes to eliminate poverty and individuals aspires to be financially independent. However, these usually come with a lot of sacrifice and investment. The question is not whether the sacrifice and investment should be made, but the amount of sacrifice and investment to be made and the likely social, economic and environmental impacts Uganda is capable of bearing as a nation.

Majority of the industries in the country are agro-based (food and wood processing) and the technologies applied are considered rudimentary and obsolete. The few non-agro-industries include metal works and chemical-based industries (paints, plastics and cosmetics). Uganda is now at the threshold of commercially exploiting her petroleum resources in the Albertine Graben that are expected to usher in a paradigm shift in the economy. The petroleum discoveries in the country have raised a lot of interest and excitement within government and the general public that Uganda is going to join the oil rich countries of the world and every citizen is going to become wealthy. While this is possible, it can become terribly bad, if the industry is poorly managed. Also, the industry is a source of many products that form precursors (raw materials) for the production of other commercially important products.

There is a growing interest the world over in the use of nanoscience and nanotechnology in medicine, drug administration, engineering, wastewater treatment and purification, energy (conversion, storage, reduction, efficiency, emission, etc), tissue engineering, metallurgy, information technology, food processing, mechanical, biological and chemical methods to cause positive social and economic transformation (internet, 1).

Apart from the imported nanomaterials (internet, 2) on the Ugandan market, nanoscience and nanotechnology is generally unknown in Uganda. There is insufficient information about the benefits and risks (safety) of nanoscience, nanotechnology and nanomaterials. Proponents suggest that nanotechnology could be the panacea to several problems and challenges in society, while opponents to nanotechnology seem to focus on the absence of sufficient information to justify its use and the potential risks to humankind and the environment. However, there is some evidence that nanotechnology has the potential of being beneficial, if properly harnessed (Berger, 2007, 2009).

The paper highlights the present status and future prospects of applying nanoscience and nanotechnology in Uganda's economy. It is based on prevailing facts and review of literature available in public domain. Some of the information was obtained through individual consultations (discussions) with persons the researcher met.

Main Body

In Uganda, nanoscience and nanotechnology is currently being applied at small-scale in fertility endoscopy, X-ray imaging, cosmetics, and plant tissue culture. There are also nanotechnology-based products on the Ugandan market such as bio-disc (a wellness product) and Ethos (a product used to improve the efficient utilisation of petroleum in vehicles). However, the popularity of these products is still low.

A review of Uganda's Renewable Energy Policy (2007) shows no explicit evidences that nanoscience and nanotechnology has been considered as a potential avenue of up-scaling renewable energy use in the country. Also, there is no deliberate mention in the other national policies and legislation of the application of nanoscience and nanotechnology in the various sectors of the economy. The policies and legislation are therefore inadequate to regulate nanomaterials in the short- and long-term. There is, therefore, need to review the national policies, legalisation and institutional provisions with a view of integrating nanoscience and nanotechnology in several applications in the country.

Although nanotechnology is in its infancy and there are dissenting voices about it, now may be the right time to explore its implication and implementation in Uganda. Many developing countries, for example, China, India, Iran, Malaysia and Thailand are investing considerable resources in research and development of nanotechnologies. However, without a long-term clear perspective, the true picture with respect to developing country engagement remains unclear. In conjunction with other technologies like information technology and biotechnology, nanotechnology will usher in social and economic transformations.

Uganda is currently facing severe challenges in many sectors *e.g.* energy (electricity), health, food processing, waste treatment and disposal, water purification, air pollution, etc. which can be ameliorated by the application of nanoscience and nanotechnology. Therefore, it is incumbent upon government of Uganda to explore avenues in which nanotechnology could be applied on a wider scale in the different sectors of the economy for the benefit of all.

What is clear is that nanotechnology applications rather than being a revolutionary new way of doing business, is evolutionary in that it offers some improvements in existing applications/ technologies and ways of doing business. Therefore, its adoption or application will depend on whether or not the technology will:

- Allow the companies to offer an improved product that could be sold at a higher price (which could or does not necessarily result in a higher margin, depending to the degree of rise in production costs) and/or gives the product a performance advantage over competition;
- Allow the companies to reduce its production cost, while at the same time improving product quality. In which case, offering the product at the same

price and achieve higher profit margin or if completion is stiff, reduce the price of the product.

In short, companies need an economic incentive to introduce nanotechnology in their applications. Before these technologies can leap from the laboratory to the market en masse, they will need to jump the hurdles of public acceptance, economic feasibility and environmental safety, which in turn would bring the price of the technology down and help spread it reach. Many of the applications are still in their infancy (Loncto *et al.*, 2007) and will require further testing to prove their reliability. Also, they will require additional capital investment to upgrade the existing equipment, systems and personnel for their effective implementation.

It is clear that there is no adequate policy, legal and institutional framework in the country to govern nanotechnology and its products.

Most of the studies that have been undertaken have focused on the potential risks nanomaterials may pose on the human body (Berger, 2009; internet, 3). Little is known about the toxicity of nanomaterials in plants and their subsequent impact within the food chain.

Conclusions

The popularity of nanotechnology and nano-products is still very low in Uganda. If nano-products can overcome the hurdle of commercialization, public acceptance and environmental safety, then they could solve many problems in society. But, for as long as the products are more expensive than their nano-free alternatives, the problems that we are facing today with the conventional approaches shall remain.

No matter how promising a new technology may be, it is hard to reap its benefits, if there are entrenched economic interests with different goals. The adoption and application of nanotechnology in developing countries is going to need a lot of political will and public support. Nanotechnology and nanomaterials will require the formulation and implementation of mechanisms for their governance.

Recommendations

There is need to review national policies, legislation and institutional provisions to identify the gaps, linkages, challenges, and opportunities regarding nanotechnology application in the Ugandan economy and make proposals for government consideration.

There is need to conduct feasibility and risks assessments on the adoption and application of nanotechnology and nanomaterials in the country, especially in areas such as commercial viability, hazard identification, dose-response assessment, exposure assessment, risks characterization and environmental impacts.

There is need to develop governance mechanisms for nanotechnology and nanomaterials in the country.

There is need to create public awareness on and appreciation of nano-applications and nanomaterials with proven social, economic and environmental viability, efficacy and safety.

References

Berger, Michael, 2007. *Water, Nanotechnology's Promises, and Economic Reality.* Nanowerk LLC. www.nanowerk.com (accessed 29-04-2009)

Berger, Michael, 2009. *Nanotechnology's Impact on Major Food Crops*. Nanowerk LLC. www.nanowerk.com (accessed 29-04-2009)

Loncto, J., Walker, M. and Foster, L., 2007. Nanotechnology in the Water Industry. *Nanotechnology Law and Business*. A Journal for Attorneys, Entrepreneurs, and Investors involved in small-scale technologies, 4(2).

Uganda Bureau of Statistics (UBOS). *Population and Social Statistics* (2005-6). Report. GoU

Internet sites

www.en.wikipedia.org/wiki/list_of_nanotechnology_application. (Accessed 24-04-2009)

www.nanowerk.com (accessed 29-04-2009)

www.ciafactbook.org (accessed 24-04-2009)

Chapter 12

Complementary Use of μRBS, μPIXE, AFM and SEM in Studying the Interaction of Germanium with Iridium and Palladium in Lateral Diffusion Couples

A. Habanyama[1] and C.M. Comrie[2]

[1]*Department of Physics, University of Zambia, P.O.Box 32379, Lusaka 10101, Zambia*

[2]*Department of Physics, University of Cape Town, Rondebosch 7700, South Africa*

E-mail: ahabanyama@yahoo.com

ABSTRACT

Existing materials and technologies in the semiconductor industry are approaching their physical limits as device sizes decrease significantly below 100 nanometers. The challenges of achieving the required lateral abruptness of semiconductor junctions and low junction leakage as well as low contact resistance require a careful rebuilding of the fundamental physics and materials platform on which to base a reliable manufacturing capability. As feature size continues to decrease, interconnects will also pose increasingly difficult challenges in terms of physics and materials. The study of lateral diffusion couples is particularly well suited for dealing with the above concerns. The Ir-Ge and Pd-Ge reactions were studied in lateral diffusion couples. The lateral diffusion couples were prepared by the deposition of Ge micro-islands on Ir or Pd thin films and then annealed at temperatures ranging from 100 to 800°C in a vacuum furnace. In both systems, the germanide phases were seen to spread out from the source region in their decreasing order of germanium content. Germinide formation, phase sequence and growth kinetics have been investigated. The techniques employed were, Rutherford backscattering spectrometry on the nuclear microprobe, *i.e.*, micro-Rutherford backscattering spectrometry (μRBS), micro-particle induced x-ray emission (μPIXE), atomic force microscopy (AFM) and scanning electron microscopy (SEM). Optical microscopy has also been employed. It is shown that μPIXE can be used in lateral diffusion studies. This technique can therefore be used to study systems of elements that are too close in atomic number to be studied by micro-Rutherford backscattering spectrometry due to excessive peak overlap. We demonstrate the complementary nature of these techniques. A notable aspect of the two lateral diffusion systems studied was that each system exhibited some interesting, unique and reproducible features. When samples of germanium micro-islands on iridium films are annealed within a range of temperatures between 600 to 800 °C, substantial lateral diffusion is observed, resulting in a number of reaction regions. Micro-Rutherford backscattering analysis indicates that the phase Ir_3Ge_7 stretches across the original island interface at all temperatures, with the phase Ir_4Ge_5 forming in the reaction region with unreacted iridium. The phase $IrGe_4$ is observed to nucleate in the middle of the island at temperatures above 800 °C. Depth information is readily obtainable from micro-Rutherford backscattering spectrometry, which is used in conjunction with atomic force microscopy data to estimate the densities of the phases formed. This demonstrates the complementary use of the techniques. Micro-RBS analysis of the Pd-Ge lateral diffusion couples indicates that the middle of the island consisted of unreacted Ge overlaying the phase PdGe at all temperatures while the growth region outside the original island interface consisted of the phase Pd_2Ge. Our SEM results showed the formation of cracks inside the source region near the original island interface. The cracks don't appear to be related to Pd diffusion into the island, as the region surrounding the growth interface does not become depleted in Pd. Pd_2Ge growth at the Pd interface was characteristically fractal. This system exhibited low temperature reaction but the extent of growth was limited because the cracks in the matrix eventually halted the reaction. At high temperatures the reaction was rapid but quickly it ceased.

Keywords: *Germanium, Iridium, Material sciences, Microspectroscopy.*

Introduction

The samples used for studying lateral diffusion reactions are composed of a thick micro-island of one material on top of a much thinner film of another material. Upon annealing the island material would react with the underlying film, going through a sequence of phases until the most island-material rich phase is formed. Since no further reaction with the underlying film is possible, the most island-material rich phase may then grow laterally until it attains a critical width, after which other phases appear and grow simultaneously. Lateral diffusion couples have been used to study the interaction between a metal and a semiconductor (Zheng *et al.*, 1982, 1983; Chen *et al.*, 1984; Ding *et al.*, 1994; Nemutudi *et al.*, 2000; Waal, 1999) or between two metals (Waal, 1999; Blanpain *et al.*, 1990; Blanpain, 1992; Joyce *et al.*, 1988, 1990). With respect to the former, the bulk of the work was done with silicon.

Micro-RBS has been used to study lateral diffusion couples (Ding *et al.*, 1994; Nemutudi *et al.*, 2000; Waal, 1999). Ding *et al.* (Ding *et al.*, 1994) used this technique to study lateral diffusion couples of the NiSi system, de Waal [de Waal, 1999] used it to study the Cr-Si and Ru-Al systems while Nemutudi *et al.* (Nemutudi, 2000) used the technique to study the Pt-Ge system. This is the first time that the technique has been used to study the Ir-Ge and Pd-Ge systems. The major advantage of this technique is its ability to give depth information. Of concern while determining the best thickness of our sample structures, was the likelihood of overlap between Pd and Ge RBS signals. This is by virtue of their having atomic numbers that lie relatively close to each other in the periodic table. This is the first time that μPIXE has ever been used in lateral diffusion studies.

Experimental Details

The lateral diffusion couples were prepared by electron beam evaporation at room temperature and a base pressure in the low 10^{-5} Pa ranges. First, a layer of Ir or Pd was deposited onto a thermally oxidized single crystal silicon wafer used as a substrate. The thickness of each layer was around 25 nm for the Ir-Ge samples and 20 nm for the PdGe samples. A Ge layer of about 100 nm in thickness was next deposited through rectangular openings (390×780 μm^2) in a silicon mask made by photolithographic techniques and selective etching on a silicon wafer. In this way micro-islands of Ge with sharp edges were formed on metal thin films.

Another SiO_2 substrate was placed adjacent to the sample structures described above, but was not masked during the deposition. Large areas of the deposited materials were therefore available for ordinary RBS investigation in order to determine their thickness. After removing the wafers from the evaporation chamber they were cleaved into small samples and furnace annealing was carried out between temperatures of 100 to 800 °C to activate solid-state interaction.

The samples were examined by optical microscopy, and representative samples selected for further analysis. For quantitative analysis, the diffusion length was measured by optical microscopy and SEM. SEM was also used to study the surface morphology. The distribution of elements as a function of lateral position was obtained by μPIXE and μRBS. The latter technique also provided information about the thickness of the films. During the μPIXE and μRBS analysis a well defined area, typically of 400×400 μm^2 was scanned using a 2 MeV α-beam focused down to 1×1 μm^2. This area was chosen to include all reaction regions observed in the optical micrographs. Sample orientation was adjusted in such a way that the interfaces of the regions of interest lay horizontally in line with the original island edge so that the microprobe beam scanned parallel to the original island interface. Both RBS and PIXE spectra were recorded along with position information. Typically, 128×128 spectra were generated in each run and once it had been established that no variation in composition was observed parallel to the interface, spectra were summed along this direction. This reduced the analysis to a one-dimensional traverse of 128 spectra perpendicular to the interface, thereby optimally exploiting beam position while improving on statistics. Finally, AFM was employed to obtain further information on the relative surface heights of the various reaction regions as a function of lateral position and the surface morphology.

Results

Figure 12.1 shows an SEM micrograph of part of a diffusion couple, which had been annealed at 800 °C for 30 minutes. The four different reaction regions observed are labeled A to D. The original edge of the island is clearly visible in region C. This sample then underwent mRBS in order to determine the composition and thickness of the various regions. An area to be scanned was chosen to include all regions observed.

A representative spectrum picked from each of the four reaction regions labeled A to D is shown in Figure 12.2. A spectrum picked from the unreacted Ir film is also included in this figure. Peak heights of the spectra taken from regions D, C and B show the phases Ir_4Ge_5, Ir_3Ge_7 and $IrGe_4$ respectively. The germanides in these three regions are seen at the surface position. On the other hand, the Ir peak from region A lies below the surface position, indicating that there was no Ir at the surface. Region A therefore consisted of a layer of unreacted Ge on top of the $IrGe_4$ phase.

The RBS data from the scanned area were analyzed to get stoichiometric information from integrated counts of the Ir and Ge peaks. Figuire 12.3(a) shows the results as a function of lateral position. Each RBS spectrum was then simulated by RUMP [Doolittle and Meth, 1986] to obtain thickness information. This information is plotted in Figure 12.3(b) as a function of lateral position. The figure shows relatively uniform thickness within each phase region (apart from the Ir_4Ge_5 region) and sharp

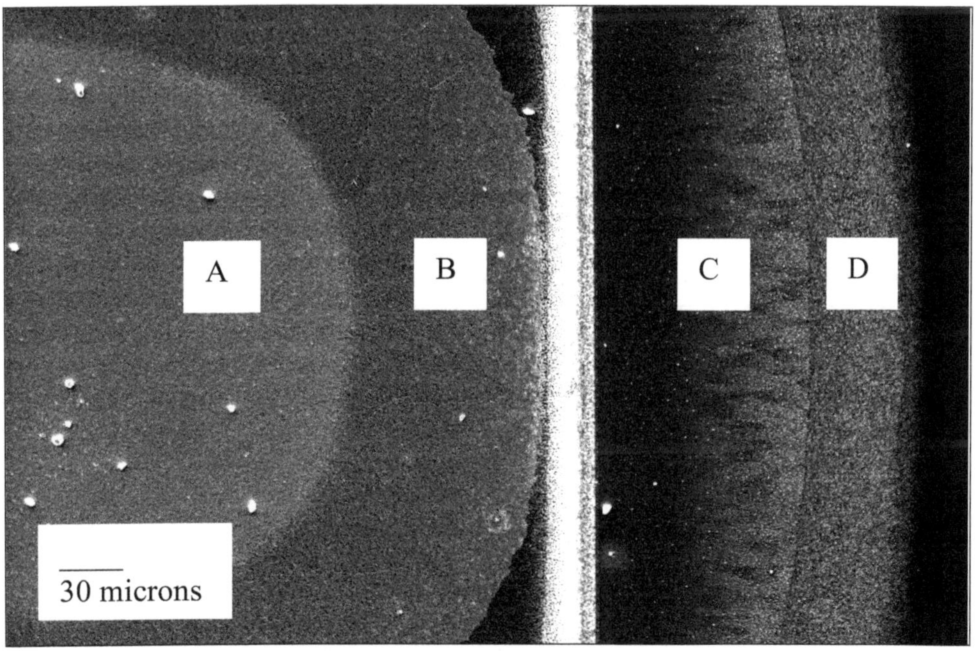

Figure 12.1: SEM Micrograph of Part of a Ge Island (250 nm) on an Ir Film (25 nm) Annealed at 800 °C for 30 Minutes Showing all the Different Reaction Regions Observed

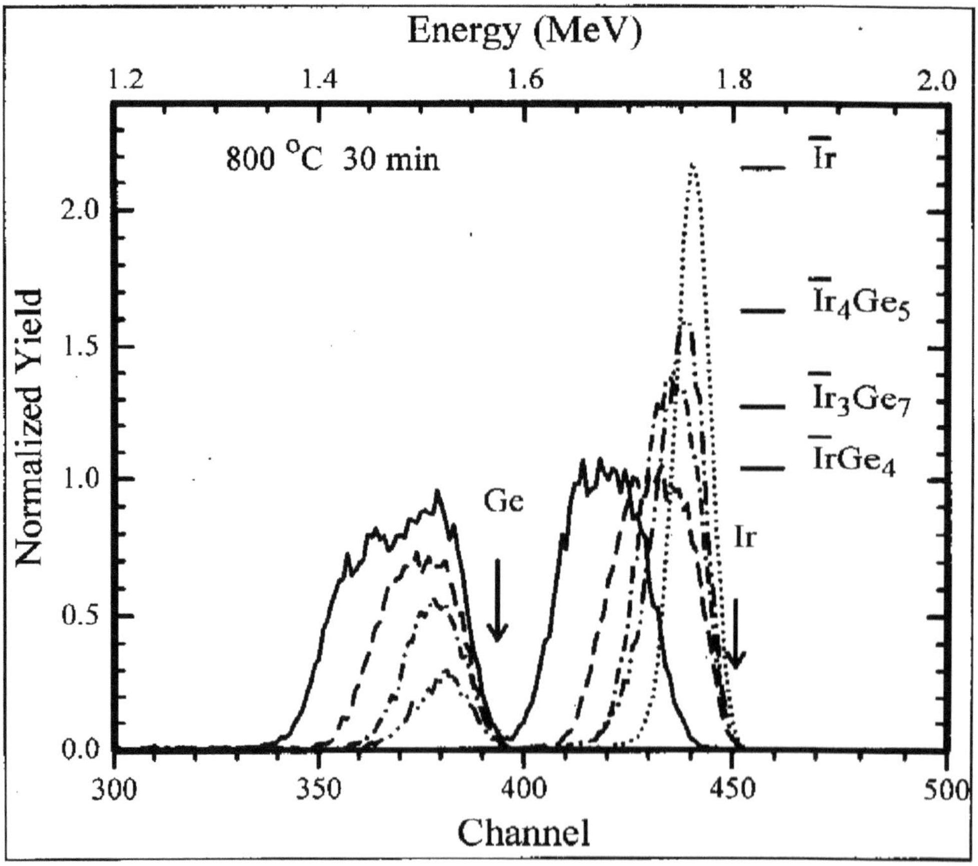

Figure 12.2: Superposition of Selected RBS Spectra from Each of the Reaction Regions and from the Unreacted Ir Film. Ir peak heights of the various phases and surface positions of Ge and Ir are indicated.

steps between regions. A narrow step is seen in the Ir_4Ge_5 at the interface with Ir, which may suggest the presence of the phase IrGe. This region was however too narrow to properly resolve and monitor.

An interesting feature seen in Figure 12.3(a) is the presence of a small region of the phase Ir_3Ge_7 inside the original island region. This region was investigated further by SEM at higher magnification as presented in Figure 12.4. In this figure, the white vertical line passing through the position labeled 'W' was identified as the original island interface. A series of darkened square spots may be visible in this micrograph; these were the result of SEM 'line scans'.

The width of this region of the phase Ir_3Ge_7 inside the original island region was observed to increase with annealing time. This suggests a gradual decomposition of the phase $IrGe_4$ into Ir_3Ge_7 by the reaction $3IrGe_4 \rightarrow Ir_3Ge_7 + 5Ge$. The vertical ridge-like structure passing through the position labeled 'V' in Figure 12.4 was identified as the reaction interface for the decomposition. A clear change in texture is seen at

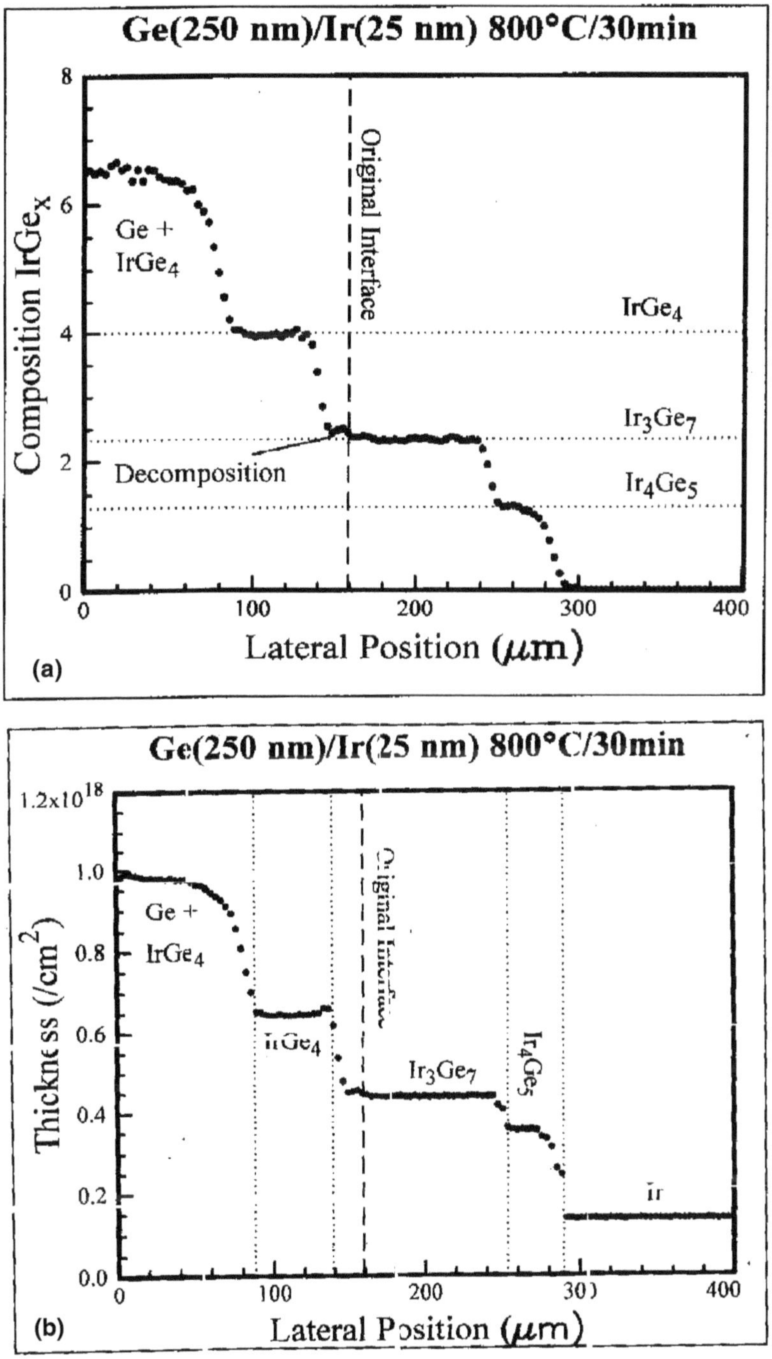

Figure 12.3: (a) Stoichiometric information of a Ge island (250 nm) on an Ir film (50 nm) annealed at 800 °C for 30 minutes, derived from integrated counts of the Ir and Ge peaks as a function of position. (b) RUMP [14] simulated thickness of the various phase regions as a function of lateral position.

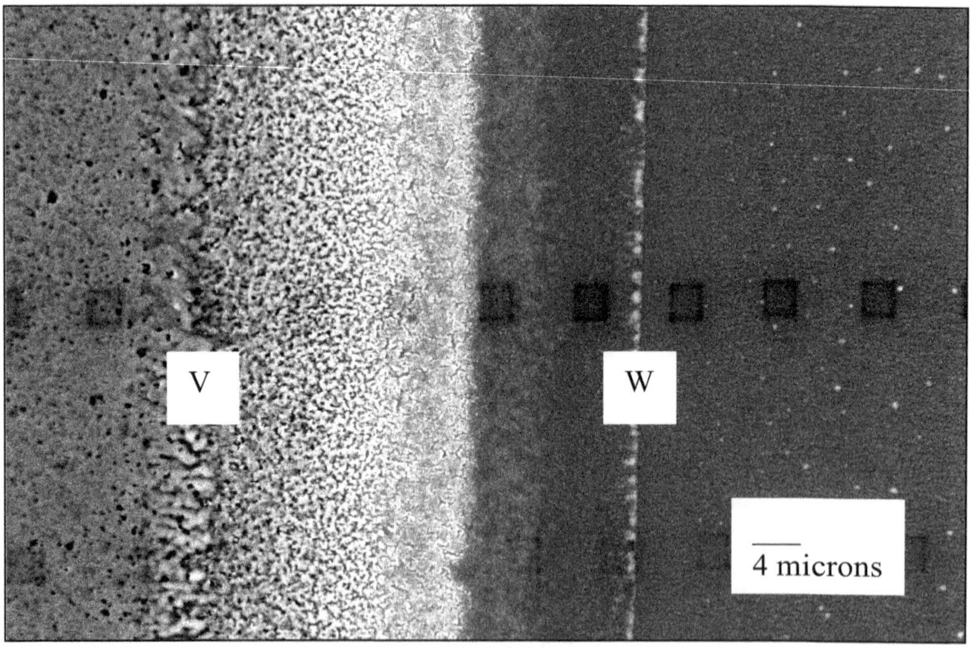

Figure 12.4: Magnified SEM Scan of the Region of Phase Ir_3Ge_7 Inside the Original Island Area, the Original Island Interface (W) and a Decomposition Reaction Interface (V)

this interface. This is a result of both thickness and phase difference (from Figure 12.3). The Ge liberated at this reaction interface joins the flux of Ge atoms from the unreacted Ge source to facilitate reactions at the other interfaces farther from the original island region. The region on the left in the SEM micrograph shows part of the exposed $IrGe_4$. The major difference between the lateral diffusion couples prepared by annealing at temperatures below 800 °C and those at and above 800 °C was the absence of the $IrGe_4$ phase below 800 °C; hence no region of decomposition was observed there. PIXE data were also extracted and added along parallel lines aligned with regional interfaces. These spectra were then analyzed to extract compositional information as a function of lateral position. The results are shown in Figure 12.5. A comparison of the PIXE results to the corresponding RBS results shows excellent agreement between the two techniques.

The sample surface was also scanned with an Atomic Force Microscope to obtain information of the relative surface heights as a function of lateral position. Figure 12.6 shows the results of a scan of region D and parts of regions C and the unreacted Ir film. A step is seen at the interface between region D and the unreacted Ir. This supports RBS observations of possible IrGe formation in this region.

The average areal density of the unreacted Ir film, from the RBS results shown in Figure 12.3(b), is found to be 140×10^{15} atms.cm^{-2}. The density of Ir is 7.0×10^{22} atms. cm^{-3}, which when divided into the 'areal density' gives the thickness of unreacted Ir as 200×10^{-8} cm. By adding the appropriate step heights from AFM results to this

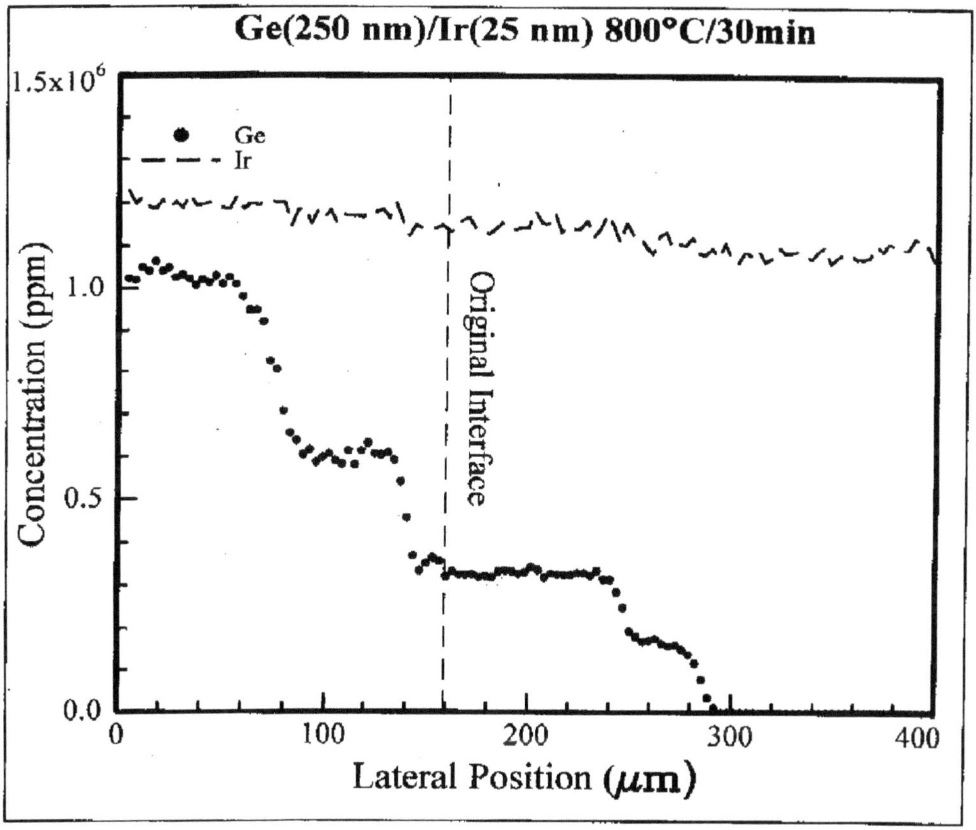

Figure 12.5: PIXE Derived Ir and Ge Concentration as a Function of Lateral Position for a Ge Island on an Ir Film Annealed at 800°C for 30 Minutes. The Ir concentration data have been offset above those of Ge for clarity but are actually lower in ppm.

base thickness, the thickness of each region was worked out. This thickness when divided into the corresponding RBS 'areal density' from Figure 12.3(b) gives an estimate of the atomic number densities of the phases formed. The results are shown in Table 12.1. The estimated densities should be treated with caution as they refer to the densities in the thin films produced in our experiments and may differ from those expected in bulk materials or single crystals.

The Pd-Ge system exhibited relatively low temperature reaction. At high temperatures the reaction in the Pd-Ge system was rapid but quickly ceased as explained later. It was therefore necessary to carry out the investigation for this system at much lower temperatures than those for the Ir-Ge system.

Shown in Figure 12.7 is an SEM micrograph of a representative sample with a Ge island on a Pd film annealed at 325 °C for 2 hours. Three distinct regions labeled A to C are seen. An area, which was chosen to include all the reaction regions observed, was scanned on the nuclear microprobe for analysis by mRBS and mPIXE. A spectrum

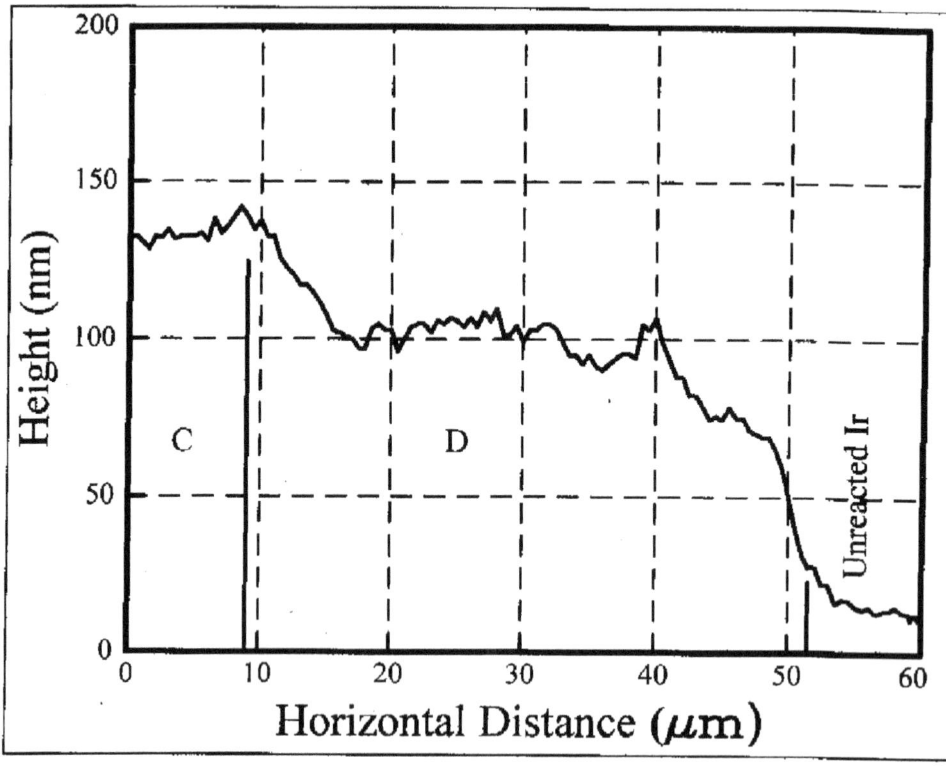

Figure 12.6: The Graph Shows the Results of an AFM Scan of Region D and Parts of Region C and the Unreacted Ir Film

picked from each of the three regions is shown in Fig. 8. The spectrum from region C shows a peak of unreacted Pd and no Ge.

Table 12.1: Approximate Height of the Steps Between Different Phase Regions as Obtained from RBS and AFM. A density estimate is also given.

Phase	*RBS 'Areal Density' (x10^{15} atoms.cm^{-2})*	*AFM Thickness (x10^{-8} cm)*	*Density (x10^{22} atoms.cm^{-3})*
$IrGe_4$ (Hexagonal)	645	2580	2.5
Ir_3Ge_7 (BCC)	440	1260	3.5
Ir_4Ge_5 (Tetragonal)	355	980	3.6
Ir	140	200 (From RBS)	7.0 (Known)

The peak height of the spectrum taken from the region B shows the phase Pd_2Ge. This phase is seen at the surface position. From the solid line in the figure, it can be seen that the region A consisted of unreacted Ge and PdGe. The Pd peak of the solid line lies at the surface position, showing that there was some Pd at the surface. The

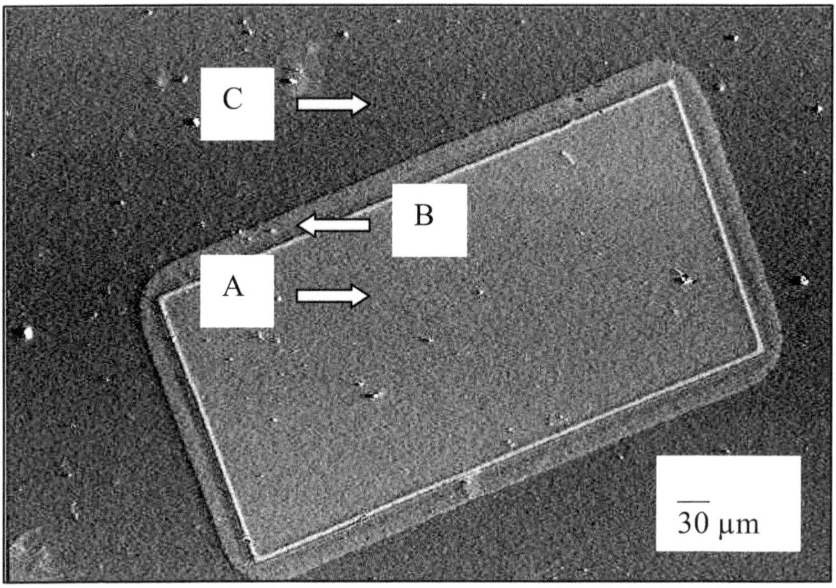

Figure 12.7: SEM Micrograph of a Ge Island (100 nm) on a Pd Film (20 nm) Annealed at 325 °C for 2 Hours Showing the Different Regions

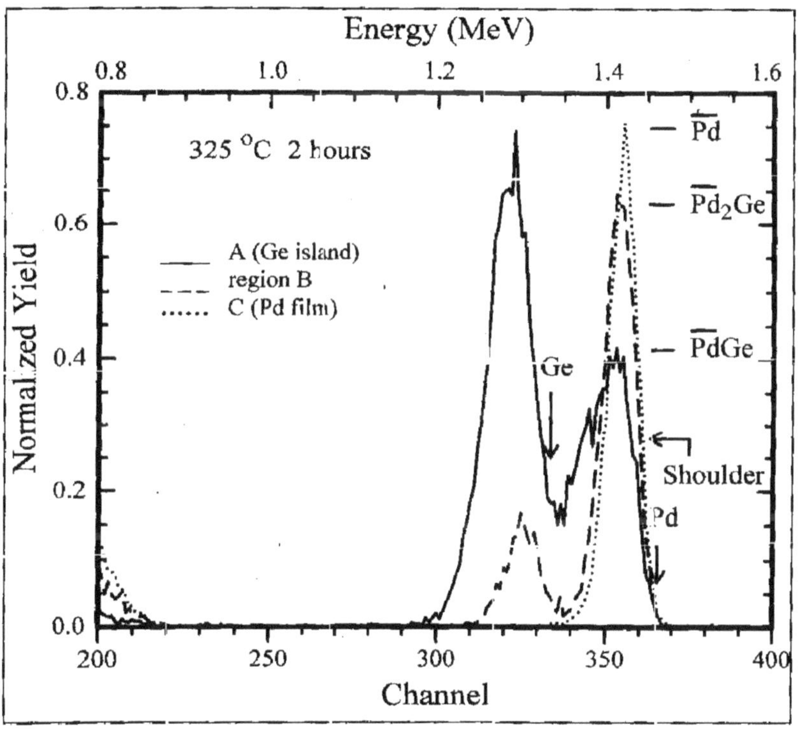

Figure 12.8: Superposition of Selected RBS Spectra from Each of the Regions of the Pd-Ge Sample. Pd peak heights of the various phases and surface positions of Ge and Pd are indicated.

'shoulder' marked in the figure indicates that there was less Pd at the surface than deeper down. Region A therefore consisted of PdGe at the bottom while at the top there was unreacted Ge intermingled with PdGe.

The elemental distribution in terms of the integrated counts of the Pd and Ge RBS peaks as a function of lateral position was obtained and is shown in Figure 12.9.

PIXE data were also extracted and the results again showed excellent agreement with the corresponding RBS results.

One feature that was consistently observed for all Pd–Ge lateral diffusion samples analyzed and seen in Figure 12.9 was the somewhat gradual drop in Ge concentration towards the interface of region B with unreacted Pd. To try and understand this, the area between regions B and C was investigated using SEM at higher magnification. An SEM micrograph of this area is shown in Figure 12.10.

We see in Figure 12.10 that the phase Pd_2Ge had a fractal type of growth, with areas of unreacted Pd clearly visible within the growth region. This was not observed throughout region B but only in the section close to region C. In Figure 12.9, the μRBS

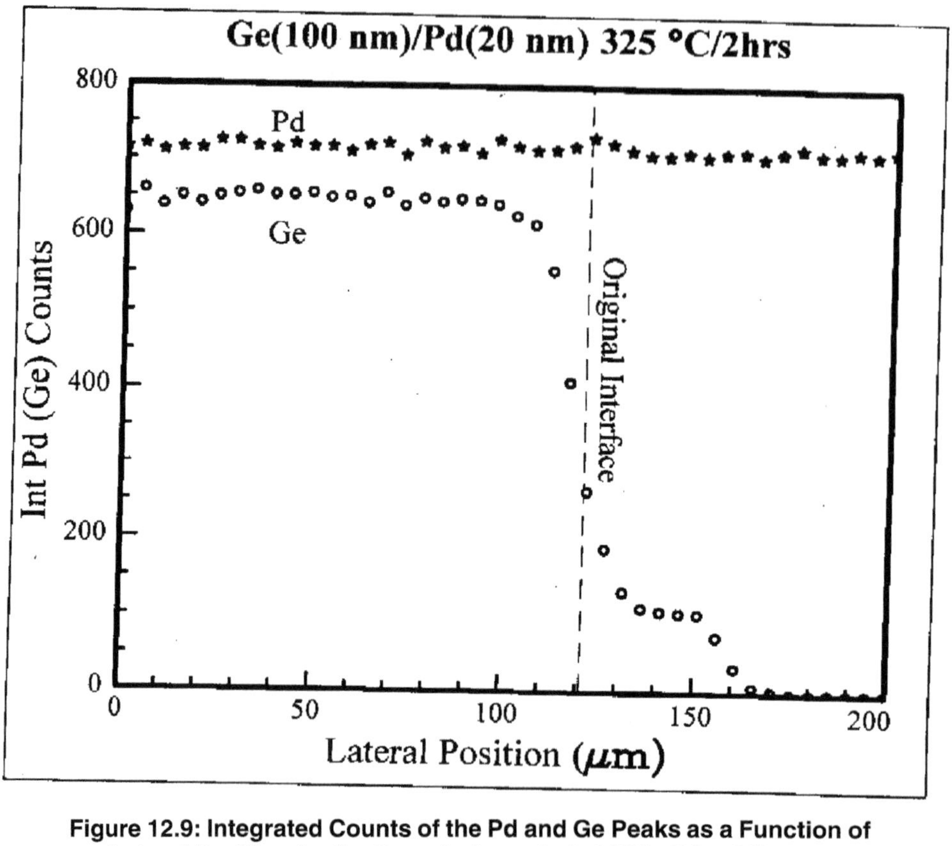

Figure 12.9: Integrated Counts of the Pd and Ge Peaks as a Function of Lateral Position for the Sample Annealed at 325 °C for 2 Hours. The data of Pd counts were offset above those of Ge for clarity.

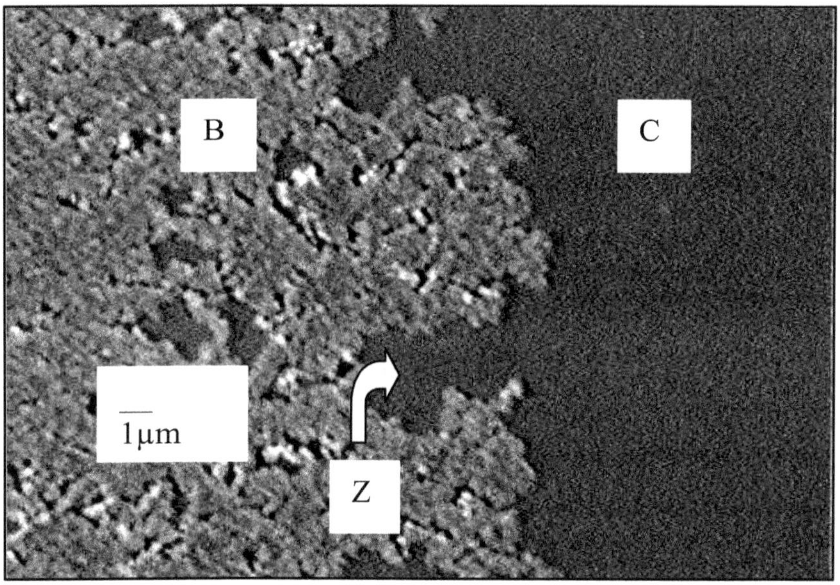

Figure 12.10: SEM Micrograph of the Area Between Regions B and C. The feature labeled 'Z' was unreacted Pd that gave a tentacle appearance to the Pd_2Ge growth.

results of this sample showed a gradual decrease in Ge concentration. It is clear from Figure 12.10 that this decrease was not as a result of a change in composition but rather the effect of summing data parallel to the interface across alternating areas of Pd_2Ge and unreacted Pd (labeled 'Z'), thereby reducing the combined Ge concentration; this reduction is greater as one gets closer to region C.

A micrograph of the area between regions A and B is shown in Figure 12.11. The RBS results of Figure 12.8 revealed the presence of Pd at the surface in region A. Looking at Figure 12.11 it would seem that the phase PdGe was mixed with unreacted Ge at the surface, in a manner that produced the 'puffy' texture observed in region A. The features labeled 'X' appear to be areas rich in unreacted Ge. A unique texture in which are features that appear to be cracks are seen in the area between regions A and B. These cracks were clearly seen as forming along the features 'X' and introduced discontinuity in the matrix of the sample.

Discussion

A schematic diagram of the different phase regions observed in the Ir-Ge system at 800 °C with various parameters used to characterize growth kinetics is shown in Figure 12.12. The subscripts α, β and γ refer to the phases $IrGe_4$, Ir_3Ge_7 and Ir_4Ge_5 respectively. Widths of the respective phase regions are represented by X_α, X_β and X_γ. Reaction interfaces between the different phases are labeled $X_{\alpha\beta}$, $X_{\beta\gamma}$ and $X_{\gamma Ir}$ respectively.

The overall lateral diffusion results are largely summed up by considering the case of a Ge island on an Ir film annealed at 800 °C. Upon annealing, interaction took

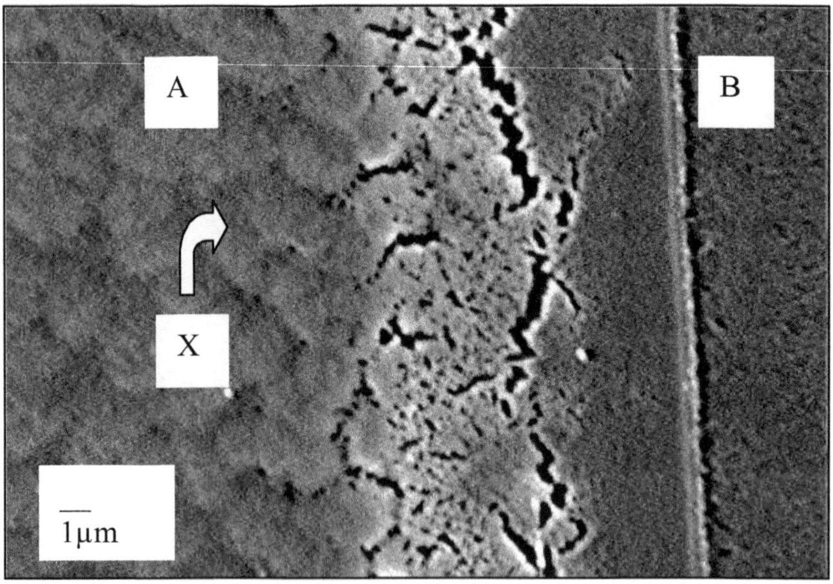

Figure 12.11: SEM Micrograph of the Area Around the Original Island Interface Between Regions A and B

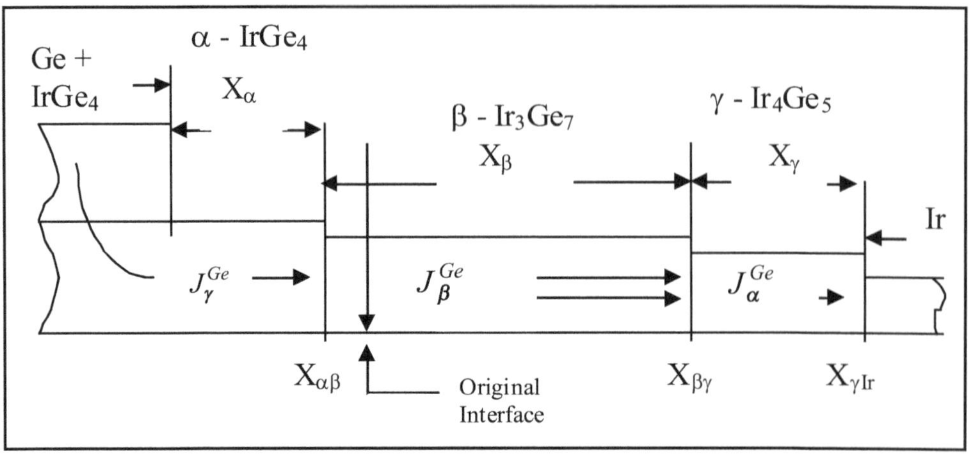

Figure 12.12: Diagram Showing Different Phase Regions and the Various Parameters Used to Characterize Growth Kinetics.

place in the source region between Ge and Ir atoms in the underlying thin film, resulting in the formation of the most Ge rich phase, $IrGe_4$. This interaction continued until the underlying Ir atoms were consumed. Atoms from the unreacted source of Ge in the middle of the island then diffused through the $IrGe_4$ with an atomic flux J_{α}^{Ge} towards the $X_{\alpha\beta}$ interface. This gradually exposed the $IrGe_4$ which decompose through the mechanism $3IrGe_4 \rightarrow Ir_3Ge_7 + 5Ge$. The heat of reaction of this decomposition is found to be approximately equal to zero (Habanyama, 2004). It is generally accepted that a nucleation-controlled reaction is characterized by a heat of reaction close to zero. This explains why the formation of $IrGe_4$, by the reverse reaction, proceeds by way of a nucleation process and exhibits sudden formation of the phase at high temperature. Ge atoms from the decomposition reaction are added to the flux from the unreacted Ge source, resulting in the flux J_{β}^{Ge} flowing through the phase Ir_3Ge_7. These fluxes 'feeds' the reaction at the $X_{\beta\gamma}$ interface were Ir_4Ge_5 is transformed into Ir_3Ge_7, a reaction that consumed Ge thereby reducing the atomic flux from J_{β}^{Ge} to J_{γ}^{Ge}.

The depletion of unreacted Ge atoms from the source region continued in parallel with the $IrGe_4$ decomposition. A gradual increase in the width of exposed $IrGe_4$ indicated that the latter process was slower. The width of the phase Ir_3Ge_7 was controlled by two processes: firstly, by the $IrGe_4$ decomposition at the interface with $IrGe_4$ and secondly, the mechanism $13Ge + 3Ir_4Ge_5 \rightarrow 4Ir_3Ge_7$ at the interface with Ir_4Ge_5. On the other hand, the phase Ir_4Ge_5 was effected by two competing processes. The above process resulted in the consumption of Ir_4Ge_5 at the Ir_3Ge_7/Ir_4Ge_5 interface and the second was the mechanism $5Ge + 4Ir \rightarrow Ir_4Ge_5$ at the growth interface with unreacted Ir. Since an increase in the width of the Ir_4Ge_5 region was observed this implies that the latter process was the faster of the two. The picture for samples annealed at temperatures below 800 °C differed only by the absence of the $IrGe_4$ phase. There was some evidence of a narrow region of IrGe from RBS [Figure 12.3(b)], AFM (Figure 12.6) and even some SEM results. However this was at the periphery of the reaction region and was not easily resolvable in most cases as it tended to be overwhelmed by the Ir_4Ge_5 formation.

The graph in Figure 12.13 is a plot of the growth width with time of anneal for a sample annealed at 800 °C. It should be pointed out that the origin in this figure does not correspond to the original island interface but refers to the interface between the $IrGe_4$ and Ir_3Ge_7 regions. The growth curve for the region labeled as Ir_3Ge_7 was a result of Ir_3Ge_7 grown from Ir_4Ge_5 and that from the decomposition of $IrGe_4$. The growth characteristics observed are parabolic with time. This is indicative of a diffusion controlled process which, as modeled by Kidson (Kidson and Mater, 1961), results in parabolic growth even in multi-phase systems.

The growth of the phases Ir_3Ge_7 and Ir_4Ge_5 was monitored at the temperatures 700 °C, 750 °C and 800 °C. Different annealing times were chosen to obtain a reasonable range of growth widths at each of the three temperatures. The squares of the growth widths were plotted against the times of anneal at each temperature and the diffusional growth constants K_{β} where obtained from the slopes. The logarithms of the diffusional growth constants were then plotted against the reciprocals of the product, $k_{\beta}T$ of the Boltzmann constant and the absolute temperatures, the resulting Arrhenius plot for the phase Ir_4Ge_5 is shown in Figure 12.14.

The activation energy, E_{α} was obtained from the slope of the straight-line fit. The average activation energy determined for the diffusion process in the Ir_4Ge_5 phase

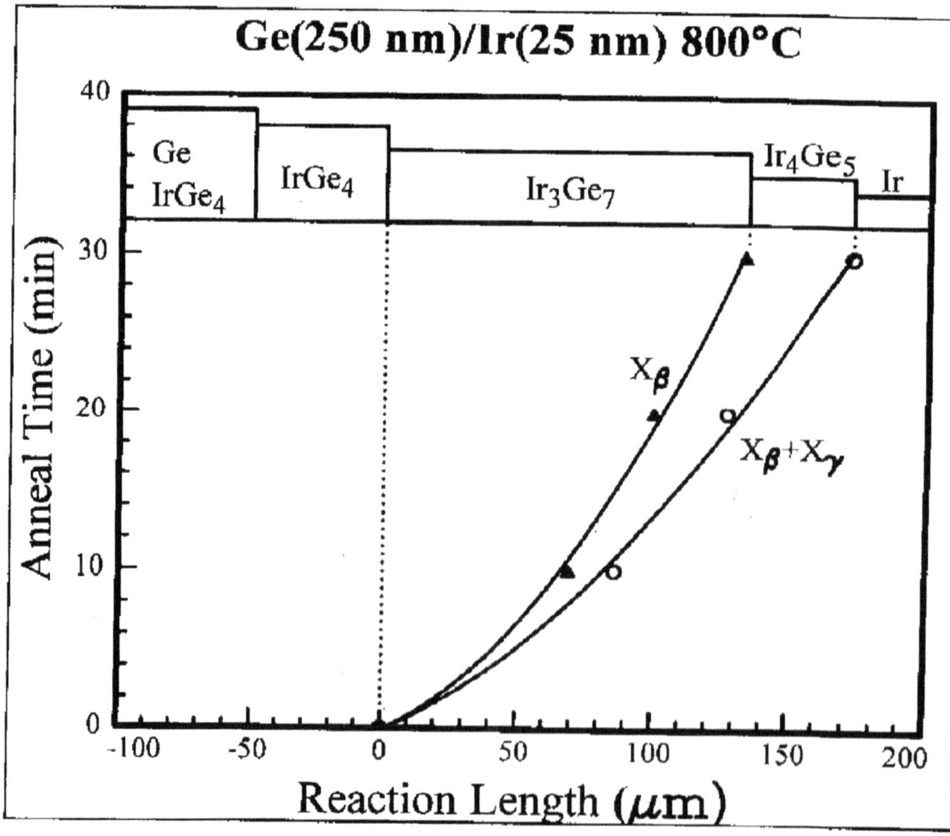

Figure 12.13: Plot of Growth Width with Time of Anneal for the Phases Ir_3Ge_7 and Ir_4Ge_5, for a Sample Annealed at 800°C

was $E_\alpha = 1.6\pm0.1$ eV. The magnitude of E_α suggests that the lateral diffusion reaction was not driven by surface diffusion of Ge atoms but rather the diffusion of Ge atoms through the interior of the lateral diffusion couple. Typical values for surface diffusion are around 0.6 eV (Eckertova, 1989).

As mentioned earlier, the Pd peak of the solid line in Figure 12.8 lies at the surface position, indicating that there was some Pd at the surface. The 'shoulder' marked in that figure indicated that there was less Pd at the surface than deeper down. This region consisted of PdGe at the bottom while at the top there was unreacted Ge intermingled with PdGe. Our SEM results further revealed the formation of cracks in the source region near the original island interface.

We now make an attempt to explain the reaction process in the Pd-Ge system: Ge in the island reacted with the underlying Pd eventually forming the most Ge rich phase PdGe which rose to the surface in columns, hence giving the surface a 'puffy' appearance as observed in the SEM. Excess Ge then diffused from the source through the PdGe to form Pd_2Ge outside the original island interface. It is believed that as the

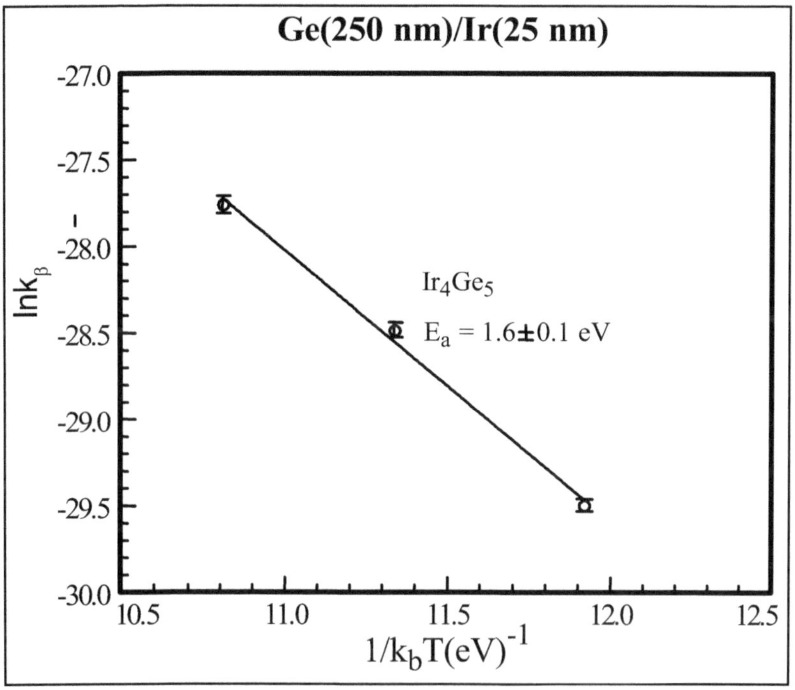

Figure 12.14: Arrhenius Plot, $\ln K_\beta$ Versus $1/k_bT$, Showing Temperature Dependence of Ge Diffusion Rate Through Ir_4Ge_5, Yielding an Average Activation Energy of 1.6 ± 0.1 eV

Ge between the columns was consumed the structure was weakened, resulting in the cracks observed by our SEM. These cracks were clearly seen as forming between the 'puffs' and they introduced discontinuity in the matrix of the sample, thereby impeding Ge flow to the reaction interface where the phase Pd_2Ge was growing. Further Pd_2Ge growth at the Pd interface was then characteristically fractal due to insufficient and inhomogenious Ge supply. The interface between PdGe and Pd_2Ge lies at the position of the original island interface and does not shift with increase in anneal time, suggesting that there was no resultant growth interaction at this interface. The exposure rate of the phase PdGe could not be monitored because the low temperature at which this investigation had to be carried out meant a slow lateral reaction and the crack formation eventually halted this reaction before the PdGe was sufficiently exposed.

Growth of the phase Pd_2Ge was monitored at the temperatures 275 °C, 300 °C and 325 °C. The periods of annealing were chosen so as to obtain a reasonable range of growth widths at each of the three temperatures; results are presented in Figure 12.15.

An Arrhenius plot of the temperature dependence of the Ge diffusion rate through Pd_2Ge was made. The average activation energy determined from this plot was $E_\alpha =$ 1.0±0.1 eV. This value is high enough to characterize a non surface diffusion process.

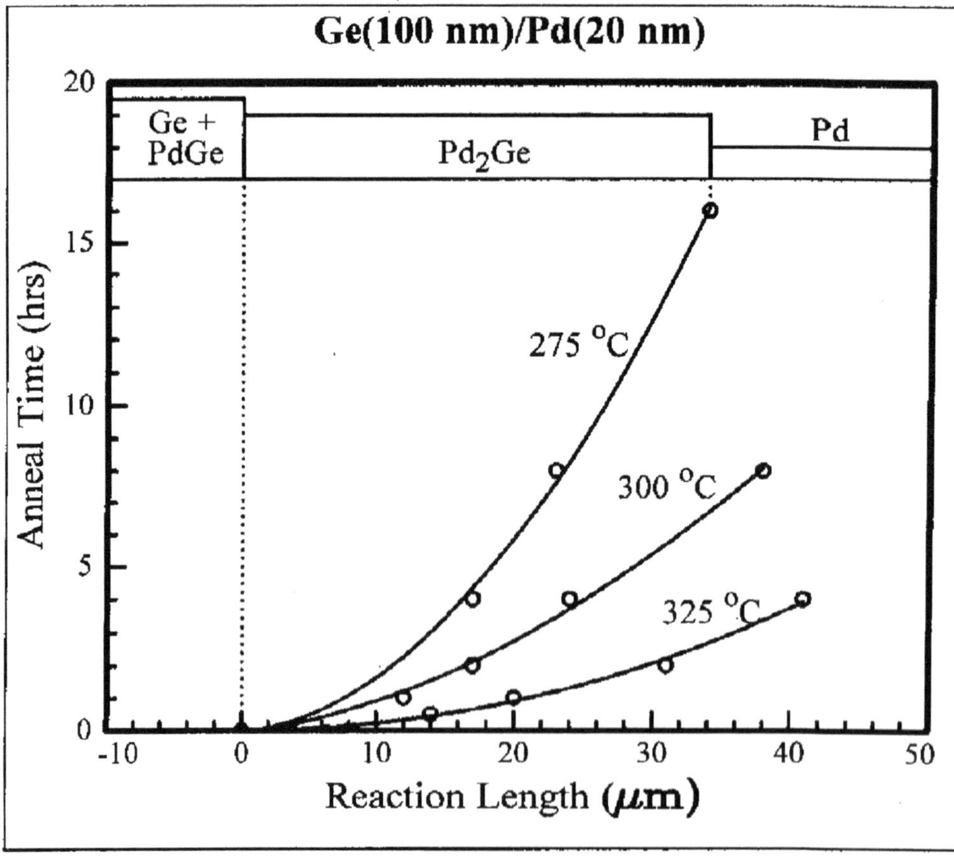

Figure 12.15: A Plot of Reaction Length with Time of Annealing for the Phase Pd_2Ge at Temperatures 275, 300 and 325 °C

Conclusion

This investigation has developed an understanding of the interactions between Ge and the metals Ir and Pd in lateral diffusion couples. This has been achieved using the established techniques RBS, PIXE, SEM, AFM and Optical Microscopy.

We have demonstrated the power of using this combination of techniques in obtaining information which is necessary in making a detailed analysis of the behavior of different diffusion systems. A few examples of how the combination of techniques was useful in this regard are presented below.

From our RBS and PIXE results of the Pd-Ge samples, the region at the outer reaction interface showed a gradual decrease in Ge concentration. By using SEM it became clear that this decrease was not as a result of a gradual change in composition but rather the effect of summing data parallel to the interface across alternating areas of Pd_2Ge and unreacted Pd in a region of fractal growth, thereby reducing the combined Ge concentration.

Another example is that our RBS and SEM results in the Ir-Ge systems indicated that Ge was consumed predominantly from around the island edges. However, a more detailed examination by AFM revealed that some material from further inside the island was also consumed. AFM was also used in conjuction with RBS to estimate the densities of the phases formed.

The origin of the 'puffs' and cracks observed in our SEM results of the Pd-Ge system was only explained after analysis of the RBS results, which revealed that some Pd atoms rose to the surface during the early stages of the reaction.

The use of Particle Induced X-ray Emission (PIXE) on the Nuclear Microprobe has been demonstrated as a useful technique in studying lateral diffusion couples. This is the first time PIXE has been used in this regard. This extends the range of possible systems like the Ni-Ge and Co-Ge systems that can be studied on the microprobe since the use of mRBS is limited to couples of elements that are not close in atomic number.

While lateral diffusion couples were first studied in the early 1980's, newly adopted techniques and apparatus have been used to study them. This work has shown that these techniques provide valuable insight, especially when they are employed in a complementary manner.

Acknowledgements

The authors wish to thank the University of Zambia, University of Cape Town, National Research Foundation and the University Science, Humanities and Engineering Partnerships in Africa (USHEPIA) for their assistance. They also wish to thank the Materials Research Group at the iThemba Labs at Faure for the use of their facilities, Miranda Waldron in the Electron Microscope unit at the University of Cape Town, Ms Terry Davies of the X-ray unit in the Geological Science Department, University of Cape Town, and the Atomic Force Microscopy team at the University of Stellenbosch.

References

Blanpain, B., 1992. Kluwer Academic Publishers, 222: 421

Blanpain, B., Mayer, J.W., Liu, J.C. and Tu, K.N., 1990. *J. Appl. Phys.*, 68: 3259.

Blanpain, B., Mayer, J.W., Liu, J.C. and Tu, K.N., 1990. *Phy. Rev. Lett.*, 64: 2671.

Chen, S.H., Zheng, L.R., Barbour, J.C., Zingu, E.C., Hung, L.S., Carter, C.B. and Mayer, J.W., 1984. *Materials Letters*, 2: 469.

de Waal, H.S., 1999. *Ph.D. Thesis*, University of Stellenbosch, South Africa.

Ding, P.J., Talevi, R., Lanford, W.A., Hymes, S. and Murarka, S.P., 1994. *Nucl. Instr. Meth.*, B85: 17.

Doolittle, L.R., 1986. *Nucl. Instr. Meth.*, B15, 227.

Eckertova, L., 1989. *Physics of Thin Films*, 2nd Edition. Plenum, New York.

Habanyama, A., 2004. *Ph.D. Thesis.* University of Cape Town, South Africa.

Kidson, G.V., 1961. *J. Nucl. Mater.*, 3: 21.

Liu, J.C. and Mayer, J.W., 1990. *J. Mater. Res.*, 5: 334.

Liu, Joyce C., Mayer, J.W. and Barbour, J.C., 1988. *J. Appl. Phys.*, 64: 651, 656

Nemutudi, R.S., Comrie, C.M. and Churms, C.L., 2000. *Thin Solid Films*, 358: 270.

Zheng, L.R., Hung, L.S. and Mayer, J.W., 1983. *J. Vac. Sci. Technol.*, A1: 758.

Zheng, L.R., Hung, L.S. and Mayer, J.W., 1983. *Thin Solid Films*, 104: 207.

Zheng, L.R., Hung, L.S., Mayer, J.W., Majni, G. and Ottaviani, G., 1982. *Appl. Phys. Lett.*, 41: 646.

– *Section III* –

NANOTECHNOLOGY: *STATUS AND PROSPECTS IN DEVELOPING COUNTRIES*

Chapter 13

Nanotechnology: Present Status and Future Prospects in Armenia

Vladimir Aroutiounian
Department of Physics of Semiconductors and Microelectronics and Center of Semiconductor, Devices and Nanotechnology, Yerevan State University, Yerevan 0025 Armenia
E-mail: kisahar@ysu.am

ABSTRACT

The paper describes investigations in theoretical and experimental nanoscience and nanotechnologies carried out at Yerevan State University (YSU), Armenia. The paper presents work on Theoretical investigations of transport, luminescence, kinetic phenomena and fractals in porous silicon, coatings for solar cells, solar thermal-photovoltaics, quantum dot technologies, photolysis, thin film based sensors to e-nose divices. Further described is the work on DNA sensors and the use of carbon nanotube-metaloxide systems for the increase of sensitivity of sensors and their work without pre-heating. Investigations of Armenian scientists and engineers were supported by many international grants. Armenia's state sponsored "semiconductor nanoelectronics" program is also presented.

Keywords: *Silicon solar cells, Solar thermal-photovoltaic conversion, E-nose, Thin film based sensors, Quantum dots, Nanoscale photodevices.*

Introduction

Yerevan State University (Faculty of Radiophysics), Department of Physics of Semiconductors and Microelectronics and Center of Semiconductor Devices and Nanotechnology (shortly YSU NT Center) has significantly contributed to Armenia's nanotechnology drive. The YSU NT centre currently has 12 Professors, Docents and Ass. -Professors, 24 full staff Researchers, 3-4 PhD students per year, 40 students per year, and the participation in teaching of 12 students per year for SYNOPSYS-ARMENIA. All semiconductor materials, photoelectrodes, cells, sensors and setups discussed below were made in the YSU NT Center. Several modern technological equipment and measurement setups were purchased in the framework of ISTC, IPP, NFSAT, and CRDF international grants. More than 50 joint papers are published and several Patents obtained in collaboration with foreign scientists. YSU NT has about 40 Patents (including inUSA, France, Russian Federation, USSR and Armenia) and are authors of several hundred publications in West countries and USSR' journals as well as in Proceedings of International conferences and meetings. More than 10 conferences were organized in Armenia. About 150 citations of local scientific papers had been made by foreign scientists in leading journals in the last 5 last years.

Investigations in YSU NT Center in last year's were carried out in the following directions:

- ☆ Investigations of high-efficient nanocrystalline photoelectrodes for photoelectrochemical conversion of solar energy in hydrogen (including titanium nanotubes) -Photolysis setups.
- ☆ Theoretical and experimental investigations as well as manufacture of different antireflection coatings for silicon solar cells.
- ☆ Gas and smoke detectors -Portable pen type hydrogen leak detector, E-nose device.
- ☆ Solar thermal-photovoltaic converters converting heat directly into the electricity using 3-5 photovoltaic diodes; other infrared devices
- ☆ Graded-bandgap photodiodes and position-sensitive detectors made of 3-5 semiconductors grown by liquid and electroliquid epitaxy
- ☆ Nanocrystalline solar cell and other nanoscale photodevices
- ☆ Formation of self-assembled InAsSbP–based strain-induced quantum dots and pits on InAs (100) substrates by liquid phase epitaxy.

A. investigations of extremely high-efficient nanocrystalline photoelectrodes for photoelectrochemical conversion of solar energy in hydrogen (including titanium nanotubes). Photolysis setups.

Oxide semiconductors (TiO_2, ZnO, $SrTiO_3$ and other oxides) and cheap technologies of the manufacturing of such materials and photoelectrodes (PEs) using such oxides or their solid solutions were developed in Armenia. Different technological methods of the preparation of ceramic, thin film and polycrystalline nano PEs were proposed and investigated. The best efficiency of photoelectrolysis of water near 2 per cent without any concentrators of solar energy was detected in metal oxide PEs, which was partially reduced and contained optimal concentration of corresponding impurities (one or two types).

An increase in the efficiency of conversion of solar energy into hydrogen is realized by the deviation from stoichiometry of stable in-water electrolytes and metal oxides (TiO_{2-x} etc.) or appropriate doping of n-type oxides with different impurities. However, even these results are not good enough for large-scale production of hydrogen. Doped PEs had the photoabsorption sensitivity range' expansion in visible part of the solar spectrum, which increased the efficiency of photoelectrochemical conversion. New type oxide compositions like high temperature superconductors and others can be also used for the preparation of rather effective photocathodes.

One way to obtain stable PEs, which will be sensitive in visible range of solar spectrum, is the synthesis of new semiconductor phases made of solid solutions of wide band gap oxide (no more than 3 eV) and rather small band gap oxides(corresponding to the maximum of the solar radiation spectrum). TiO_2-MnO_2, ZnO-CdO, and Fe_2O_3-Nb_2O_5 etc. were synthesized and investigated. The method of anodic or electrochemical oxidation of industrial titania films or plates in water solutions of electrolytes at DC condition was used for manufacturing of nanoporous films of n-TiO_2. Discussion on prospects for the use of porous and nanocrystalline oxide PEs in photoelectrochemistry was carried out..

Several versions of the photoelectrolysis setups were realized and tested in Armenia. Setups contained matrices made of very cheap ceramic, thin film and nano PEs. In first equipment, hydrogen had been obtained on PEs with small area (1.5–2 cm^2) at high concentration of the solar radiation (up to 50 W/cm^2). The equipment had a concentrator in the form of a parabolic mirror with the area of 0.2 m^2, having the photoelectrolysis cell in its focus. A matrix of PEs with common area up to 550 cm^2 was used in other three photoelectrolysis' equipment. First setup had not any concentrator, in second–the solar radiation double concentration had been released using the parabolic-cylindrical concentrator (Figure 13.1). Other technologies proposed by us allow preparing large area electrodes by the use of metallic sheet covered by powder of TiO_2 by using of conductive glue. These systems did not require a complicated tracking system. The output of this equipment supplying by parabolic-cylindrical concentrator and the rutile PEs' matrix with the area of about 100 cm^2 was equal to 0.046 l/h of solar hydrogen.

The weight of this solar–hydrogen equipment was about 80 kg. Mean efficiency of solar energy conversion to hydrogen was less than 3 per cent. Cheap nickel electrodes were used.

The SEM image of titanium oxide films prepared is shown on Figure 13.2. TiO_2 nanoporous nanotubes made are shown in Figure 13.3.

B. Theoretical and experimental investigations as well as manufacture of antireflection coatings for silicon solar cells, made of nano-grain films of porous silicon, diamond-like carbon and silicon oxynitride

Multilayer anti-reflection coatings for silicon solar cells. Si-based solar cells reflect approx. 36 per cent of the incident light in the spectral range where Si is photosensitive. One of methods of the increase in efficiency of cells is a reduction of reflectance of sunlight from the solar cells, which can be realized by the use of the different types of antireflection coating layers (ARCs). Such an approach leads to an increase in the absorbed part of solar irradiation in the whole optical and near IR region when corresponding ARC having appropriate refractive index (n) and layer thickness (d) is achieved.

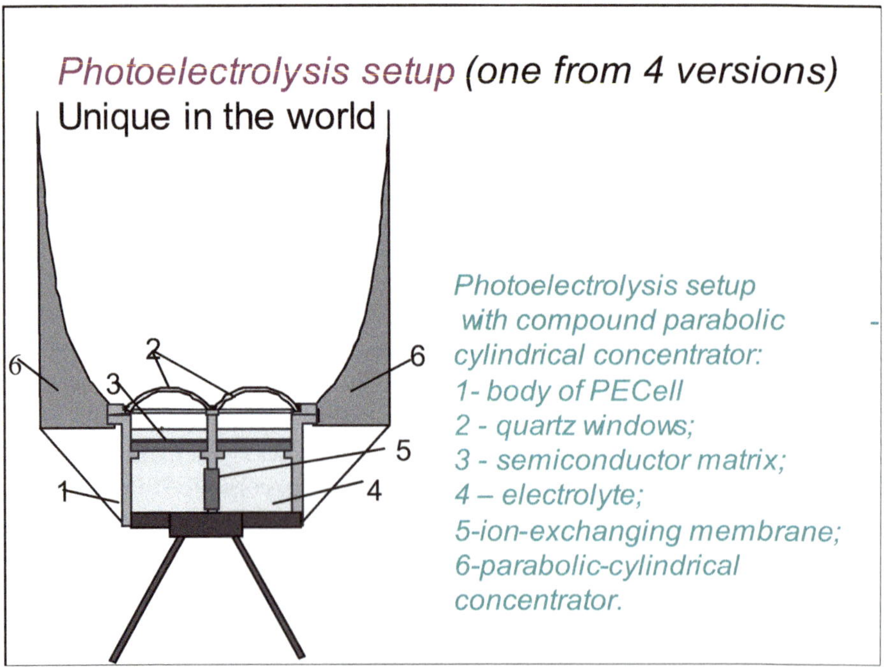

Figure 13.1: Output 0.028 l/h–Without Concentrator

In the equipment with the Freznel lens and rutile PEs, the output of hydrogen was 7 l/h; the size of the equipment (in mm) was 3600x2300x300 size of one PE–1 cm^2.

Investigations and the development of porous silicon layers for their use as antireflection coatings of solar cells, as well as for semiconductor gas sensor applications are being successfully carried out at YSU. Promising results have been obtained after investigations of commercial solar cells with standard ZnS ARC and the same cells after removal of the ZnS layer and realization of the porous silicon ARC. The measurements of optical, photoelectrical and photovoltaic characteristics of such solar cells, performed under the same conditions, have shown that manufactured solar cells with porous silicon ARC by their main characteristics (short-circuit current I_{sc}, open-circuit voltage Uoc, low reflectance) are much better in comparison with the industrial cells with the ZnS ARC. The advantage of solar cells with the porous silicon ARC is that they are stabilizing even after some degradation of photoelectrical parameters in time. The saturation of the degradation characteristics was confirmed in multiple testing measurements of current-voltage characteristics.

Today ZnS, Al_2O_3, Ta_2O_5 layers are used in industry as single-layer ARCs, SiO_2/TiO_2, Ta_2O_5/ZnS and MgF_2/ZnS layers are used as double-layer ARCs, and MgF_2/Al_2O_3 /ZnS coating is used as a triple-layer ARC. Recently Si_3N_4 layers are used as ARC also. A new theoretical approach to the calculation of the reflectance of using silicon solar cells with different multilayer coatings with a corresponding choice of optical length, refractive indexes (n) and thickness (d) of layers manufactured had been presented. Powerful optical matrix approach for the simulations, allowed the property calculations of different new types of anti-reflection coatings using simple

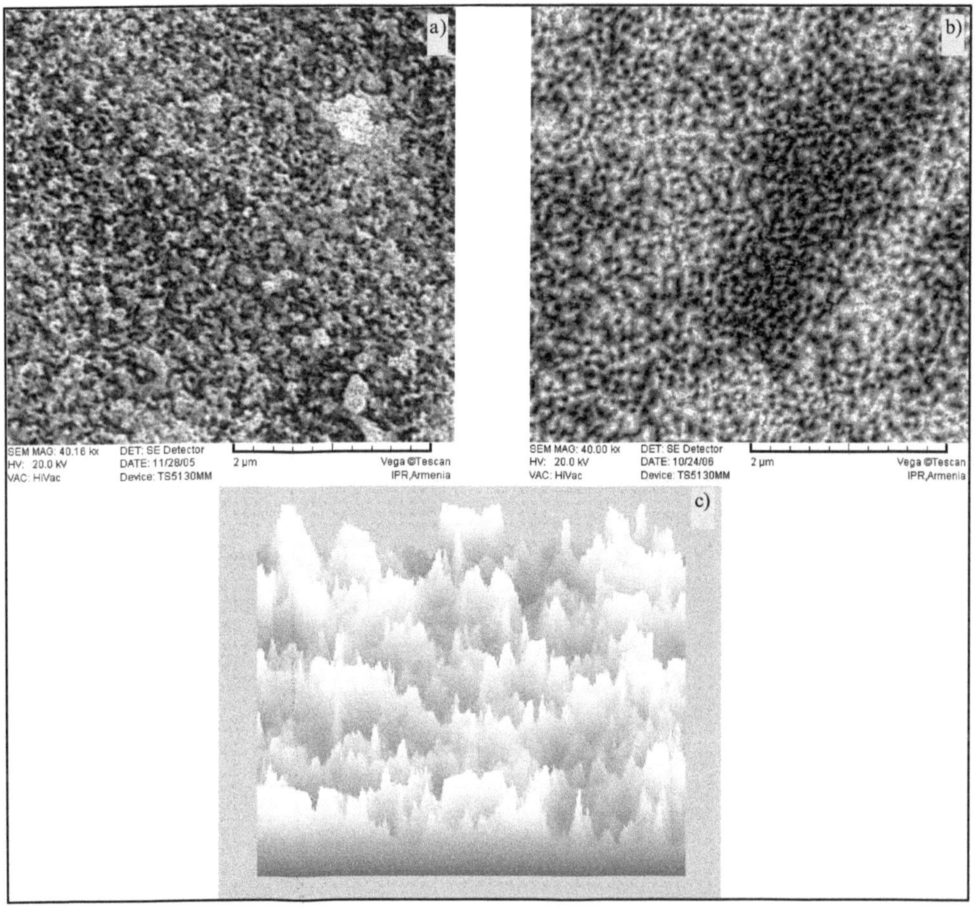

Figure 13.2: SEM Images of Titanium Oxide Films Prepared at 30 V Before (a) and After (b) Treatment in the Air; 3-D View of the Samples (c).

software "Mathematica-5". The reflectance spectra of frequently used MgF_2/ZnS and SiO_2/TiO_2 double-layer coatings calculated coincided with experimental results for these coatings.

For this purpose, the reflectance spectra of different double-layer ARCs had been determined and compared with corresponding experimental results. MgF_2 /ZnS and SiO_2/TiO_2 double layer ARCs found large application in silicon solar cells. The reflectance spectrum of all coatings have been determined and compared with the results of calculations from experimental data. As can be seen below, the calculations and experimental results are in very good agreement (Figure 13.4).

Coatings made of porous silicon (PS) layers are of special interest, as the surface of porous silicon layer is highly textured which leads to reduction of reflectance losses. It is possible technologically to control the band gap of porous silicon layer for optimization of solar irradiation absorption. Such PS layer can be easily formed on large area silicon solar cells by simple electrochemical anodization. Moreover,

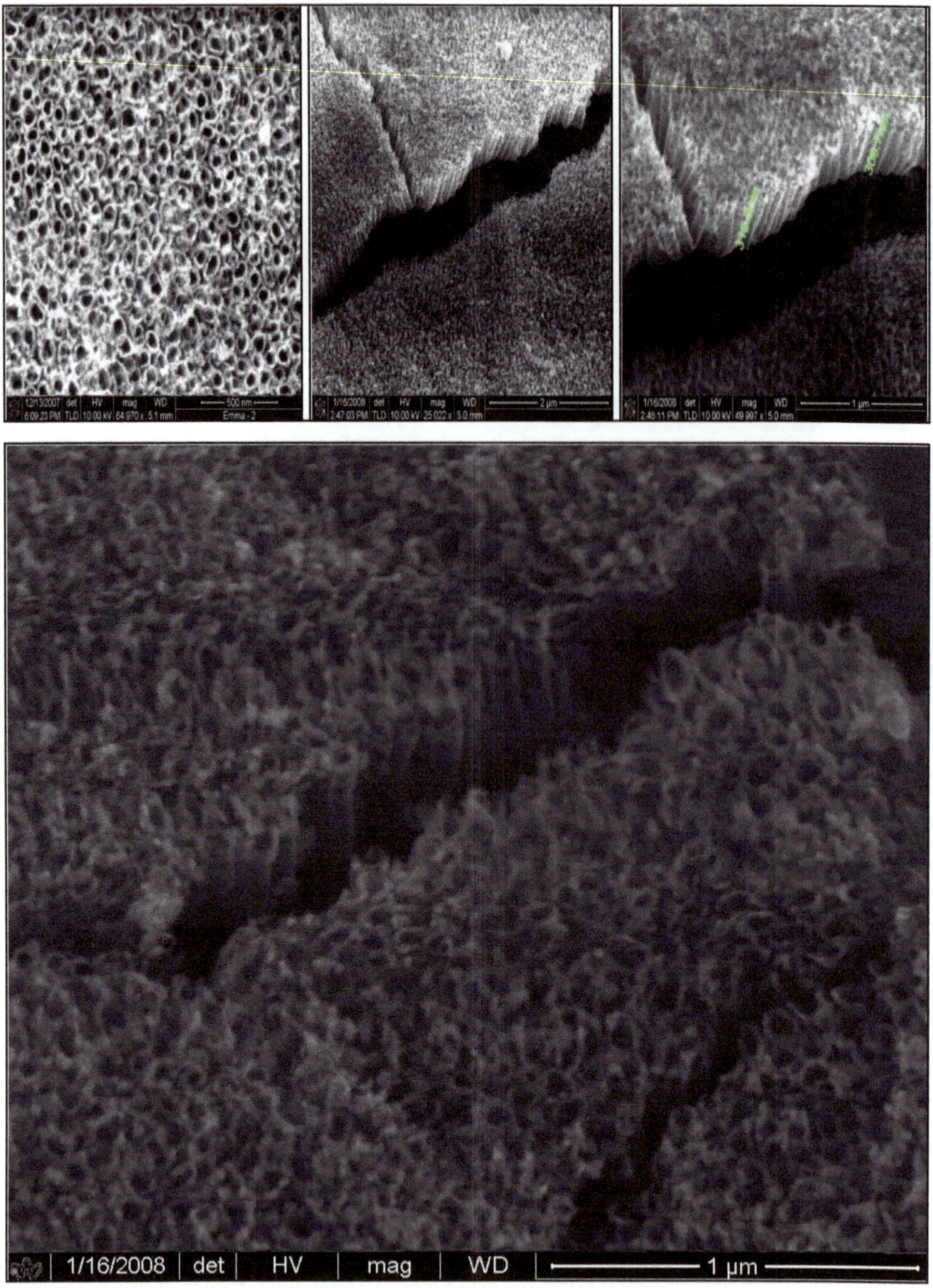

Figure 13.3: TiO_2–Nanotubes. D~ 70-100 nm, H~ 270-320 nm

application of PS layers leads to the reduction of surface recombination velocity, enlargement of spectral response of silicon solar cells into short wavelength region,

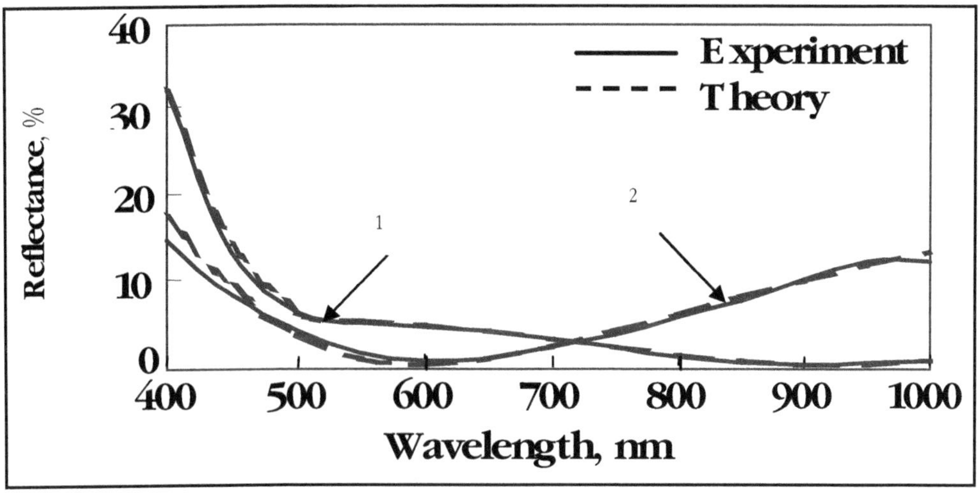

Figure 13.4: Experimental and Calculated Reflectance vs. Wavelength of the MgF_2/ZnS (1) and SiO_2/TiO_2 (2) Double-layer Coatings

improvement of photogeneration of carriers. Because of its porous nature, the PS refractive index varies from 1.25 to 3; therefore such layer can be used as ARC. On the other hand, by varying the current density during electrochemical anodization in one technological cycle, it is possible to obtain layers with different refractive indices and distribution profile and therefore to obtain multilayer ARC. However, it is known that PS degrades with time if special methods of preservation are not used. It is necessary to protect the PS layer from ambient conditions by other layers. In this context, diamond-like carbon (DLC) and silicon oxynitride (SiO_xN_y) films are of great interest. Advantages of each layer could be considered separately.

The advantages of DLC layer in silicon solar cells as ARC are: such layers have a very high hardness, a high stability to hostile environments such as aggressive chemical attacks, high dose irradiation and ambient condition fluctuation. The technology of DLC film growth is well established using CVD from hydrocarbon precursors and sputtering of graphite target. In addition by changing growth parameters, it is possible to change the refractive index of such a material from 1.5 to 3.1. Note that the band gap of DLC layers can reach up to 4 eV, therefore it can be used as a large bandwidth optical window and it has a high transparency in the UV region, making it especially promising in solar cells for space application.

The advantages of silicon oxynitride layers are: silicon oxynitride layer promising candidate for replacing pure SiO_2 in microelectronics. SiO_xN_y films can be manufactured by direct oxynitridation of Si surface or a nitridation of the SiO_2 layer. It leads to a remarkable decrease in the surface states concentration and to extremely low surface recombination velocity and leakage current. Besides, SiO_xN_y film is stable to different chemical influences to high-energy irradiation, more resistant to hydrofluoric acid and to the diffusion of different impurities and ions. Generally SiO_xN_y can be considered as a nitrogen-doped silicon dioxide. In such layers the properties of silicon nitride (high stability to chemical influences) and silicon oxide (silicon surface state passivation) are combining. A very important feature of the SiO_xN_y films is that they can have energy gap and refractive index values intermediate between silicon dioxide (SiO_2) and stoichiometric silicon nitride (Si_3N_4) films depending on the x and y parameters of SiO_xN_y. Therefore, it is possible to mediate the refractive index of such material from 1.45 up to 2.

The use of single-layer ARCs based on silicon oxynitride or diamond-like carbon is well known. However, the combination of PS with silicon oxynitride or diamond-like carbon layers for solar cells does not realized experimentally. The possibility of application of double-layer and triple layer ARCs based on porous silicon layer in combination with diamond-like carbon and silicon oxynitride layers have been investigated. The best simulation results have been obtained with the following parameters of layers, when the refractive indices were n(SiO_xN_y)=1.5, n(PS)=2.8, and thicknesses were d (SiO_xN_y) =84 nm, d(PS)=52 nm in the case of silicon oxynitride over layer and refractive indices were n (DLC)=1.6, n (PS)=2.9, and thicknesses was d (DLC)=88 nm, d (PS)=50 nm in the case of diamond-like carbon over layer. They are shown in Figures 13.5 and 13.6 respectively. As can be clearly seen, it is possible not only to preserve low reflectance in the mid-IR and visible region of solar irradiation but also to expand it to short-wavelength (UV) region of spectrum. Therefore, application of double-layer ARCs proposed in silicon solar cells will be effective.

Theoretical and experimental investigations of double- and triple ARCs were carried by scientists from YSU NT Center and Prof. P. Soukiassian (CEA, Saclay and University of Paris-Orsey France) and are covered by 2 French Patents.

"SOLAREN" Co. in Yerevan (V. Afiyan) and "Viasphere-transistor" (A. Vardanyan) produced in small scale solar panels using silicon solar cells made out of Armenia. STUA are working on covering of Si solar cells with diamond like carbon ARC (J. Panosyan).

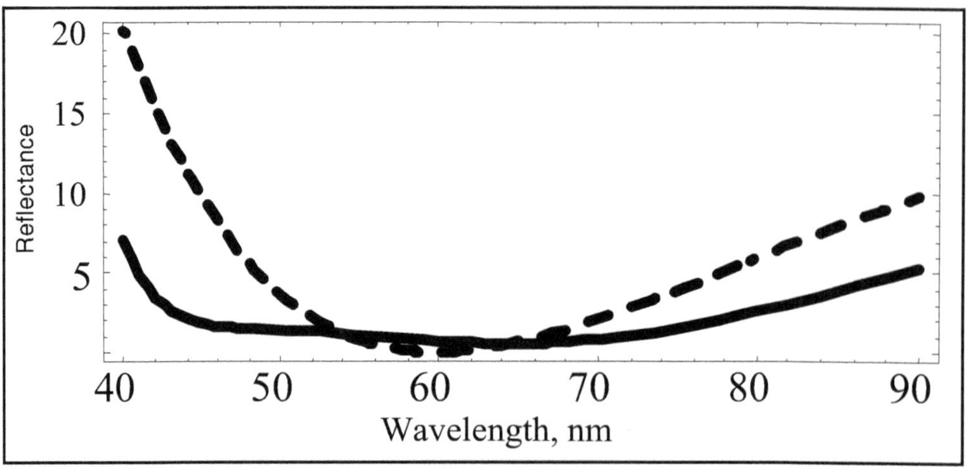

Figure 13.5: Reflectance Spectrum of SiO_xN_y/PS (Solid line) and SiO_2/TiO_2 Double Layer ARCs (Dash-dotted line)

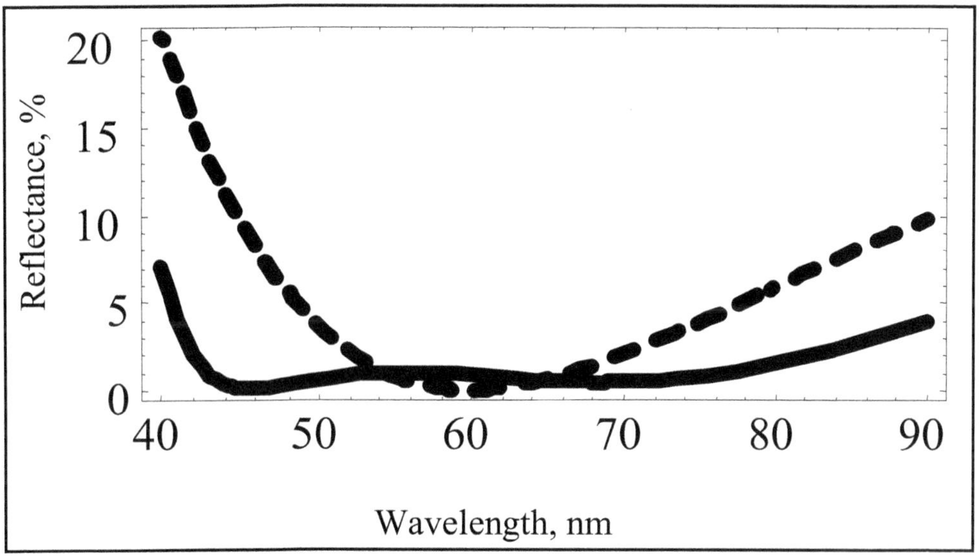

Figure 13.6: Reflectance Spectrum of DLC/PS (Solid line) and SiO_2/TiO_2 Double Layer ARCs (Dash-dotted line)

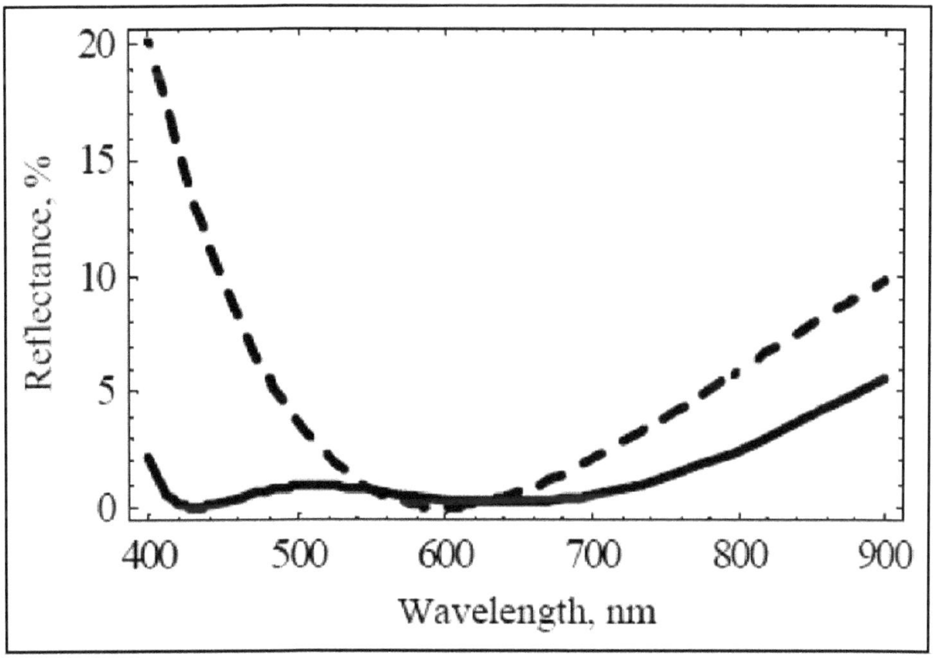

Figure 13.7: Reflectance Spectrum of Triple-layer ARC with DLC Film (Solid line) and SiO_2/TiO_2 Double Layer ARC (Dash-dotted line)

Chemical Sensors and Gas Detectors

Besides residential applications, gas and fire detectors are used under operating conditions with liquid fossil, gas media, fuel cells as well as in nuclear power plants, in order to prevent different type of incidents. It is very important to have information about the concentration of hydrogen, methane, carbon monoxide and dioxide, and other gases as well as humidity in order to prevent a false alarm. Research and development of world class semiconductor metal-oxide adsorptive sensing elements (with micro-and nano sized sensitive grains) for different gases made of different metaloxide semiconductors as well as integral components of new type sensors and detectors for non-expensive hybrid multisensor home Security system are carried out at YSU.

Development and investigations of corresponding technologies and regimes for obtaining of relatively thermally stable nano-sized SnO_2 particles of very small dimensions, thin films on its base hydrogen and CO sensors are carrying out by Drs. A. and Z. Adamyan. It is established that the sensitivity of gas sensors strongly depends on nanocrystal's granularity. Weak dependence of grain size on films on calcination temperature is found. Low operating temperatures and consumed electric power are typical for our sensors. The sizes of nanoparticles measured using TEM are shown in Figure 13.8, dependence of the sensitivity on working temperature on Figure 13.8. The technique offered ensures more stable grain sizes (in comparison with known techniques). The SnO_2 grains remained less than 5-6 nm even after calcinations at 750°C.

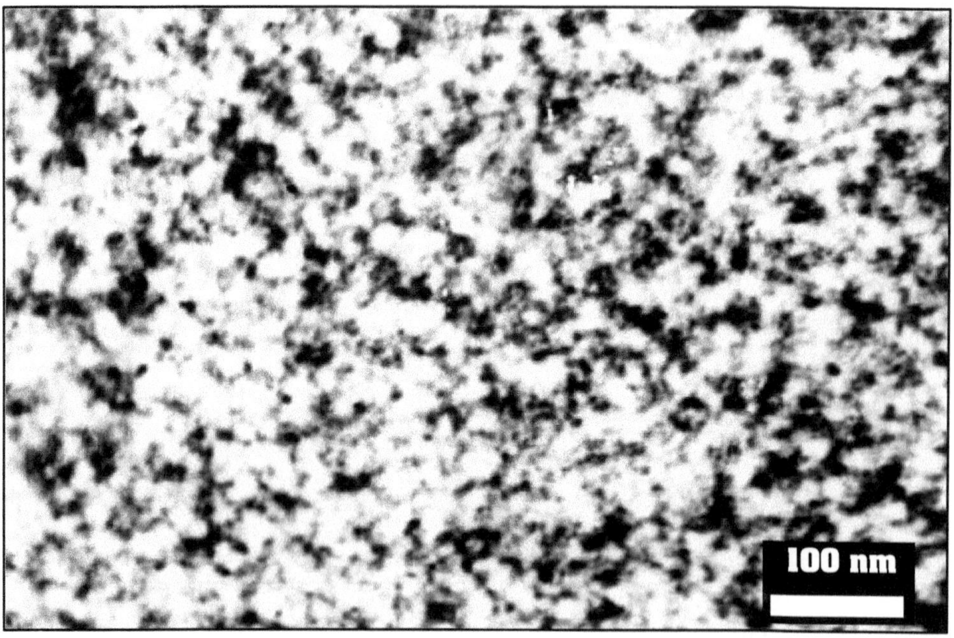

Figure 13.8: TEM Image. Minimal grain size is equal to ~4 nm. Solid-core pictures of the SnO_2 nano-sized particles.

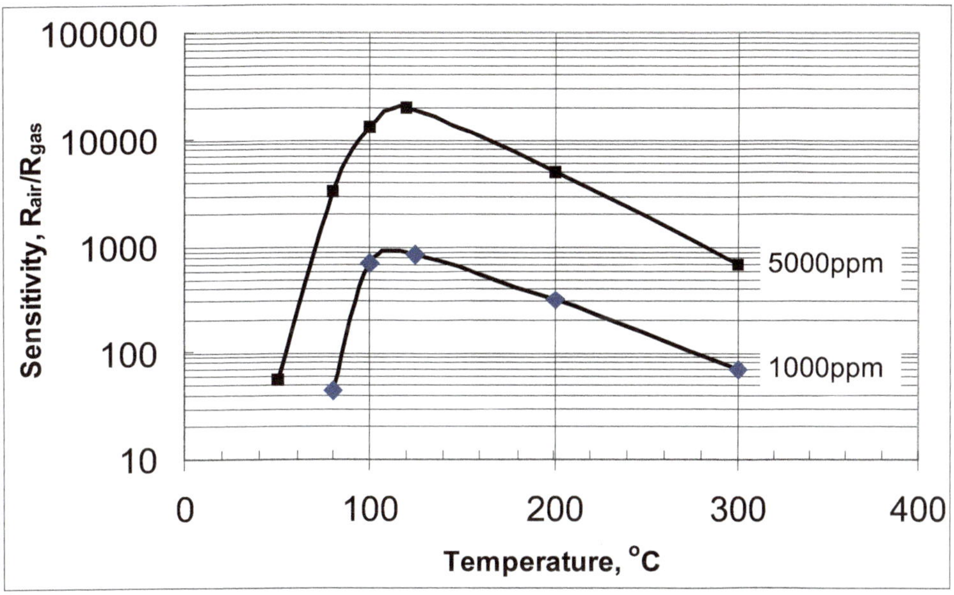

Figure 13.9: H_2 Sensitivity of Sol-gel Sensors vs. Operating Temperature

Different semiconducting oxides ($Sn0_2$, ZnO, TiO_2, Fe_2O_3), catalytic oxides (V_2O_5, MoO_3, CuO, NiO), metals deposited on oxide supports (Pt/SnO_2, Pt/ZnO, Pd/SnO_2), and mixed (or complex) oxides exhibit different physical properties on exposure to different gas species.

The following nanosensors are ready for innovation in YSU NT Center:

Bi_2O_3 adsorptive type smoke and NO_x sensors for multi-sensor system solutions (Patented in Armenia, the first modification was patented also in USA) Nanocrystalline SnO_2 thin-films based sensors for hydrogen and CO detection (Patented in Armenia) Figure 13.10 shows an example.

The key advantage of the product is workability of the sensing element at near room temperatures and high sensitivity to H_2 and CO gases, ensures high security at hydrogen detections, and exhibits selective-sensitivity to CO at other operational temperature regime. Fuel cells are main applications of H_2-sensors.

Portable Pen Type Hydrogen Leak Detector (Figure 13.11) is another development in this area.

The detector operates in autonomous regime from battery (14 hours of run-time) as well as can also be run continuously from the battery charger (mobile phone charges is acceptable).

Porous silicon–metal oxide tandem structures for Hydrogen and other hazardous gases sensing including CO (Patented in Armenia). Hydrocarbon gases sensors, ZnO based nano sensor is another development.

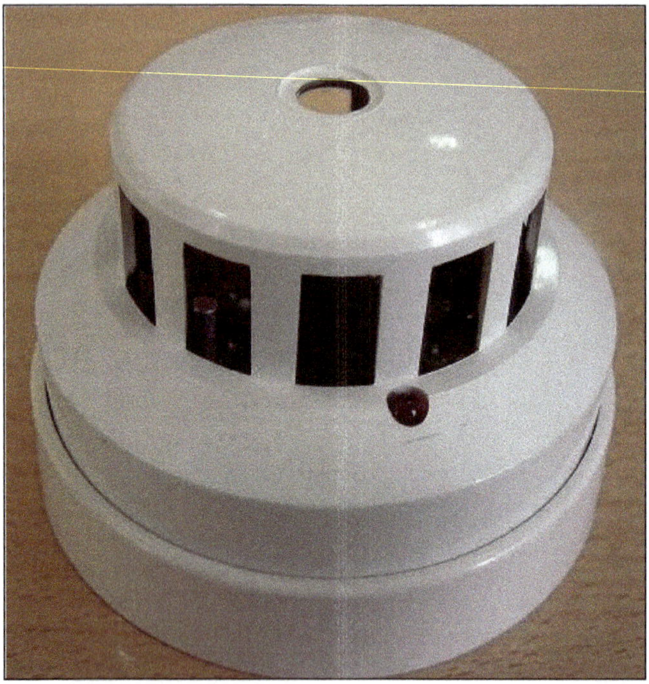

Figure 13.10: An Example of a Sensor Based on Nanotechnology

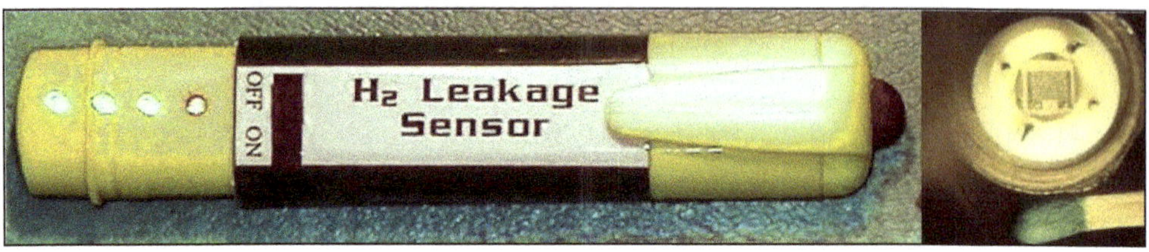

Figure 13.11: Pen Type Hydrogen Leak Detector

Electronic nose (Figure 13.12) was another development.

The E-nose device for home safety allows measuring low concentrations of methane, carbon monoxide and hydrogen simultaneously. Other gases sensors can be added by request.

The sensitivity of above-mentioned sensors was earlier measured by classical method-comparison the resistance of sensors in gas media and air. Measurements of noise spectral density of sensors were also carried out. A comparison of data for several Figaro sensors is given in Figure 13.13.

As well as sol-gel H_2 and porous silicon sensors produced in the lab. Higher sensitivity of spectral dependence of noise (SDN) on gas concentration in comparison with the method of the measurement of the change in the Ohmic resistance of samples allows using such powerful method for determination of gas concentration in the air or environment (see Figures 13.13–13.14).

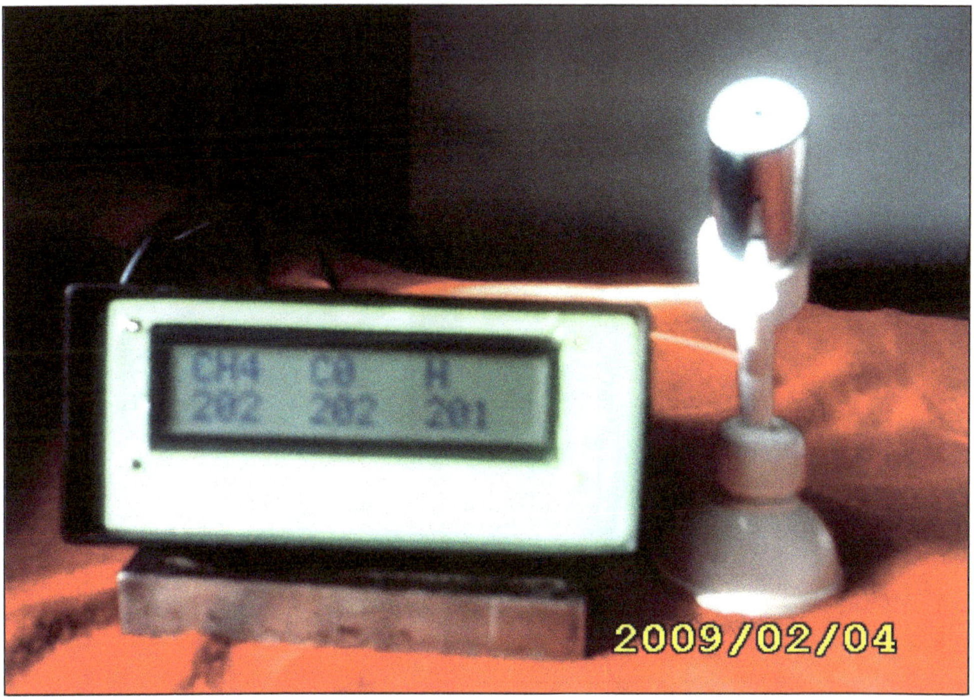

Figure 13.12: E-nose Device

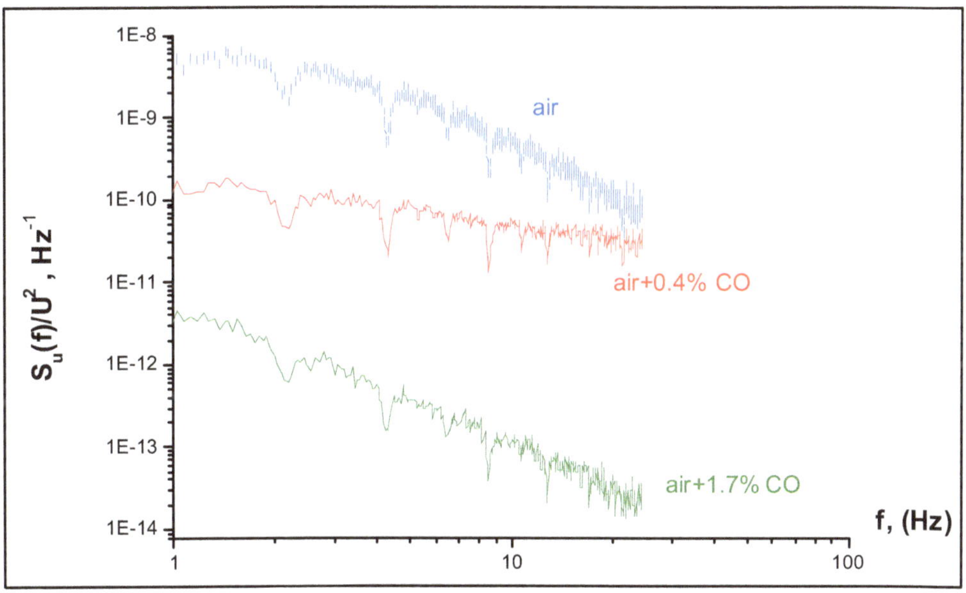

Figure 13.13: Noise Spectroscopy of Gas Sensing

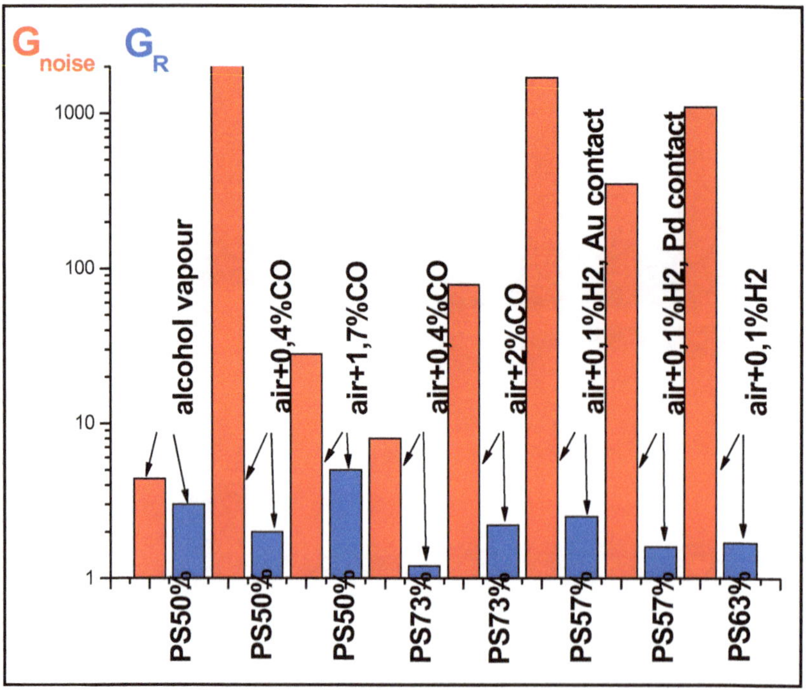

Figure 13.14: Comparison of the Sensitivity of Different Sensors Obtained by Measurements by Classical and Noise Methods

Locally developed sensors are working without remarkable pre-heating of work body, consume very low energy and ready for large-scale production. Their cost and size are much lower (Figure 13.15) in comparison to photoelectrical and radioactive ionization detectors, which are today in the market. Recently the detectors were successfully tested at Southwest Research Institute (San Antonio, USA).

Quantum Devices and Materials

Quantum dot cells and graded-bandgap photodiodes and position-sensitive detectors made of 3-5 semiconductors Quantum dot p–i- n nanocrystalline solar cells (QD cells) were proposed in YSU in 2000 and realized in Japan in 2003. Other nanoscale photodevices were proposed in YSU and Institute for Radiophysics and Electronics of NAN Armenia. Graded-bandgap photodiodes and position-sensitive photodetectors made of 3-5 semiconductors were grown in YSU by the method of liquid and electroliquid epitaxy. Thermophotovoltaic cells were proposed and grown in YSU. Recently YSU realized self-assembled InAsSbP–based strain induced islands on InAs substrates and saw their evolution from pyramids to QDs.

Solar thermal-photovoltaic (TPV) converters convert heat directly into the electricity using 3-5 photovoltaic diodes grown and developed in YSU NT Center.

The input power to the TPV cell is provided by a radiant heat source ("emitter")- see Figure 13.16. The typical emitter temperature available is T=1200°-2000°C. The

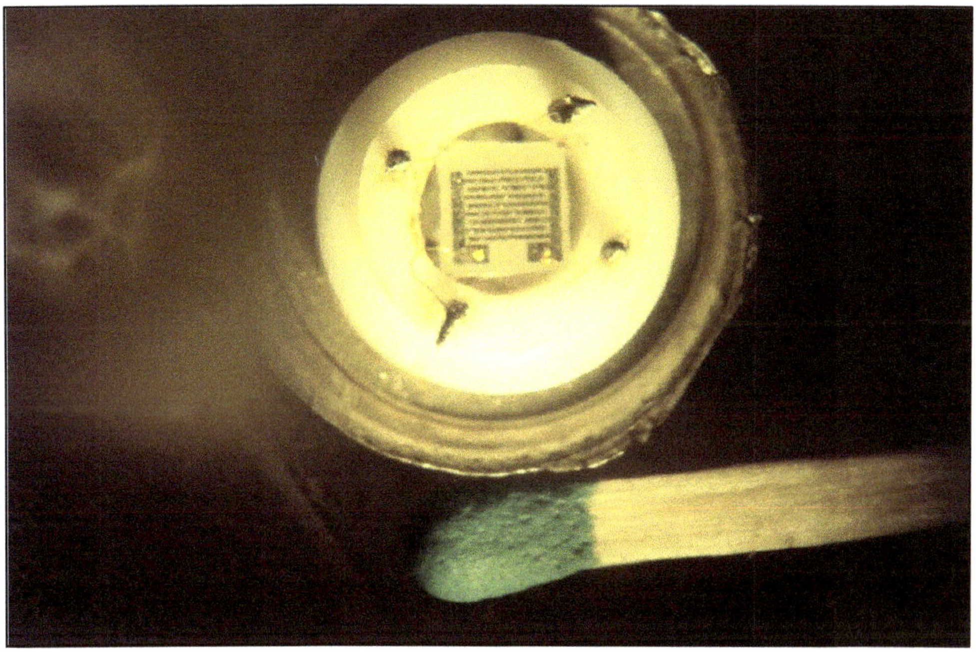

Figure 13.15

main difference between solar cells and TPV converters is a selective optical filter used to reflect radiation, which cannot be absorbed by the TPV cell, directing it back to the emitter to reheat it. Due to this approach, the efficiency of a TPV solar cell (see Figure 13.16) dramatically increased in comparison with PV using direct solar irradiation. Solar cells were made of InP-InAs system. Good matching between the emitter radiation spectrum and diode characteristics increases the efficiency of the system; the topic of the present technology is novel 3-5 materials involved quantum dots grown in YSU NT Center for optimizing this "matching" efficiency (see Figure 13.16). TPV cells can significantly increase the efficiency limit of existing solar cell technology and reaching up to 35–40 per cent in efficiency.

Formation of self-assembled InAsSbP–based strain-induced quantum dots and pits on InAs (100) substrates by liquid phase epitaxy is investigated by Dr. K. Gambaryan at YSU.

The Liquid Phase Epitaxy (LPE) technique is used in YSU NT Center also for the formation of self-assembled InAsSbP–based strain-induced quantum dots (QD) and pits (QPs) on InAs(100) substrates (Figure 13.17).

YSU NT research has looked at the technological approach and methodology for the growth of quantum dots-pits cooperative structures from the quaternary composition liquid phase as well as the energetic task and dots–pits competing nucleation mechanisms in strain-induced InAsSbP system. The critical strain when the mechanism of nucleation is changed from the growth of dots to the nucleation of pits is calculated. The concentration regions for the growth of InAsSbP quaternary

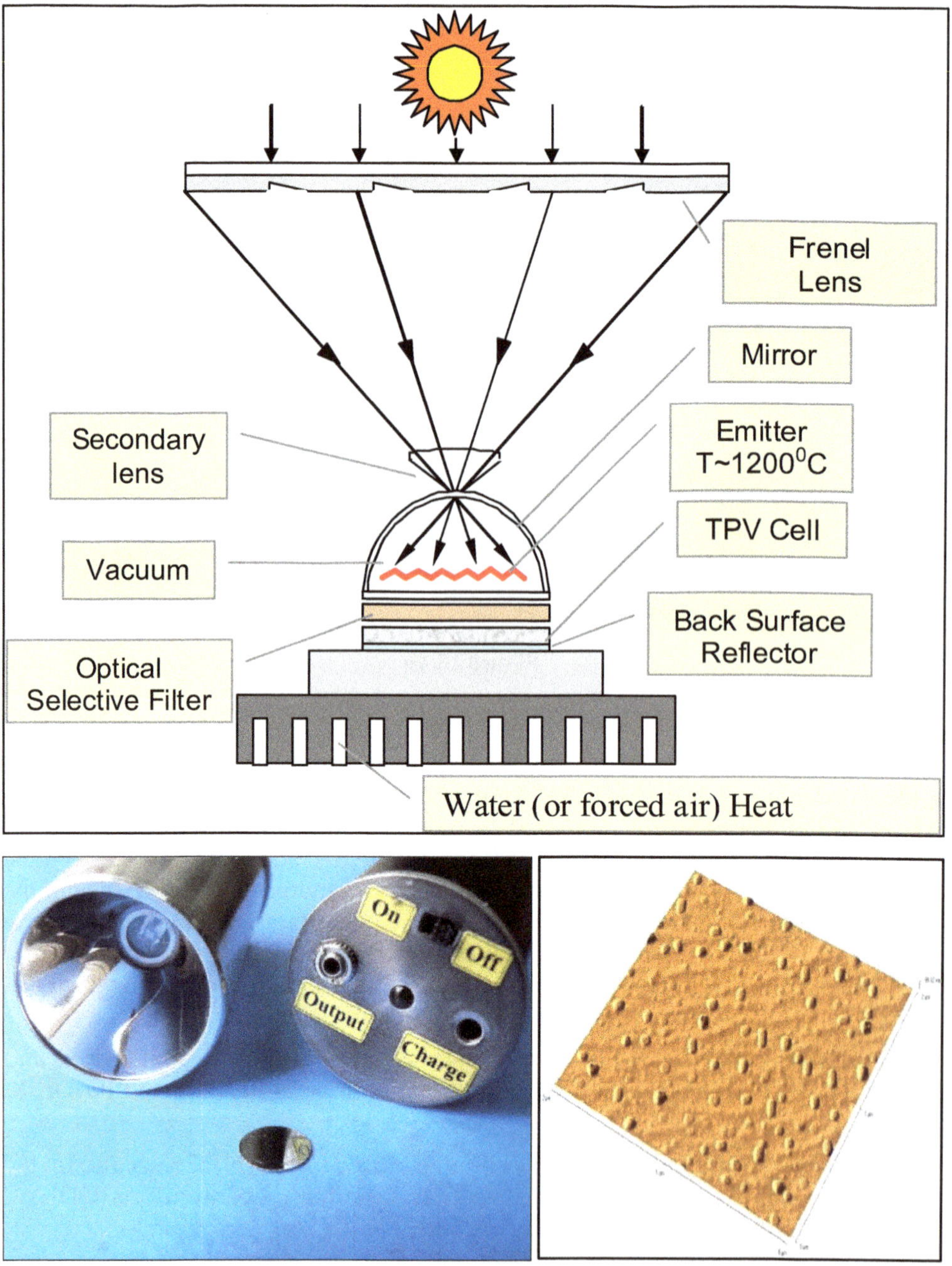

Figure 13.16

dots and pits are denoted. The morphology, dimensions and distribution density of grown dots and pits are investigated by high resolution Scanning Electron Microscope (SEM–EDXA). The QDs average density ranges from 5 to 7×10^{9} cm^{-2}, with heights

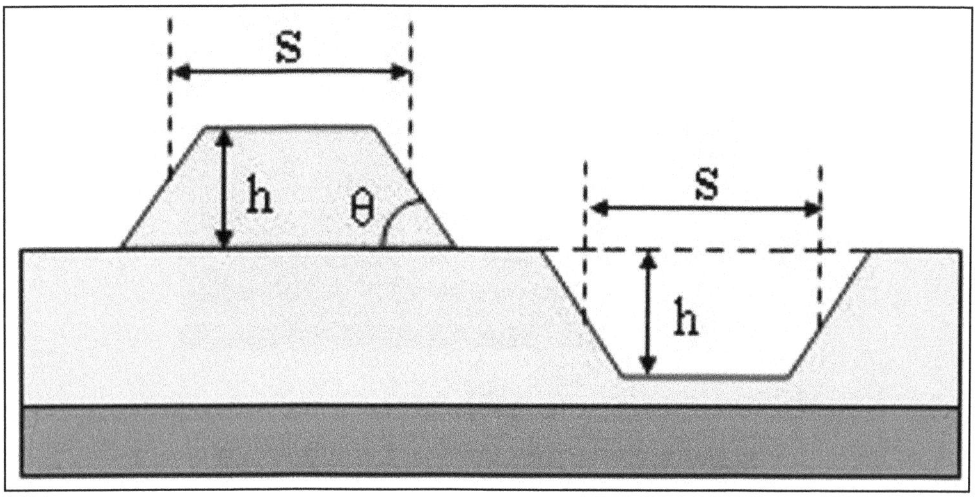

Figure 13.17: Schematic View of the Cross Section of Shape Assumed for Dot and Pit

and widths dimensions from 5 nm to 15 nm and 10 nm to 40 nm respectively. The average density of the QPs is equal to (2-6) $\times 10^{10}$ cm^{-2}, with dimensions of 5-30 nm in width and depth. A distribution for both quantum dots and pits versus their average diameter is experimentally detected. The mathematical approximations of these distributions are carried out and the corresponding adjustable parameters are calculated. The Fourier-Transform Infrared Spectrometry (FTIR) measurements allowed detecting a displacement of the absorption edge towards the long wavelength region as well as enlargement towards the short wavelength region.

To conclude, the study has opened up interesting fundamental and applied prospects including the underlying physics and the influence of quantum dots and pits for InAs-based diode heterostructures and device properties and performances for several mid-infrared applications.

Theoretical investigations in the field of nanophysics and nanotechnology are carrying in Physical-Technical Faculty at Armenian-Russian (Slavonic) University in Yerevan. Chair of general and theoretical physics headed by Prof. Eduard Kazaryan involve 2 Professors, 3 Docents working with 2 PhD and 10 students per year.

Scientists at ARU have been engaged in the following areas of research:

- Investigation of electronic properties of quantum nanostructure with nontrivial geometry (ellipsoidal quantum dots, quantum lens, spherical and cylindrical quantum layers, quantum rings).
- Investigation of coulomb (exciton, impurity) states in quantum well, wires and dots.
- Consideration of different mechanisms of absorption in quantum well, wires and dots (direct and indirect, interband and intraband)

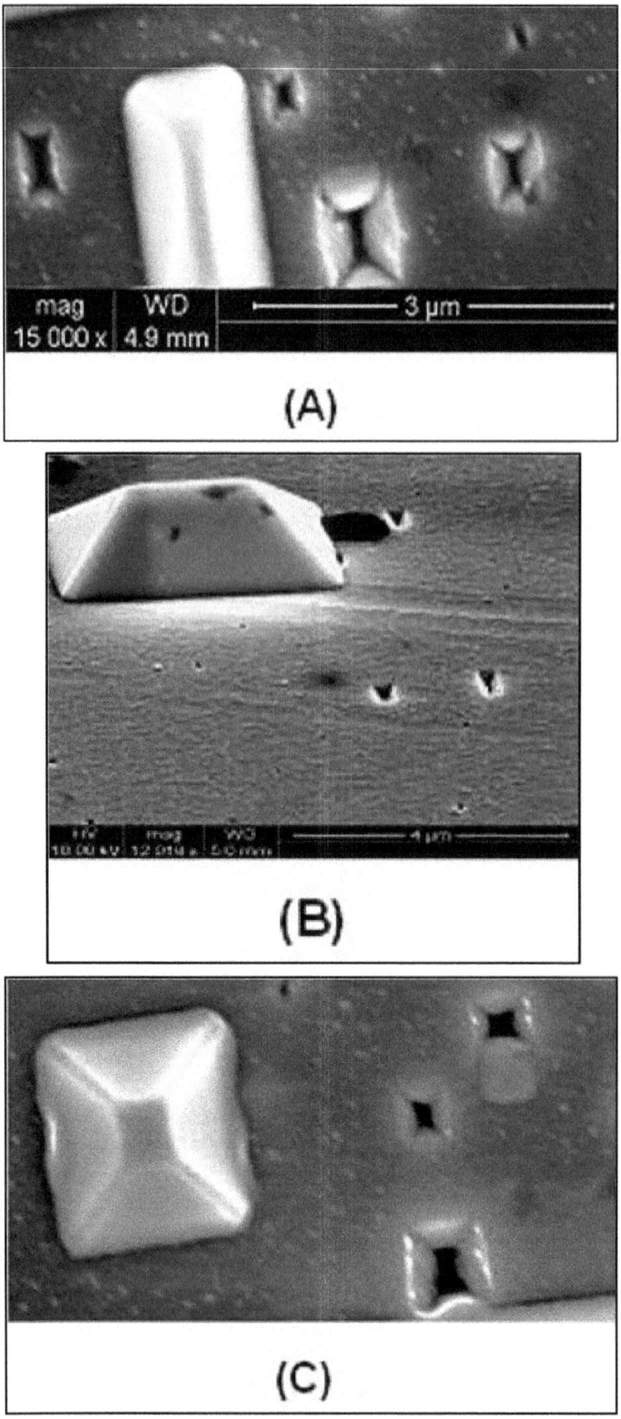

Figure 13.18: High Resolution SEM Images of the InAsSbP Strain-induced Islands–pits Cooperative Structure

Methods of adiabatic approximation as well as theory of perturbation are widely used for investigation of single electronic states in ellipsoidal quantum dots. It is important to mention results obtained while electronic states of quantum dots with weak ellipsoidality has been investigated for finite height of confining potential. Adiabatic approximation widely used for cases when geometry of the nanostructure allows split the Hamiltonian of the system into two parts, fast and slow. For example, such situations take place for the strongly oblate and strongly prolate ellipsoidal quantum dots and quantum lens of small width. In the case of layers with small width the model of rigid rotator can be used, because in this situation the radial motion of the particle is much faster than rotational motion.

The behavior of the coulomb systems in quantum wells, wires and dots has been investigated for the problems of optical absorption in such structures. The threshold of the absorption can be changed by influence of electron-donor interactions or presence of the impurity centre. This is the reason of investigation of low dimensional coulomb problems. It is important to mention that for some compounds the dispersion law of charge carriers is essentially non-parabolic. Particularly for compounds of InSb the dispersion law for the electron and light hole is relativistic one (two band Kane model). The researchers have shown that in case of InSb the two-dimensional coulomb problem is unstable and ground state of the system is decayed. One-dimensional coulomb states take place in quantum wires and in that case the consideration of quasi one-dimensionality eliminates the instability in opposite to pure one-dimensional case.

The other class of the physical problem connected with the investigation of different mechanism of light absorption in the nanostructures. In particular, the absorption of light in an ensemble of non-interacting cylindrical quantum dots in the presence of a magnetic field is discussed. Theoretical results are compared with experimental data on magneto-luminescence in an $In_{0.53}Ga_{0.47}As/InP$ cylindrical quantum dot system, and it is show that there are good agreement between theory and experiment. The same problem was solved for narrow-gap InSb cylindrical quantum dots.

Interesting class of problems connected with the impurity and inter-impurity absorption in the quantum wells. The optical absorption related to donor acceptor pair transitions in GaAs quantum wells had been investigated and had been shown, that there is a blue shift in the absorption spectra. Comparison with available experimental data had been made.

Activity of scientists in State Engineering University of Armenia and other chairs at Yerevan State University is carrying in the fields of nanophysics, microwave and photovoltaic devices.

Note that activity of Armenian scientists in nanotechnology and nanophysics is highly appreciated. Z.N. Adamyan, V.M. Arakelyan, Kh. S. Martirosyan, G.E. Shahnazaryan, and V.M. Aroutiounian of Yerevan State University got the Prize of President of Armenia for investigations in 2004-2006. ARU' scientists E.M. Kazaryan, H.A. Sarkisyan, K.G. Dvoyan and L.S. Petrosyan got the Prize of President of Armenia for investigations in 2005-2007.

References

Adamyan, A.Z., Adamian, Z.N., Aroutiounian, V.M., Arakelyan, A.H., K.J. and Touryan, J.A., 2007. Turner. Sol-Gel derived thin-film semiconductor hydrogen gas sensor. *Intern. Journal of Hydrogen Energy*, 32(16): 4101–4108.

Arakelyan, V.M., Aroutiounian, V.M., Shahnazaryan, G.E. and Khachaturyan, E.A., 2008. Thin film n-titanium oxide photoanodes for photoelectrochemical production of hydrogen. *Renewable Energy*, 33(2): 299–303.

Aroutiounian, V. M., 2007. Semiconductor metaloxide hydrocarbon gas sensors. *Intern. Scientific Journal for Alternative Energy and Ecology*, N3(47): 33–42.

Aroutiounian, V. M., Martirosyan, Kh. S. and Soukiassian, P., 2006. Almost zero reflectance of a silicon oxynitride/porous silicon double layer antireflection coating for silicon photovoltaic cells. *J. Phys. D: Appl. Phys.*, 39: 1623–1625.

Aroutiounian, V., Arakelyan, V., Galstyan, V., Martirosyan, Kh. and Soukiassian, P., 2009. Hydrogen sensor made of porous silicon and covered by TiO_{2-x} or ZnO<Al> thin film. *IEEE Sensors Journal*, 9(1): 9–12.

Aroutiounian, V., Martirosyan, Kh. and Soukiassian, P., 2004. Low reflectance of diamond–like carbon/porous silicon double layer antireflection coating for silicon solar cells. *J. Physics D: Applied Physics*, 37: L25–L28.

Aroutiounian, V., Petrosyan, S. and Khachatryan, A. Photocurrent of quantum dot p–i–n solar cells. *Ibid*, pp. 59–64.

Aroutiounian, V.M., 2007. Metal oxide hydrogen, oxygen, and carbon monoxide sensors for hydrogen setups and cells. *Int. Journal of Hydrogen Energy*, 32(9): 1145–1158.

Aroutiounian, V.M., Arakelyan, V.M. and Shahnazaryan, G.E., 2005. Metal oxide photoelectrodes for hydrogen generation using solar radiation-driven water splitting. *Solar Energy*, 78(5): 581–592

Aroutiounian, V.M., Arakelyan, V.M., Shahnazaryan, G.E., Hovhannisyan, H.R., Wang, H. and Turner, J.A., 2007. Photoelectrochemistry of tin-doped iron oxide electrodes. *Solar Energy*, 81: 1369–1376.

Aroutiounian, V.M., Arakelyan, V.M., Shahnazaryan, G.E., Stepanyan, G.M., Khachaturyan, E.A., Wang, H. and Turner, J.A., 2006. Photoelectrochemistry of semiconductor electrodes made of solid solutions in the system Fe_2O_3–Nb_2O_5. *Solar Energy*, 80(N9): 1098–1111.

Aroutiounian, V.M., Petrosyan, S.G. and Yesayan, A.E., 2004. Multiple quantum well photodiode with lateral p-n-junction. *Thin Solid Films*, 451–452: 389–392.

Atayan, A.K., Kazaryan, E.M., Meliksetyan, A.V. and Sarkisyan, H.A., 2008. Magneto-absorption in cylindrical quantum dots. *The European Physical Journal B*, 63: 485–492.

Atoyan, M.S., Kazaryan, E.M. and Sarkisyan, H.A., 2004. Direct interband light absorption in a cylindrical quantum dot in quantizing magnetic field. *Physica E*, 22: 860–866.

Buniatyan, V.V. and Aroutiounian, V.M., 2007. Wide gap semiconductor microwave devices. *Journal of Physics. D: Applied Physics*, 40: 6355–6385.

Buniatyan, V.V., Aroutiounian, V.M., Zekentes, K., Camara, N. and Soukiassian, P., 2004. Silicon carbide TUNNETT diodes. *Solid–State Electronics*, 48(9): 1569–1577.

Dvoyan, K.G., Hayrapetyan, D.B., Kazaryan, E.M. and Tshantshapanyan, A.A., 2007. Electron states and light absorption in strongly oblate and strongly prolate ellipsoidal quantum dots in presence of electrical and magnetic fields. *Nanoscale Research Letters*, 2: 601–608.

Dvoyan, K.G., Hayrapetyan, D.B., Kazaryan, E.M. and Tshantshapanyan, A.A., 2009. Electronic states and light absorption in a cylindrical quantum dot having thin falciform cross section. *Nanoscale Research Letters*, 4: 130–137.

Dvoyan, K.G., Kazaryan, E.M. and Petrosyan, L.S., 2005. Electronic states in quantum dots with ellipsoidal symmetry. *Physica E*, 28: 333–338.

Galstyan, V.E., Martirosyan, Kh.S., Aroutiounian, V.M., Arakelyan, V.M., Arakelyan, A.H. and Soukiassian, P.G., 2008. Investigations of hydrogen sensors made of porous silicon. *Thin Solid Films*, 517: 239–241.

Gambaryan, K.M., Aroutiounian, V.M., Boeck, T., Schulze, M. and Soukiassian, P., 2008. Strain-induced InAsSbP islands and quantum dots grown by liquid phase epitaxy on a InAs(1 0 0) substrate. *J. Phys. D: Appl. Phys.* 41, art. no. 162004, p. 162004–1–162004–1–5.

Gevorkyan, V.A., Aroutiounian, V.M., Gambaryan, K.M., Kazaryan, M.S., Touryan, K.J. and Wanlass, M.W., 2004. Liquid-phase electroepitaxial growth of low band-gap p–InAsPSb/n–InAs and p–InAsP/n–InAs diode heterostructures for thermo–photovoltaic application. *Thin Solid Films*, 451–452: 124–127.

Harutyunyan, V.A., Êazaryan, E.Ì., Êostanyan, À.À. and Sarkisyan, H.A., 2007. Interband transitions in cylindrical layer quantum dot: influence of magnetic and electric fields. *Physica E*, 36: 114–116.

Juharyan, L.A., Kazaryan, E.M. and Petrosyan, L.S., 2006. Electronic states and interband light absorption in semi-spherical quantum dot under the influence of strong magnetic field. *Solid State Communications*, 139: 537–540.

Kazaryan, E.M., Kostanyan, A.A. and Sarkisyan, H.A., 2007. Optical absorption in GaAs quantum wells caused by donor–acceptor pair transitions. *Journal of Physics: Condensed Matter*, 19, id 046212(9pp).

Kazaryan, E.M., Meliksetyan, A.V. and Sarkisyan, H.A., 2007. Interband transitions in an InSb narrow-gap cylindrical quantum dot. *Technical Physics Letters*, 33: 49–56.

Kazaryan, E.M., Petrosyan, L.S. and Sarkisyan, H.A., 2008. Hidden symmetry and excitonic transitions in the quantum well. *Physica E*, 40: 536–541.

Tshantshapanyan, A.A., and K.G. Dvoyan, E.M., 2009. Kazaryan, light absorption in coated ellipsoidal quantum lenses. *J. Mater. Sci. Mater. Electron*, 20: 491–498.

Chapter 14

The Cuban Approach Towards Nanobiotechnology: A New Step on Scientific Development of the Country

Ariel Felipe
Scientific Advisor Office, Council of State, Republic of Cuba
E-mail: ofascience@enet.cu

ABSTRACT

The philosophy behind the development of the Center for Advanced Studies of Cuba (CEAC) project is based on, and the most relevant achievements of the Cuban Biotechnology are presented. Projects based on bionano in initial phases are undertaken by different Cuban Scientific Institutions, which are the foundation of the approach toward the Bionanotechnology. Bionanotechnology deals with the properties at molecular scale and the applications of biological nanostructures; and such as, it is presented as an interface among the chemistry, the biology, and the physical sciences and engineering. It is using new nanofabrication techniques and processes of molecular auto assembling that the Bionanotechnology allows the production of new materials and devices including fabrics and cellular engineering scaffolds, molecular motors, and biomolecules for the delivery of drugs, sensors of wide spectrum, and other medical applications. The political vision and will of undertaking the constant scientific consolidation and development of a society based on science people, and knowledge; will turn this Project an unquestionable reality.

Keywords: *Nanotechnology, Bionanotechnology, Developing countries, Intellectual property rights.*

Introduction

Nanotechnology is a potentially enabling technology with applications in water filtration, energy, computers, telecom, medicine, agriculture, etc. (Roco *et al.*, 2002).

The goal of the challenge is to discover new nanotechnologies that will ameliorate, or provide sustainable alternatives to some of the human activities that are most severely affecting the social and economical balance of the developing counties (Juma and Yee-Chong, 2005).

The application of nanotechnology is not without its downsides, due to our incomplete knowledge. Although we are able to make totally new, nano-derived materials, especially nanoparticles and carbon nanotubes, how they will behave in living systems and in the environment is incompletely understood. Some new nanoparticles appear to offer huge benefits for healthcare–in the targeting of drugs to the disease site, and as contrast agents in the early identification and treatment of disease. It is therefore critical to understand how these nanoparticles work, and what their long-term effects in the body might be, to realise their real benefit.

Industrialization has led to the generation of huge 'wealth' for western society, based on asset-stripping the natural world, and exacerbating the divide between the rich and poor nations. The rate of consumption and waste has increased so dramatically and is now so vast that it is threatening the very existence of all humankind on the planet. Across the globe 1.2 billion people live in absolute Poverty (WHO, 2005). Of these, almost 300 million live in sub-Saharan Africa. The majority of these people are small-scale farmers who depend either entirely, or to a large extent, on agriculture and the management of natural resources for their livelihoods.

Nanotechnology is one highly important facet of a spectrum of technologies that may offer some solutions to creating a sustainable future. Its key selling point has been heralded as offering benefits for the environment through 'more for less'; the concept of being able to fit more features into smaller, smarter products. This should translate into products that use less resources (reducing waste and pollution) to produce, and are more energy efficient. Nanotechnology, in conjunction with other technologies, may provide alternatives to activities that are resource intensive; and also may offer new opportunities for a satisfying way of life in this future world, more just and secure, without jeopardizing our environment.

Fabio Salamanca- Buentello *et al.* (2005) lists ten nanotechnologies considered in a survey of leading nanotechnologists to be the most important to the developing world. These are:

- Energy storage, production and conversion
- Agricultural productivity enhancement
- Water treatment and remediation
- Disease diagnosis and screening
- Drug delivery systems
- Food processing and storage
- Air pollution and remediation
- Construction
- Health monitoring
- Vector and pest detection and control

However, as J. Schummer stated (Schummer, 2007) -The Impact of Nanotechnologies on Developing Countries, (2007): -"at least 1.1 billion people lack access to safe water, and this results in several million deaths per year, mostly children in poor Asian and African countries. Many authors have suggested that nanotechnology will provide the crucial remedy. To assess that promise, we need to have a closer look at the pollutants that are particularly important for unsafe water in non-industrialized areas, which are microbes that are usually from human sources due to insufficient sanitation and heavy metals dissolved from minerals, then it is difficult to see where nanotechnologies could help here other than by competing with established and simple but efficient water treatments, like chlorination, that are still lacking in many areas. The main problems are not of technological nature, but a lack of basic infrastructure, facilities, and education". Invernizzi and Foladori, (2005) cite the ability of China and Vietnam to significantly reduce malaria in the last century without the use of emerging technologies.

The above-mentioned alerts us that the selected areas should be assessed carefully by the developing countries themselves considering their realities and needs. It is crucial to keep in mind the possible technological infrastructures to be developed with the potential human capacities of each country. Furthermore, Brown notes that, "within development circles there is a suspicion of technology boosters as too often people promoting expensive, inappropriate fixes that take no account of development realities" (Brown, 2001).

J. Schummer in the same paper (Schumer, 2007) states-" in the earliest state, when decisions about governmental R&D funding are made, negative impacts particularly result from wrong decisions that are misled, for instance, by unfounded hopes, hype, unclear concepts, or wrong information. While richer countries can afford the waste of research money to a certain degree, the effect of big and misguided projects on poorer countries with small research budgets can be disastrous, because their relative investments, for instance for new instrumentation, is much higher. It is particularly important, therefore, that they do not simply copy industrialized countries but instead focus their research efforts on well-defined projects tailored to specific needs rather than on such vague projects as nanotechnology overall. Informed science policy decisions would also be cautious about hype-words such as 'novel' and, instead, rely on careful patent researches that proof whether a research field is already claimed by patents or not. If such a new and promising field is identified and focused on, developing countries could benefit not only from meeting their specific needs but also from becoming leaders in a new technology"-.

Intellectual Property's Rights and Nanotechnologies

Although it is not a specific problem of nanotechnology, they are emerging in a time that the rights on the intellectual property have changed in the Western world. Such changes have a negative effect on the developing countries. Two negative tendencies can be observed in this situation. Of one of them, the approach of what is patentable and what is not has followed a tendency evidently liberal; in such a way, that even the basic knowledge and the databases of basic knowledge now can be patented. On the other hand, there are new regulations for the Universities on their scientific magazines publications imposing them a report to the administration of the University before the topic to be published is sent. This is called "disclosure report", it approves the publication or decides its protection with patent. Due to these tendencies, the knowledge generated in the academic institutions, initially accessible in scientific magazines, is now another problem for the developing countries (Lage, 2000).

This politics has triggered the quantity of patents, with the rising increment of profits for the use of the knowledge protected licensing for the developed countries–particularly the USA. For example, the earnings for this concept went from 200 Million dollars in the nineties to 1.4 billion dollars in 2004. The annual growth from 1996 at the 2002 was approximately of 15 per cent, reaching a sustained increase pace from that moment on. On the other hand, although the public expenditure in the 2006 increased to 6.4 from 4.6 billion dollars in 2004, the private expenditure increased at 6 from 4 billion dollars in the same year. That is to say, the private sector increased in the same proportion the public sector did, which points toward a private bigger domain of the knowledge generated by this high level of investments in investigation (Igami and Okazaki, 2007).

While this tendency has promoted the technological transfer from the University to small local companies (Start ups) that open new income sources to the University; the access to the technological development for third parties gets complicated and increases its price notably, since every bit of basic engineering has to be bought now. The developing countries are then less benefited with these policies that require the payment for basic knowledge licenses, free up to now. Therefore, what brings benefits for the developed countries increases dramatically the technological gap between the developing and developed countries.

In most of the Latin-American countries, the researchers prefer the scientific publication world–with its own rules and rewards–than the ones of patents and the commercial exploitation of the knowledge. Although, the researchers say that the incentives for patenting are continually increasing, is not easy to change the traditional

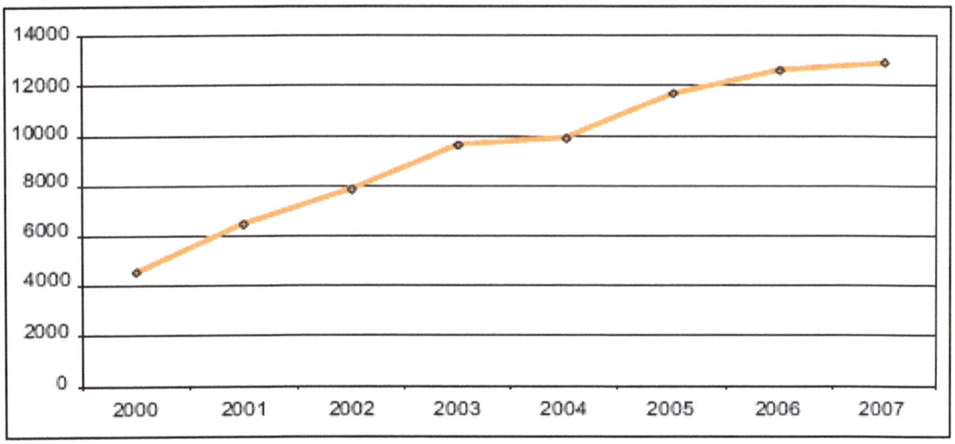

Figure 14.1: Growth of Patents in Nanotechnology

orientation of the investigations and convince the scientific community of not publishing–abandoning their common way of communicating among them, in order to defend the novelty of their invention that can be in future a patent worth (La Nanotecnología en Iberoamerica: situación actual y tendencias, 2007).

Early analyses of patent distribution have shown ownership concentration among a select group of countries, led strongly by the United States (US), Germany and Japan and strong private sector influence in patenting within the U.S. (Heines, 2003). Based on current investments and patents' information of the sector it is foreseen that for the 2020 the nations that will dominate the world Market will be the United States, Germany, and China.

It is also the idea that knowledge is different from other economics resources and that it can be freely appropriate by anybody it's an extremely naive belief. The use of knowledge require substantial, competent human resource, access to information channels, organizational resources, functional links between academia and industry, negotiation skills and many other elements, whose scarcity in south constitutes a formidable barrier far more powerful than any that a customs system could impose.

One could describe knowledge and science as two sides of the same coin. In today's world, the ability to use knowledge is increasingly linked to the ability to generate it. What we need to do is to promote scientific capability, because the models of knowledge transfer and technology no longer work, because change now happens too fast (Castro-Díaz, 2004).

Cuba–Background

In order to understand the current development of the sciences and technologies in Cuba, it is necessary to approach the background that made it all possible, starting with education.

The materialization of a scientific development in Cuba originated from a political decision that goes back to 1959, which as a result brought up dramatic transformations in the fields education, science, and culture.

In 1958, there were only three public universities–one of them the University of Havana, founded in 1782–with a registration of 15600 students. By the middle of the 50s, the Cuban population was about 6 million inhabitants. A million of them were illiterate and only the 28 per cent between the ages of 13 and 19 years continued their studies in the secondary level. Of about 17,400 graduated only 3 per cent were women. In 1961, a national campaign was carried out with the objective of eradicating the illiteracy that existed in the country that affected more than 30 per cent of the population. This campaign was followed by another supplementary campaign dedicated to drive those recently alphabetized students to reach the sixth grade and later on the ninth. Many of them reached higher levels of education. The budget dedicated to the education has been increased in more than 20 times, by the state, in regard to the year 1958.

The genesis of the infrastructure and the human potential that the biotechnical Cuban industry counts today, goes back to the creation in 1965 of the National Center of Scientific Investigations (CNIC) that became in initiator of multiple advances of the

Cuban science and in formation's Center for many the of scientific outstanding leaders in the field of biomedicine and biotechnology. Nowadays, Cuba counts with 65 Centers for Higher Education–53 universities and more than 10 independent faculties. It also counts for more than 500,000 students in higher education, and 82 curricula programs in 28 specialties (all without exception, public).

One can affirm that the Cuban University become more and more centre's for research. 77 per cent of the university closely participate in investigation activities in national and regional projects and more than 40 per cent of the students of pre-grade years; participate in curricular activities of research work.

Cuba has gone through several organizational models of scientific policies–included by the middle of the 90s, a national system for the scientific and technological innovations, revitalizing the national technologies adapting science and imported technologies. The approach became especially important in the adaptation of the electric plants of the country during the year 2003, a process that allowed Cuba, to generate near 95 per cent of its electricity, using the local fuels. On the other hand, nickel industry, with the modernization of the technology and the computerized management techniques, has obtained a productive record of 800,000 tons from the year 2002, which transforms it into the fourth largest producer globally (Castro-Díaz, 2001).

It is important to point out that from that from 2002 Cuba invests more than 1.6 per cent of its GDP in activities related with sciences. The total figure of workers linked to the spheres of the sciences climbs to more than 64,000, of which 52 per cent are women. Our country counts today of an infrastructure of 200 Research Centers in most varied areas of the scientific activity of the economy and of the humanistic studies. Of the economically active population of the country, 15 for each 1000 inhabitants work in the scientific sector. This figure is comparable with some European developed countries (Castro-Díaz, 2004).

In so early dates such 1982, a team of Cuban scientists–after being trained in advanced international centers–were able to produce in Cuba the alpha leucocitary interferon, event only achieved in a smaller number of developed countries at present. This led to the idea that biotechnology could represent a strategic line in the scientific development for a small country as Cuba, lacking natural resources.

With this objective and as a government priority decisive concepts that allowed the success of the idea gradually settled down (Herrera, 2004):

- ✰ Closed cycle organizations: The most important institutions were structured in research, production and commercialization organizations; in which the whole cycle, from the money invested by the state in the investigation, the development and the production, until the recovery of the investments in the market, is subordinated to the same management with the same level of responsibility.
- ✰ Organizations integrated to the national system of health–although it is not subordinated to these organizations–and guided toward the exports: Once the internal market–which is small–is satisfied; the continuous

development of the products are made with an eye toward the global market, this strategy created the drive to improve quality levels, according to the established parameters globally, and certified internationally, that allow the production and the supply of vaccines and other products according to international requirements.

- ✰ The intellectual property, although it is a polemic topic, arising out of the effort to develop highly innovative products for the most important markets, needs to be protected by patents. The effort to create a particular platform for intellectual property increased after the Cuba's entrance in the World Trade Organization (WTO) in 1995. Currently, the main organizations of the Cuban biotechnology have more than 700 registered patents abroad. This also matches the decision of not limiting the biotechnical products "me-too" or "biogenerics", but increase the competitiveness in the field of the innovative products. The existence of this intellectual property platform allows us to aspire in these moments to a fair compensation for our investments through the negotiations with industrialized countries' companies. (Lage, 2000)

Nowadays, the main development area, which contains more than 45 scientific research institutions, production and commercialization, is the so called Scientific Western Pole of Havana that has also extended its capacities to other provinces of the country. Out of the 10 most broadly marketed biotechnical products for human use at international level, 8 are produced in Cuba. In addition to that, out of the 13 vaccines with which all the children in the country are immunized, 10 are Cuban. The development of the biotechnology still have a great impact in the public health, which has resulted in the production of an impressive number of new products, vaccines and own bimolecules are included, as well as patented technologies of production.

In others of the branches of the pharmaceutical industry, about 70 Cuban drugs have been registered at international level. In Cuba, about 900 drugs are produced at the moment. It possesses a system of national health that applies this new pharmaceutical generation to the whole population; it also counts with more than 68,000 doctors–1 for each 165 inhabitants. In addition to all that, its infantile mortality rate is of 6.1 for each 1000 born alive and an expectancy of life more than 77 years (Castro Díaz, 2006).

The results achieved by the Cuban biotechnology match up with the planned objective of using the new technologies to boost economic development. The fact that this sector represents a considerable per cent of the GDP of Cuba at present time, illustrates it.

Bionanotechnology and their Perspectives for Cuba

The achievements and the infrastructure reached in the development of the biotechnical sector and medicine, offer a favorable framework, at the same time, supports Cuba's quest of achieving a global position in bionanotechnology, that guarantee the continuity of the biotechnical Cuban products in the international market.

In an analysis, in the 21st century, the biggest challenge is to be able to determine the exact relationship between biomarkers and clinical pathologies of the illnesses, as well as to develop non invasive technologies able to detect the illness in very early stage.

The technical imaging, extraordinary tools in the life sciences (RMN, TAC, and PET) will benefit notably with the advances in the nanoparticles development that will increase this techniques' capacity. New contrast agents, biomarkers quantum dota and intelligent nanoparticles will be able to extend under the cellular' level–to a molecular level–the selectivity and signaling of anomalous characteristic of illnesses.

In near future, the medical therapies, will respond to the challenge of the drug delivery, and controlled dose in the affected cells of the organism, avoiding the exposition of healthy tissue to the toxicity of the medications, as well as excesses of drug in an organism.

All these envisage this new development area (Castro Díaz, 2008)–which concern new properties to molecular scale and nanostructure applications and their interaction with biological systems–as such, it is presented as an interface among the chemical, the biological, and the physical sciences, medicine and engineering.

It is not about large scale biological materials' production, such as proteins or the specific genetic modification of plants, organisms or animals to increase interest properties for humans–characteristic of the biotechnologies. It is about the use of new nanofabrication techniques and processes of molecular self-assembling, that Bionanotechnology allows the production of new materials and devices, including fabrics and cellular engineering scaffolds, molecular motors, and biomolecules–for the drug delivery, sensors, and medical applications.

When embarking on Bionanotechnology, the Cuban state, not only makes it from the point of view of the current knowledge, but also as an important area to reach for competitiveness and to assimilate this technology for future development. The state and its institutions' role will consist on developing the interdisciplinary and institutional teaching; to promote the needed multidisciplinary cooperation for the Bionanotechnologies' advance and to support the creation of strategic networks in the field of Nanociencias and the Nanotechnologies. (Castro Díaz, 2008)

Nanotechnology in Cuba

Since late 90s, several universities (UH, IPSJAE, INSTEC, UCLV, UMCC), and investigation centers (CNIC, IMRE, CQF, CIM, CIGB) began to develop limited investigations on designing, obtaining and characterization of nanomaterials and nanosensors, and on biomedical applications of the nanotechnologies. Essentially, all these investigations have been framed as nanosciences without advancing in characteristic technological developments of the nanotechnologies. It means papers, thesis of masters and doctorates with few patents and no technological transfer to the industry.

Cuban Institutions Related with Nanosciences

1. Scientific Advisor Office, Council of State Republic of Cuba
2. Ministry of Science, Technology and Environment (CITMA)
   - Direction of Science
   - Cuban Observatory of Science and Technology (OCCyT)
   - Higher Institute for Applied Science and Technology (INSTEC).
   - Institute for Cybernetics, Mathematics and Physics (ICIMAF)
3. West Scientific Pole.
   - Centre for Genetic Engineering and Biotechnology (CIGB)
   - Centre for Chemistry Biomolecular (CQBM)
   - Centre for Molecular Immunology (CIM)
   - National Centre for Scientific Research (CNIC).
   - Centre for Neurosciences (CNEURO)
4. Ministry of Higher Education (MES).
   - University of Matanzas (UMCC).
   - Higher Polytechnic Institute José Antonio Echeverría (ISPJAE).
   - Central University (UCLV)
   - University of Havana (UH).
     - — Faculties of Chemistry, Physics and Biology. Institute of Pharmacy and Food Science
     - — Centre for Biomaterials (BIOMAT) and Institute of Materials Science and Technology (IMRE)
   - Other Ministries–such as the Ministry for Public Health and the Ministry for Food–are related to the problem of Nanosecurity through the CIDEM (Centre for Drug Research and Development, and the CECMED (State Center for Drug Quality Control).

In the Universities, it is necessary to mention, the Biomaterials' Centers of Havana University, with their results in the adhesive tissues and the bioceramics for closing wounds in skin and mucous; as well as fillers in bone defects and tissue regeneration. The Center of Studies of Proteins of the same University has searched of proteases inhibitors (IP) in marine organisms, which has allowed them to identify biological nanostructures potentially important. In Faculty of Chemistry, they have worked on self- assembly of small molecules and biomolecules on metallic surfaces and electrodes, as potential biosensors and nanopartículas with biological purposes, both in the areas of supramolecular chemistry.

Also forming part of the Havana University is IMRE which researches on Zeolites and nanoporous solids focused to the development of applications in molecular separation and adsorption (filters); theoretical works in hydrogen storage in nanocavities; and developments of different types of solar cells (Sensitized and Multi-

Quantum Well). The IMRE and the CIIQ also work in the nanoencapsulation of pharmaceutical active principles (medications) in polymeric matrixes.

The Centers of Biotechnologies of Matanzas' University and the Center for Bioactives of Las Villas University also stand out for their investigations on the enzymes and the chemical-physical principles of bioproducts with potential applications in health and agriculture. The NanoCIME of the José Antonio Hechevarría University (CUJAE) is active on Molecular Electronics for sensor design as well as the preparation of introductory courses of nanotechnologies at under graduated level. From the Ministry of Science, Technology and Environment (CITMA)– the INSTEC, the ICIMAF, and the CEADEN–whose Groups, due to their solid formation in physical sciences, mathematics and chemistry, approach essential thematic of nanociencias; examples are, the surfactants self- assembling in solutions with applications in biotechnology; pharmacy and new materials; the modeling of the quantum dots; and the design of scientific instruments respectively.

From the Ministry of Public Health (MINSAP) the CBQF and the CIDEM, research in the field of the medicine; and the CECMED (State Center for Drug Quality Control) in the establishment of rules and regulations of the nanotechnological products' income to the Cuban market.

From the West Scientific Pole of Havana the CIGB and CIM, their developments in biology and molecular immunology, therapeutics vaccines and new technologies for drug delivery. Also forming part of the West Pole is the National Center for Scientific Research (CNIC), which had introduced the new concept for the antibiogram's implementation that allows nanostructured kinetic biomaterials as support for controlled delivery of antibiotics, which makes possible a rapid microbiologic diagnosis.

In spite of these efforts, in this moment, the critical mass of investigators with the equipment and necessary resources to achieve a real competitiveness at international level in nanotechnologies is not enough.

About the Creation of a Centre for Advanced Studied Directed to Bionanotechnology

The existence of a common centre to gather the necessary expensive infrastructure for the development of bionanotechnology and the formation of multidisciplinary groups is important. It would face the new challenges and implies this new stage of the development for the sciences in Cuba (CEAC Creation Proposal, 2006).

Among the characteristics and expectations, it is important to highlight:

- ☆ The investigations in this field are characterized by the multidisciplinary nature. For its activities, the center will be OPEN to scientists from different national and international institutions.
- ☆ At the same time, it will be endowed with the services, infrastructure, and necessary capacities to excel at an international level.

Vision

To become into an international advanced level scientific institution, with a tangible impact in Cuba's development and Latino American relationships.

Mission

"To assimilate, generate and transfer the new knowledge and state-of-the-art, to contribute to the scientific progress of Cuba and other countries of the region"

- ☆ The facilities will be open for scientist from high level biotechnology centers and universities favoring the integration of knowledge from all over the country.
- ☆ It will integrate two national scientific priorities (Biotechnology and Informatics) in an area of special development, in between of the two fields. *i.e.;* biotechnology and medicine.
- ☆ It will facilitate the scientific cooperation offering the right framework for multinational projects.

Centre Facilities

- ☆ Nanostructured materials obtaining
- ☆ Composition and structure analysis with nanometric resolution
- ☆ Visualization and manipulation in a nanometric scale
- ☆ Development of nanodevices
- ☆ Advanced computing capacities

Scientific Activities of the Centre

- ☆ *Obtaining and Functionalizing* nanomaterials through physic, chemist biological methods
- ☆ *Characterization* (*nanoanalytic*) of nanomaterials and molecular biological systems.
- ☆ *Computing Modeling and simulating* of meso- scale phenomena.
- ☆ Development and application of new nanostructures and nanodevices (*Nanoengineering*).
- ☆ Manipulation and study of biological material in the meso- scale (*Nanobiology*).

It would have the possibility of associated laboratories to the center, even though they are not physically located in close proximity.

Main Research Activities on Bionanotechnology

- ☆ Nanoencapsulation of proteins and bioactive compound for drug delivery
- ☆ Nanostructure supports for diagnostic and other applications
- ☆ Visualization, manipulation and characterization of nanostructures

- Nanosecurity: Health and environmental impact of nanoparticles. Characterising the Potential Risks posed by Engineered Nanostructures.

Currents Research in Progress

- PEGylation enhances pharmacokinetic properties of IFN-α
- Profiles of BSA delivery by nanoparticles
- Development of nanoemulsion based formulation for vaccines and drug delivery
- Development of nanodispersions with compressed fluids as co-solvents for drug encapsulation
- Use of Cryomicroscopy of viral capsides for understanding the work of the virus
- VSSP: a rational design of an nanoadjuvant for Cancer Immunotherapy
- Studying the molecular mechanism of action of the anti-tumor antibody Nimotuzumab by Atomic Force Microscopy (AFM).
- Optical Tweezers: Real time detection of a single DNA Helicase at work
- The study of the behavior thermo- magneto of iron oxide nanopartículas and their functionalized Ferro fluids with the antibody monoclonal Nimotuzumab for therapies
- Characterization by means of atomic force microscopy of damaged mechanisms of and the DNA reparation
- Obtaining, characterization and study of nanoparticles properties of functionalized magnetite for medical application.

Human Resource Development

Another essential pillar for this new Center's success is the generation of human resources with necessary expertise from high level foreign Institutes, suitably identified. It is about preparing recent graduates, in foreign laboratories; which have the "analytic new tools" that the development of the bionanotechnology requires. Although it is true that mastering and the possibility of the use of the classic analytical techniques (TEM, SEM, DRX, XPS, MS, RMN, FTIR, etc) are a starting point for the Nanobio investigations.

Besides established methods for studying matter with nanoscale lateral resolution, such as atomic force microscopy (AFM), scanning tunnelling microscopy (STM), and scanning electron microscopy (SEM), typically give very little or no chemical information. Some exceptions exist–for example derivative AFM tips that can be used for chemical recognition. In general, however, only the shape of a nanostructure, its surface topography, and its local electric properties can be measured, albeit in many cases with excellent spatial resolution, down to the atomic scale. However, with the advent of nanotechnology and nanoscience, the need for chemical and in particular molecular analysis has become urgent.

Several new approaches to spectroscopic observation and imaging with nanometer-scale resolution have appeared recently (Zenobi, 2007), and will continue to set trends in the fields of nanotechnology, is in this new direction that we need to prepare the young scientific minds to acquire the knowledge and experiences necessary to work in the challenge that bionanotechnology presents.

Conclusions

Cuban experience demonstrates that a key factor in the application of science and technology to national economic development was the ability of a large numbers of workers with a sound basic education, able to learn new task quickly and the political decision to advance in a prioritised direction.

Cuba has also promoted skill creation and the introduction of several research and educational institutions of excellence, which have promoted the development of basic science and technology especially in life sciences.

The main conclusion derived and consolidated from the Cuban's decision in *The Cuban approach towards Bionanotechnology: a new step on scientific development of the Country* is that each of our countries has to find–in accordance to its reality–their own way to develop nanotechnologies. The attempt to copy the developed countries experiences in the development of nanotechnologies, in facing the relevant challenges in developing countries, is a wrong strategy. Sadly however, this is the basic strategy implied in many systems for international cooperation.

In fact developing country engagement and interest in nanotechnology R&D, itself, will not promote a more equitable engagement with global science. Considerable barriers exist for less-developed countries seeking to engage with nanotechnology R&D on a national level. Early signs are that developing countries will direct nanotechnology R&D to 'value-add' to export markets, rather than use nanotechnology to promote sustainable development.

If existing nanotechnologies are less able to address the specific needs of developing countries that is because the specific socio-economical contexts have been ignored in framing the problems to which the technologies should be the solution. Smarter engineers would consider these contexts from the very beginning and develop their products accordingly, without caring about the arbitrary nano-label.

Cuba looks forward to advances on its development bases in biotechnology toward a new stage of the Cuban science, we look forward to the future with optimism when basing the development strategy on the knowledge and the intellectual productions, sustained by an expressed government political will.

In 1960, when the Revolution was barely beginning its work, Fidel Castro did express: "the future of Cuba must necessarily be a future of men of Science and Thought." Recently, in the inauguration of the University of Informatics Sciences (UCI) he said: "Cuba will depend in the future of its intellectual productions".

Acknowledgements

I would like to thank Dr. F, Castro Díaz Balart for his critical reading of the manuscript and his contributions for make it. Without them it would have been impossible its fulfillment. I also thank BA. Lázaro G. Díaz Váldes for his support during the work preparation. I am grateful for the financial support for my participation and presentation of this work provided by the Centre for Science and Technology of the Non-Aligned and Other Developing Countries (NAM S&T Centre) and Iranian Research Organization for Science and Technology (IROST), and the Iranian Nanotechnology Initiative Council (INIC) in Tehran.

References

Brown, M.M., 2001. Foreword in Human Development Report: Making new technologies work for human development, United Nations Development Programme (Ed.), Oxford University Press, New York.

Castro Díaz-Balart (Coordinator Editor), 2006. Science, Technology and Innovation: Challenges and Uncertainties for South. Ediciones Plaza, Habana.

Castro Díaz–Balart F., 2008. Opening Lecture, *2nd International Seminar for Nanoscience and Nanotechnology*, Havana.

Castro Díaz–Balart F. (Finlay's Statement), 2008. *Meeting at the Science Academy of Cuba.*

Castro–Díaz, Balart F., 2004. Ciencia, Tecnología y Sociedad: Hacia un desarrollo sostenible en la Era de la Globalización, Editorial Científico–Técnica, La Habana.

Castro–Díaz, Balart F., 2001. Ciencia, innovación y futuro. Cuba, Barcelona, Editorial Grijalbo–Mondadori (2nd edition).

Castro–Díaz, Balart F., 2004. Ciencia, Tecnología y Sociedad: Hacia un desarrollo sostenible en la Era de la Globalización, Editorial Científico–Técnica, La Habana.

CEAC Creation Proposal, 2006. Experts' Ad Hoc Group, Internal report.

Heines, H., 2003. "*Patent Trends in Nanotechnology*". Available: http://library.findlaw.com/2003/Nov/4/133136.html.

Herrera, L., 2004. *Biotechnology in Cuba*, presentation at the Global Biotechnology Forum, Chile, March.

Invernizzi, N., Foladori G., 2005. "*Nanotechnology as a Solution to the Problems of Developing Countries*?" Available: http://cspo.org/ourlibrary/documents/NanoSolut.pdf.

Igami, M. and Okazaki, T., 2007. Capturing nanotechnology's current state of development via analysis of patents, STI Working Paper, 2007/4, OECD, París. Disponible en http://www.oecd.org/dataoecd/6/9/38780655.pdf.

Juma, C. and Yee-Chong, L., 2005. "*Innovation: Applying Knowledge in Development*". Available: http://bcsia.ksg.harvard.edu/BCSIA_content_stage/documents/TF–Advance2.pdf.

Lage, A., 2000. Propiedad y expropiación en la economía del conocimiento, Ciencia, Innovación y Desarrollo, Vol 6, No 4, 25–37, La Habana, CITMA.

La Nanotecnología en Iberoamerica: situación actual y tendencias, 2007. Observatorio Iberoamericano de Ciencia, Tecnología e Innovación del Centro de Altos Estudios Universitarios de la OE.

Lage, A., 2000. "Las biotecnologías y la nueva economía: crear y valorizar los bienes intangibles", *Biotecnología Aplicada*, 17(1): 55–61. La Habana.

Roco, M.C. and Bainbridge, W.S. (eds.), 2002. *Converging Technologies for Improving Human Performance:Nanotechnology, Biotechnology, Information Technology and the Cognitive Science*, Arlington, VA: National Science Foundation.

Salamanca-Buentello F. *et al.*, 2005. *Nanotechnology and the Developing World*, April, www.plosmedicine.org

Schummer, J., 2007. *The Impact of Nanotechnologies on Developing Countries*, Hoboken, NJ: Wiley.

Schummer, J., 2007. *Nanoethics: Examining the Societal Impact of Nanotechnology*, Hoboken, NJ: Wiley.

The Royal Society and Royal Academy of Engineering, 2004. "*Nanoscience and Nanotechnologies: Opportunities and Uncertainties*", The Royal Society and Royal Academy of Engineering, London.

WHO, 2005. Water, sanitation and hygiene links to health: Facts and figures. Available at http://www.who.int/entity/water_sanitation_health/factsfigures2005.pdf. UN Millennium Project. 2004.

Zenobi Renato, 2007. Analytical tools for the nano world. *Anal. Bioanal. Chem.*, Springer–Verlag.

Chapter 15

Nanotechnology: Present Status and Future Prospects in Indonesia

Silvester Tursiloadi
Research Center for Chemistry, Indonesian Institute of Sciences, Kawasan PUSPIPTEK, Serpong, Tangerang 15314, Indonesia
E-mail: tursiloadi@yahoo.com

ABSTRACT

Indonesia is located in the crossroads between Australia and Asia and between the Pacific and Indian Oceans. It is a string of over 17,000 islands that stretches 5,152 kilometers from West to East and 1, 770 kilometers from North to South along the equator. Indonesia is just a new player in nanotechnology, Indonesia has comparative excellent in the form of various natural resources as raw material for product making and energy resources, moreover there are floral and fauna biodiversity in a great extent. That is why, nanotechnology development should be directed to manage and increase the added value of Indonesian natural resources significantly. Indonesia is a tropical country, which palm oil and essential oil are one of the important commodities, which contributes to economic development in Indonesia, especially in agro industrial sector. In mining sector, Indonesia has high natural resources, contain 5 big major minerals namely; copper, gold, silver, nickel and cobalt. Without ignoring other metal materials, the 5 major materials are potential to be developing with technological touch. One of the latest technologies which are able to increase added value of the mining materials dramatically is nanotechnology. There have been several nanotechnology development on metal materials were applied. Especially on gold and silver nanoparticles can be applied for sensor and pharmaceutical application. Whereas for copper, nickel and cobalt could be developed as nanocatalyst materials for applied to chemical industries. While for iron sand material, application of nanotechnology could be applied to pharmacuetical and chemical industries. For titanium, the nanotechnology is used in chemical, cosmetic, electronic and transportation industries.

Keywords: *Present status of nanotechnology in Indonesia, Indonesian natural resources, Mining sector, Nanotechnology application in industrial world.*

Introduction

Indonesia is located in the crossroads between Australia and Asia and between the Pacific and Indian Oceans. It is a string of over 17,000 islands that stretches 5,152 kilometers from West to East and 1, 770 kilometers from North to South along the equator. It is inhabited by more than 230 million people, from 300 ethic groups, speaking about 200 languages, but possessing one national language, Bahasa Indonesia.

Indonesia is a tropical country, which palm oil and essential oil are one of the important commodities, which contributes to economic development in Indonesia, especially in agro industrial sector. Palm oil estate covered 5.6 million ha (2004) and produced up to 13 million tons of crude palm oil in 2006. Indonesia and Malaysia produce about 90 per cent of the world's palm oil (The Jakarta post Dec. 19, 2007). Indonesia's output overtook that of Malaysia 2006 as it has more land for expansion. Indonesia has record harvest of about 17 million metric tons in 2007, and about 18.4 million tons for 2008 (Indonesian palm oil association). Indonesia is the highest producer of palm oil in the world. Palm oil has been used as edible or cooking oil for many years. Now, palm oil is also consumed as an industrial raw material. Oleochemical industries, for example consume palm oil as one of their raw materials. On the other side, essential oil is a natural organic product obtained from distillation of plant body parts, which includes the distillation of leaves, roots, flowers, seeds, or others. Essential oil is very potential to be developed in Indonesia and is a major commodity for the incensement of the country revenue. From 40 essential oils in the world market, Indonesia contributes 12 essential oils, such as patchouli oil, citronella oil, vetiver oil, cananga oil, clove oil, and nutmeg oil. In 2005, Indonesia received a surplus of US$ 2.531 million, where exports are reported as US$ 4248 million and import of US$ 1.370. Nanocatalysts have become increasingly important for the efficient synthesis of active compounds from certain herbs available in Indonesia especially from clove oil, nutmeg, and cinnamon, in particular for the production of fine chemicals and pharmaceuticals. High level of selectivity and efficiency can be achieved using nanocatalysts leading to processes with lower energy level. A unique aspect of nanotechnology is the vastly increased ratio of surface area to volume present in many nanoscale materials, which opens new possibilities in surface-based science, such as catalysis.

In mining sector, Indonesia has high natural resources as shown in Table 15.1, contain 5 big major minerals namely; copper, gold, silver, nickel and cobalt. Without ignoring other metal materials, the 5 major materials are potential to be developing with technological touch. One of the latest technologies which are able to increase added value of the mining materials dramatically is nanotechnology. There have been several nanotechnology development on metal materials were applied. Especially on gold and silver nanoparticles can be applied for sensor and pharmaceutical application. Whereas for copper, nickel and cobalt could be developed as nanocatalyst materials for applied to chemical industries. While for iron sand material, application of nanotechnology could be applied to pharmacuetical and chemical industries. For titanium, the nanotechnology is used in chemical, cosmetic, electronic and transportation industries.

Table 15.1: Potential Natural Resources and National Metallic Mineral Reserve

Sl.No.		*Mineral Resources (Ton)*			
		Hypothetic	*Inferred*	*Indicated*	*Measured*
1.	Primary Gold	1,394,919	465,430,000	111,050	366,990,000
2.	Alluvial Gold	52,500,000	30,000,000	81,000,000	66,070,880
3.	Silver	-	31,630,000	14,310,000	7,000,000
4.	Copper	18,900	321,550,250	48,000	191,500,000
5.	Tin	-	3,200,000	224,193	509,097
6.	Nickel	-	347,300,000	417,566,200	385,350,000
7.	Manganese	809,525	714	2,379,763	437,040
8.	Iron Sand	3,406,770	250,000	61,241,872	459,772
9.	Lateric Titan	15,694,843	3,700,000	828,673,796	-
10.	Placer Titan	994,193	28,153,713	12,036,998	136,420
11.	Lead	120,825	17,525,935	6,120,000	1,760,000
12.	Zinc	318,900	17,440,935	13,495,000	-
13.	Platinum	-	30,000,000	32,250,000	52,500,000
14.	Monazite	-	-	179,712	-
15.	Molybdenum	-	681,000,000	-	-
16.	Placer Chromite	3,239,590	250,000	1,382,471	891,813
17.	Primary Chromite	975,075	292,000	234,000	10,000
18.	Cobalt	-	424,800,000	422,300,000	625,920,000
19.	Lateric Iron	-	416,856,930	574,597,784	-
20.	Primary Iron	3,721,000	1,132,000	-	65,339,600
21.	Bauxite	-	3,356,606	-	97,843,757

Source: National resources and reserve, Dirjen Geologi and Sumber Daya Mineral, 2004.

Technology development has significantly increasing industrial competitiveness of nation. In 21 century there have been several paradigm changes in viewing the technology, where properties and performance materials cold be modified in such different way more effectively and efficiently. In nanometer scale (10^{-9} m) infect, the materials have properties and performance with unique phenomena more superior compare with micrometer scale (10^{-6} m). Using nanotechnology, materials could be designed and arranged in atomics or molecules order. By rearranging or modifying material structure in nanometer level, there will be obtained a certain excellent material. That why, nanotechnology has become center of attention in the world concerning with its application on industrial world. According to *Technology Transfer Centre* report during 2006-2010 period, Japan has allocated more than US$ 6.5 billion, followed by America, Europe, and China around US$ 6.0, 4.8 and 3.2 billion, respectively. These numbers indicate that they have a great hope toward nanotechnology application for developing their industries.

Table 15.2: National Resources and Metallic Minerals Reserve

Sl.No.		Mineral Reserve (Ton)		Number of Location
		Prediction	Explored	
1	Primary Gold	401,982,000	2,977,504,000	45
2	Alluvial Gold	-	-	17
3	Silver	399,932,000	2,877,504,000	24
4	Copper	1,060,000,000	3,712,162,000	26
5	Tin	307,512	148,753	7
6	Nickel	486,690,000	112,170,000	25
7	Manganese	-	300,000	17
8	Iron Sand	500,000	200,000	12
9	LatericTitan	-	-	9
10	Placer Titan	-	-	14
11	Lead	-	-	10
12	Zinc	-	-	7
13	Monazite	-	-	9
14	Cobalt	19,510,000	318,620,000	22
15	Lateric Iron	2,050,000	2,470,000	16
16	Primary Iron	-	-	24
17	Bauxite	-	2,770,000	11

Source: National resources and reserve, Dirjen Geologi and Sumber Daya Mineral, 2004.

The experts have predicted that nanotechnology revolution will give a big impact comparable with four industrial revolutions before (Revolution of steam engines, Railways, automobiles and computers) which required 2 centuries. The same impact of four this revolution industry will be overcome in several years only as shown at Figure 15.1.

According to European expert, the potential of developing nanotechnology will accelerate industrial products. Figure 15.2 shows that up to 2005, nanotechnology opportunity in industrial market has not given significant impact. This indicates that research and nanotechnology development is still relatively new. A long with procession time during 2010 to 2020 period, there will be a great acceleration of

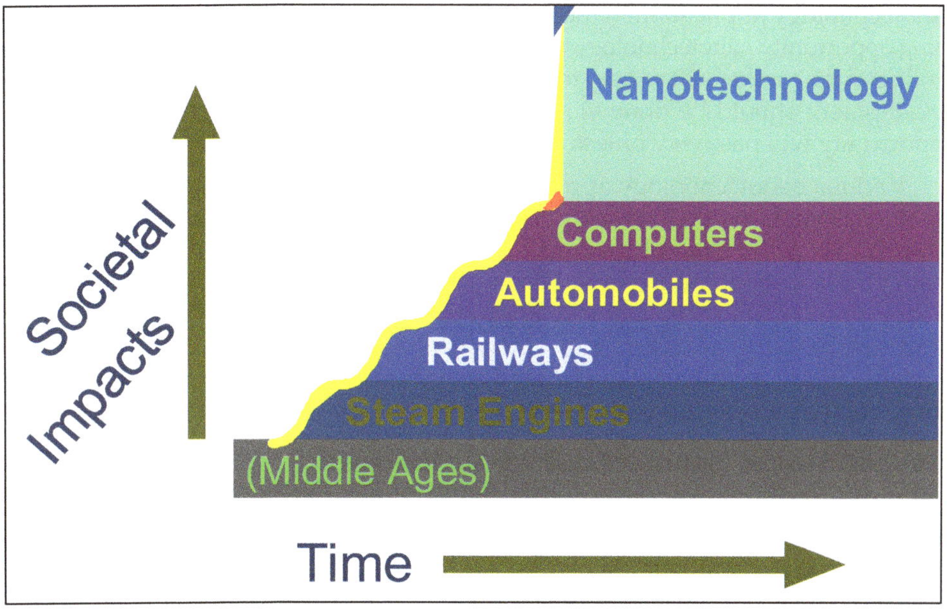

Figure 15.1: Comparison of Impacts on Society from Several Industrial Revolutions and Nanotechnology (*Source*: Center for Responsible Nanotechnology www.CRNano.org)

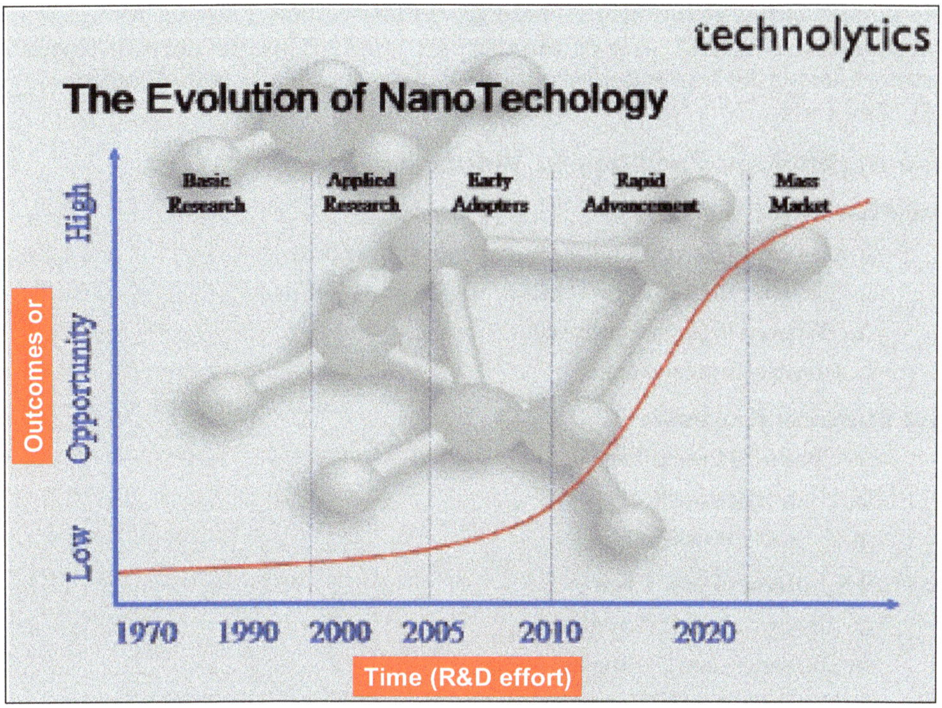

Figure 15.2: Nanotechnology Opportunities in Industrial World (*Source*: Technolytics)

nanotechnology application in industrial world. Furthermore in next few years the using opportunity nanotechnology will be surfeited, where at that time nanotechnology product in the market will be very massive in number. For that reason, nanotechnology development should be done vastly at present. Development of nanotechnology opportunity will pass way unless we take action now.

Various nanotechnology applications on present products have been applied on electronicall, medical, pharmacuetical, cosmesticall, food industries, textile, ceramics etc. As an example nanotechnology development in computer world has changed not only computer size become more compact but also has boosted its capability and capacity so that enable giant programs solution in short time. Many products as nano textiles, nano ceramics, nano coating, nano film, nano pharmacueticals etc, have spread to our live through social status boundary of human being. For that reason, nanotechnology has become moving power for new business.

Present Status Nanotechnology in Indonesia

Indonesia has comparative excellent in the form of various natural resources as raw material for product making and energy resources, moreover there are flora and fauna biodiversity in a great extent. That is why, nanotechnology development should be directed to manage and increase the added value of Indonesian natural resources significantly. On the other hand, the research and development nanotechnology Indonesia have been started in same research institutes (Indonesian Institute of Sciences (LIPI), National Nuclear Energy Agency for Indonesia (BATAN), Agency for Assessment and Application of Technology (BPPT), National Agency for Space and Aeronautics (LAPAN), etc) or universities. Table 3 shows current status of nanotechnology R&D in Indonesia. (NM: NanoMaterial, NB: NanoBiotechnology, ND: NanoDevice, NC: NanoChemistry, NE: NanoEducation).

Nanotechnology Facilities in Indonesia

Gas-Phase Processing

- ☆ Gas Condensation with Thermal Evaporation
- ☆ Vacuum Evaporation on Running Liquids (VERL)
- ☆ Thermal Plasma Synthesis
- ☆ Combustion Synthesis

Wet Chemical Processes

- ☆ Chemical Precipitation
- ☆ Hydrothermal Processing
- ☆ Sol-Gel Processing

Sol-Gel Synthesis Flow Chart

- ☆ Thermo chemical Synthesis
- ☆ Sonochemical Synthesis
- ☆ Hydrodynamic Cavitations

Table 15.3: Current Status Nanotechnology R&D in Indonesia

Name of Institute	Group	Research Area				
		NM	NB	NO	NC	NE
Indonesian Institute of Sciences (LIPI)	Research Center for Physics	X		X	X	X
	Research Center for Chemistry	X			X	
	Research Center for Calibration, Instrumentation, and Metrology	X		X		
	Research Center for Electronics and	X		X		
	Research Center for Metallurgy	X	X			
	Research Center for Biotechnology		X			
	Research Center for Electrical Power and	X		X		
National Nuclear Energy Agency for Indonesia (BATAN)	R&D Center for Nuclear Fuel and Recycle	X				
	Center for Isotopes Application and Radiation	X	X			
	Technology Center for Nuclear Materials Industry	X		X		
Agency for Assessment and Application of Technology (BPPT)	Center for Pharmaceutical and Medical	X	X		X	
	Center for Polymer Technology	X			X	
	Center for Materials Technology	X		X	X	
National Agency for Space and Aeronautics (LAPAN)	National Aeronautics and Space			X		
University of Indonesia (UI)	Department of Physics	X		X		X
	Department of Metallurgy and Material	X				
	Department of Electronical Engineering			X		
	Institute of Human Virology and Cancer Biology		X			
	Faculty of Public Health		X			
	Department of Chemical Engineering				X	

Contd...

Table 15.3–Contd...

Name of Institute	*Group*	*Research Area*				
		NM	*NB*	*NO*	*NC*	*NE*
Bandung Institute of Technology (ITB)	Department of Physics	X		X	X	X
	Department of Physical Engineering	X		X		X
	Department of Chemistry	X			X	
Gajah Mada University (UGM)	Department of Chemistry	X			X	
	Department of Chemical Engineering	X			X	
	Department of Industrial and Mechanical Engineering	X	X			
	Faculty of Pharmacy	X	X			
	Department of Physics	X		X		X
Sepuluh November Institute of Technology (ITS)	Department of Physics	X		X	X	X
	Department of Material Engineering	X			X	
Andalas University (Unand)	Department of Chemistry	X			X	X
Surabaya University (UBAYA)			X			
Airlangga University (UNAIR)	Department of Chemical Engineering		X			
Yogya Muhammadiyah University		X			X	
Atmaiaya Catholic University			X			
ISTA	A. Yani Institute of Science and Technology				X	
Soedirman University	Department of Physics		X			X
Padiaiaran University	Department of Physics	X			X	

Contd...

Table 15.3–Contd...

Name of Institute	*Group*	*Research Area*				
		NM	*NB*	*NO*	*NC*	*NE*
State Polytechnic of Bandung	Department of Chemical Engineering				X	X
State University of Jakarta (UNJ)		X				
Gunadharma University		X				
Islamic National University		X				
Muchtar Riady Center		X	X			
Eijkmann Institute			X			
University of Sebelas Maret	Department of Chemical Engineering	X				
University of Syiah Kuala	Department of Physics	X				
University of Diponegoro	Department of Chemical Engineering	X			X	

Solid-State Processes

- ☆ High-Energy Milling
- ☆ Mechanochemical Synthesis

Indonesian Society for Nanotechnology

- ☆ Established on 2005
- ☆ Indonesian International Focal Point on Nanotechnology Issues
- ☆ Constructing the nanoscience and nanotechnology consortium of Indonesia that would integrate and coordinate interests from academics, industry, and government.

Present Indonesian Governmental Policy

Ministry of Research and Technology of Indonesia (Ristek) take a certain attitude toward the new global nanotechnology development by giving guidance in certain activity with a great theme: Research and development on science and technology (PENELITIAN DAN PENGEMBANGAN ILMU PENGETAHUAN DAN TEKNOLOGI) with sub theme development on Nano Science and Technology [document Ristek 2006]. The activity has been issued from 2005 to 2009. The content of the document is:

Objective and Target of Activity

Objective

Research Development of Nanoscience and Nanotechnology Program has some objective;

- ☆ Strengthening and developing researcher ability in framework of mastering nanoscience and technology.
- ☆ Giving focus direction on developing research activity based on natural resources.
- ☆ Giving synergy upon national researcher's ability.

Target

All the activity of Nano science and technology programs are focusing to support the achievement of.

- ☆ Capability of fulfilling society need at large by nanotechnology.
- ☆ Capability of fulfilling products at domestic and international market.
- ☆ Optimal natural resources management.
- ☆ Achievement of nano science and nanotechnology in national scale.
- ☆ Arrangement of nano science and nanotechnology development policy in national scale.

Platform of National nano science and nanotechnology development shown in Figure 15.3.

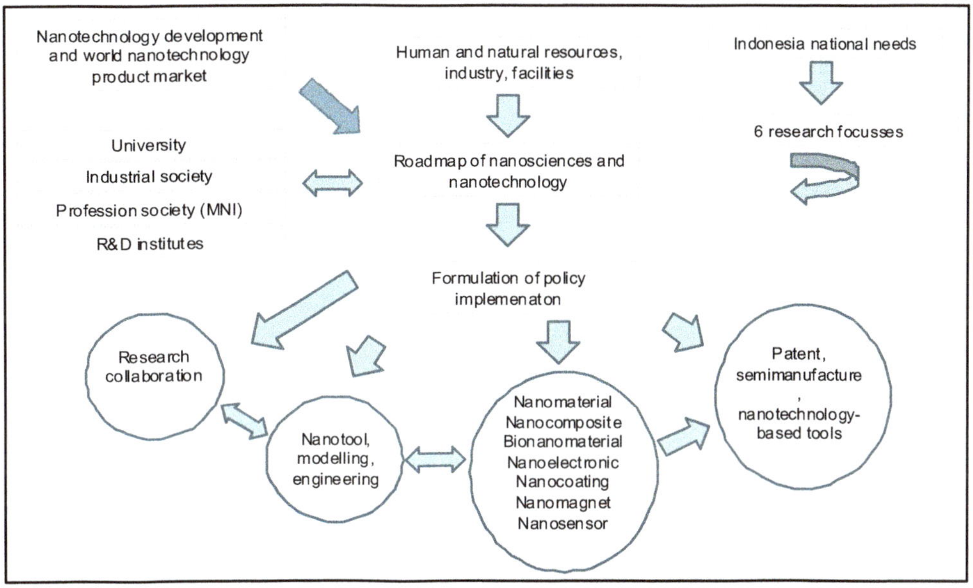

Figure 15.3: Platform of Nanotechnology
Ministry of Research and Technology, Republic of Indonesia

The goverment take very active role to improve nanotechnology in Indonesia by intervention through various policy and financial support for R&D and make implementation of nanotechnology in national industries. There have been tendency of nanotechnology application on European industries based on potential a marked need namely: 1) Health and Biology, 2) Information and Telecomunication Technology and 3) Chemistry.

The shown trend is almost the same as seen in Asia Pacific industries that is: 1) Chemistry 2) Health and Biology, 3) Daily needs and 4) Automotive dan transportation.

Figure 15.4 show the roadmap of development of nanotechnology based on local natural resources to support national industries through technology push–market pull approach. Combination of technology push and market pull approach is very relevant strategy becaouse it has not only consider diversity and the search of natural resources opportunity and possessed ability but also demand of the consument.

If the trend is annllized more detailed with considring recent trend, present R&D status, opportunity and challange, cappability, core competency, and national industry excellence and also nanotechnology application in industrial world, every country in the world as their own focus in nanotechnology field. Generally, at least there are three main development focus on nanotechnology namely 1) nanomaterial, 2) pharmatical and health 3) energy.

Research and development nanotechnology in Indonesia is dominated by nanomaterial with focus method and technical preparation, which applied extensively

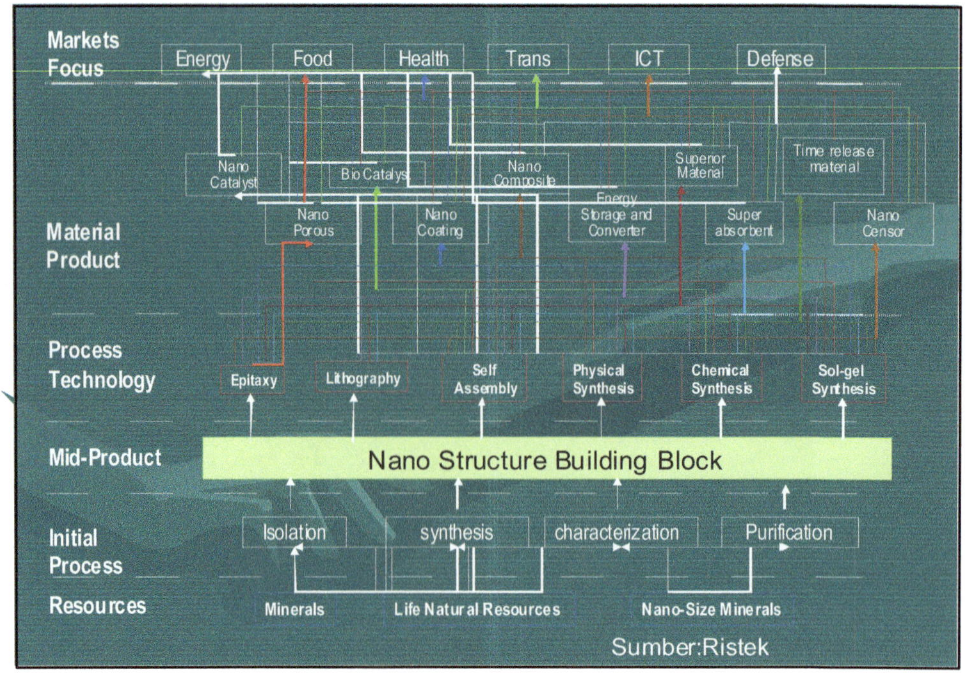

Figure 15.4: Roadmap Nanotechnologies with Supply Technology Approach (*Source*: Ristek)

for cosmetic industry, health, coating, ceramics, electronics, energy, etc. Nonobiotechnology development has drawn some attention for healthy medicine and improvement of agricultural. Some research on energy field has been conducted with environmental friendly energy as the target.

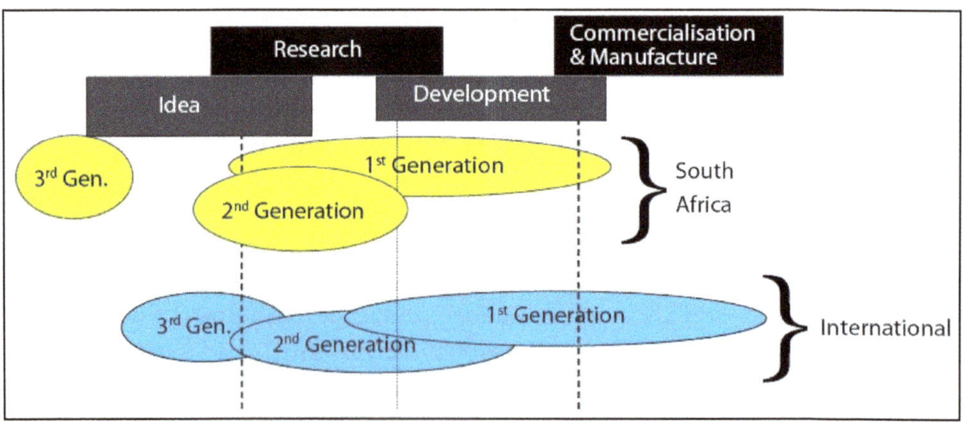

Figure 15.5: Schematic of Nanotechnology Development in Indonesia

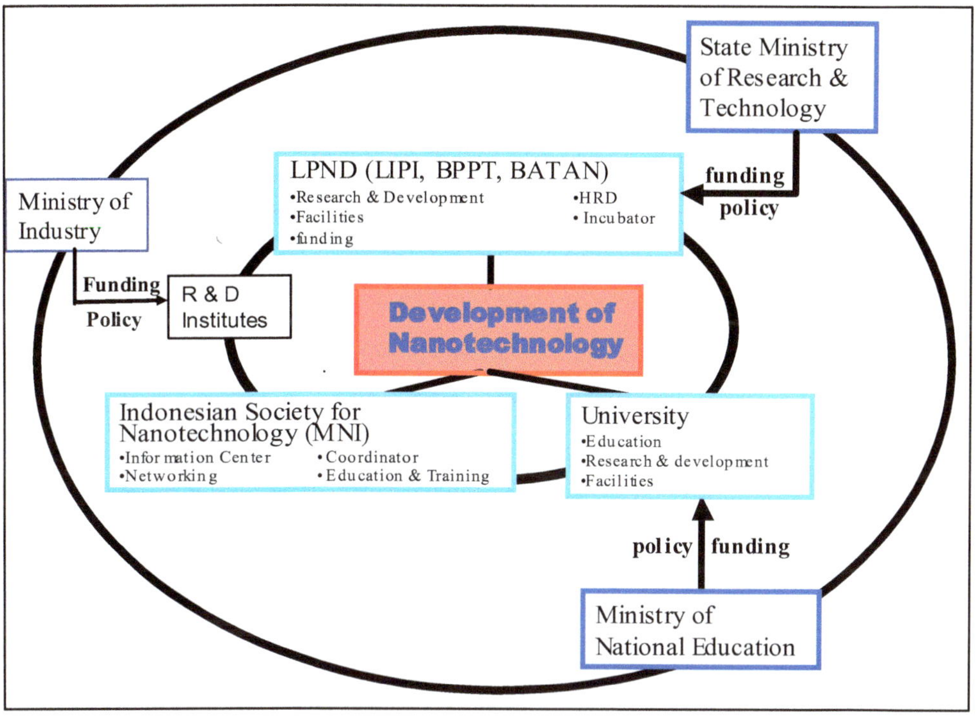

Figure 15.6: Strategy to Socialize Nanotechnology in Indonesia

Those mentioned about, will be useful and significant factor as a reference of consideration in the framework of arranging strategy of nanotechnology based industries development.

Future Prospects in Indonesia

Nanotechnology as trend technology in the future, require some attention from various sides for its development. Concern that could be done urgently is by making nanotechnology as a national program. Nanotechnology as national program has been done by developed countries but also by ASEAN countries such as Malaysia, Thailand, and Vietnam.

In Indonesia at least involvement of various institutions which directly connected with nanotechnology development is needed. These institutions are ministry of industrial affairs, ministry of commerce, ministry of energy and mineral resources, Indonesian Institute of Sciences (LIPI), Agency for Assessment and Application of Technology (BPPT). Besides that, universities also have some involvement concerning research development in academic world and preparing potential human resources. Industry as the side that uses also has become the most important partner in nanotechnology development. Some activities that have been done among others are research of nanostructure/ nanoparticle / new nanomaterial, determining potential application, product commerzation and market entry. Moreover, it is needed to establish processes product of standardization process and nanotechnology product, improvement of natural resources, end-user education, and improvement of infrastructure and readiness supply energy. Other important things are participating in international forum and focus group discussion forum; these activities will strengthen national industries competitiveness.

Conclusion

Indonesia has comparative advantage on natural resources, and amount of population. Combined with nanotechnology this can be the added value for Indonesia natural resources

- ☆ Good synergy between Academic-Business/Industry-Government will support the acceleration of Nanotechnology in Indonesia
- ☆ Participating in international forum and focus group discussion forum, these activities will strengthen national industries competitiveness.
- ☆ Industrial side need to be supported with steadinees in nanomaterial field, due to raw material has become centre of attention in application of nanotechnology in industrial activity.

References

Center for Responsible Nanotechnology www.CRNano.org.

Dirjen Geologi dan Sumber Daya Mineral, 2004. National resources and reserve.

Document Ristek 2006

Kozo Obara, The role of Nanotechnology and Spin in New Electronic Energy Revolution, Proceeding of the 10th Industrial Electronic Seminar, October 30th 2008, EEPIS-ITS Surabaya.

Mikrajuddin Abdullah, Yudistira Virgus, Nirmin, Khairurrijal, 2008. Review: Sintesis Nanomaterial. *Journal Nanosains dan Nanoteknologi*, 1(2): 33-57.

Nurul Taufiqu Rochman dkk, 2006. Current Status of Nanotechnology in Indonesia. *Indonesian Nanoletter*, 1(1): 13–36.

The Jakarta post Dec. 19, 2007

Yateman Arryanto, Siti Amini, MF Rosyid, Arif Rahman, Pedy Artsansi, Iptek Nano di Indonesia, Terobosan, Peluang dan Strategi, Kedeputian Perkembangan Riptek, Kementerian Negara Riset dan Teknologi, Mei 2007.

www.technolytics.com.

Chapter 16

An Overview on Nanotechnology Activities in Iran

***Saeed Sarkar*[1] *and A. Beitollahi*[2]**

[1]*Director of Iran Nanotechnology Initiative Council (INIC), and Director of Research Center for Science and Technology in Medicine (RCSTIM)/Tehran, University of Medical Sciences (TUMS)*
E-mail: sarkar@sina.tums.ac.ir
[2]*Department of Metallurgy and Materials Engineering, Iran University of Science and Technology (IUST) and Director of Infrastructure Development, Iran Nanotechnology Initiative Council (INIC)*

ABSTRACT

During the last decade, nanotechnology has gained remarkable attention by international scientific and industrial societies. In this respect, various countries have launched their short and long term programs to benefit from the huge market expected by utilization and implementation of this technology in their industrial/technological infrastructures. Islamic Republic of Iran, considering the undeniable impact of this technology on her nation's wealth and economy, initiated national nanotechnology development plan referred to as "Future Strategy" in 2005. The strategic plan outlines countries major action plans for ten-year period up to 2015 with the target to be amongst the top 15 countries active in this very important area. This paper, while reviewing briefly the objects, targets of the plan also considers the current status and achievements obtained by implementation of this strategic plan. Further, the impact of various developed infrastructures on the success of the program is also discussed.

Keywords: *Nanotechnology, Iran Nanotechnology Initiative Council (INIC), Nanotechnology Development Plan.*

Introduction

Nanotechnology is a field of science and technology that involves the control of matter at the atomic, molecular and macromolecular level. It deals with understanding and control of matter and processes at the nanoscale (~1-100 nanometers) in one or more dimensions where the onset of size-dependent phenomena usually leads to novel applications. At such scales, various properties of materials are modified remarkably giving rise to new, exciting and different characteristics that can generate a vast array of novel products. Such a technology is considered both as enabling and disruptive that alone, or in combination with conventional technologies, can lead to high performances and capabilities. According to U.S. National Science Foundation (NSF) estimates, by 2015 the annual global market for nano-related goods and services will top US$1 trillion highlighting the immense impact of this technology on worldwide economy. The last two decades has noticed highly growing interest and support for nanotechnology internationally and there are numerous studies and reports detailing this growth.

Considering the huge impact of nanotechnology on global market and national economies of various countries, most of developed and developing economies have established strategies and encouraging commercialisation plans. Islamic republic of Iran also by recognizing the future impact of this very important field on her nation's wealth and to accomplish her previously implemented national knowledge-based economy programs launched her national nanotechnology development plan referred as "Future Strategy" which passed the approval of cabinet in 2005. The plan outlines countries major action plans for a ten year period starting from 2005 up to 2015 with the target to be amongst the top 15 countries active in this field in all ring's of value chain by the end of this period. *i.e.*: human resources, science generation, technology development, production and commercialization. The plan was devised by Iran Nanotechnology Initiative Council (INIC) established at 2003. INIC is composed of six ministers and five experts. INIC is not only responsible for policy making in this field but also supporting, supervising and evaluating the activities, and outcomes as well.

Iran's national nanotechnology development plan (INNDP) is composed of three major divisions: outlook, mission, main goals and consists of 53 executing programs with each having quantitative goals acts as a road map for all of the responsible bodies involved. The main goal in the strategic plan was chosen to be raising people's quality of life and generation of wealth.

According to INNDP, the countries priorities are as the followings:

1. Energy (such as: oil, gas, petrochemical, solar cells)
2. Health (such as: targeted and smart drug delivery, diagnostic kits, nanobiosensors)
3. Water and Environment (such as: water desalination, water purification, air pollution)
4. Materials (such as: nanoparticles, nanocomposites and nanocatalysts)

The strategic plan is mainly composed of four sections as the followings:

1. Advocacy and public awareness
2. Human resources development
3. Research and development infrastructure.
4. Technology development and production

In this respect, in order to facilitate the implementation of the devised programs four responsible committees were organized. The aims and activities of these committees will be briefly reviewed here.

Advocacy and Public Awareness Committee (APAC)

This committee started to introduce the current and future potentials of nanotechnology for policy makers and top decision making organizations in the country. Publication and distribution of more than 130 newsletters and establishing special websites for experts, public and even school children was proved to be very effective for familiarizing these groups of society.

Human Resources Development Committee (HRDC)

Human resources are one of the basic requirements for any development plan. HRDC by providing a package of supports for students, researchers, and academic staffs have been able to attract the attention of academia and postgraduate students towards nanoscience and nanotechnology. At the moment more than 50 universities and research institutes are engaged in these areas amongst which 12 universities are running masters programs and 5 universities deliver doctoral level courses in this field. Due to the widespread interest towards this field amongst Iranian postgraduate students, more than 2000 MSc and 400 PhD theses are completed or currently running at various universities/research institutes. Furthermore, HRDC also have other supporting packages for nanotechnology related job training as well as sabbaticals. These supporting schemes have had a remarkable impact on the developing status of Iran in terms of science generation. Taking the number of ISI publications as an indicator of science production one can monitor the improvement of Iran in this field. In 2003, when INIC was established the number of nanotechnology related ISI papers in Iran was only 30 which put Iran in position of 51 in the world ranking. At the end of year 2008 Iran was ranked 19 having published 771 ISI papers in this field. Since 2003 up to now Iran has enjoyed the highest rate of growth every year. Figure 16.1 displays the rate of increase in the number of ISI papers and Iran's global ranking in this field within the period of years 2000 to 2008.

The growth in the quantity of the ISI papers may not fully describe the scientific achievements of Iran in this field. The quality of the published papers is normally implemented in the impact factor of the journals. Though, this factor varies rather significantly for different areas of science and technology.

Figure 16.2 indicates the impact factor (IF) distribution profile of the nanotechnology related ISI papers published by Iranian scientists.

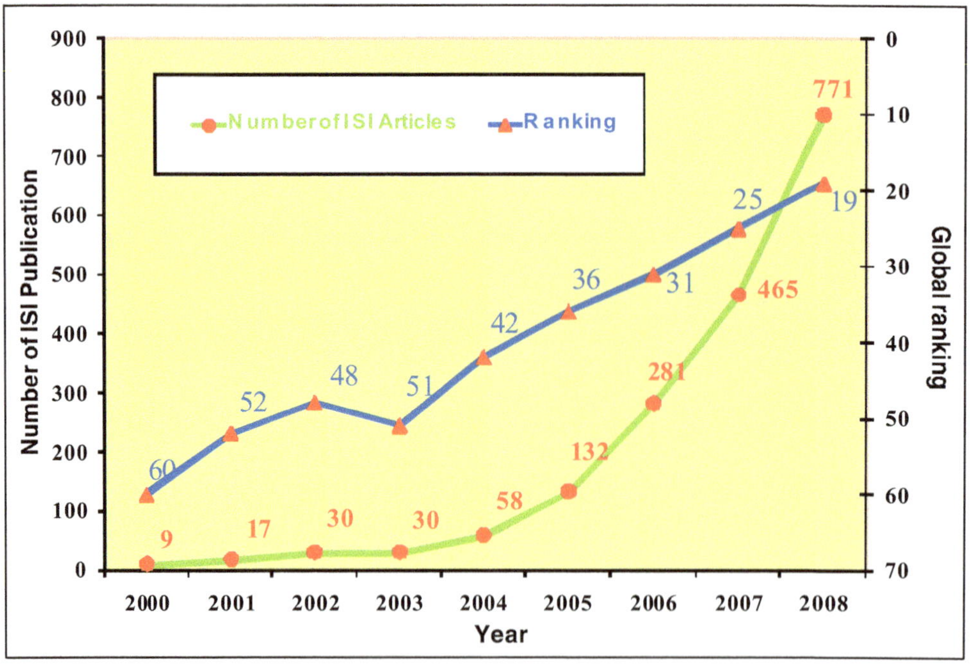

Figure 16.1: The Rate of Increase in the Number of ISI Papers and Iran's Global Ranking in the Field of Nanotechnology

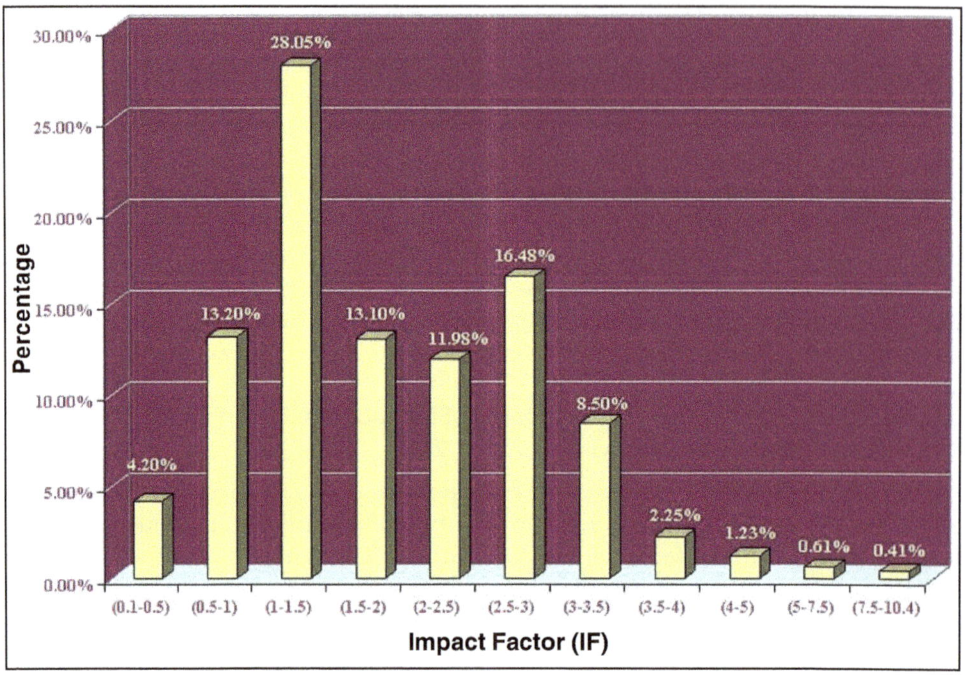

Figure 16.2: The Impact Factor (IF) Distribution Profile of the Nanotechnology Related ISI Papers Published by Iranian Scientists

Research and Development Infrastructures Committee (RDIC)

The major aim that RDIC follows is the development of the necessary infrastructures for the support of R&D in academic and industrial institutions. The establishment of Iran Nanotechnology Laboratory Network (INLN) in 2004, Iranian Nanotechnology Standardization Committee (INSC) and more than 10 patent offices in 2006 can be considered as the required steps towards the achievements of the goals of this committee. INLN is a virtual network currently embracing 40 well-developed labs. It provides various services in all major fields of nanoscience and nanotechnology. The member labs are ranked based on human resource capabilities, facilities and the services they offer. The homepage of the INLN provides comprehensive and updated information on the facilities, expertise and available services of member labs (www.nano.ir). Moreover, INLN aims to set up collaborations with other regional/international lab networks for the exchange of services and expertise.

The issue of standards is very crucial for the development of any product including nanotech products. As an active member of the international nanotechnology standardization committee of ISO (ISO/TC229 committee), INIC has organized her related mirror committees with the close collaboration of institute of standard and industrial research of Iran (ISIRI). The establishment of a nano-metrology center is another important task that INIC is following.

Technology Development and Production Committee (TDPC)

Iran has realized that in the immediate future Iranian enterprises possess no option other than to utilize novel technologies to increase its share of the future global market and to improve the level of nation's wealth. Nanotechnology, in this respect lends itself as the most promising opportunity for Iran. TDPC is in charge of promotion of research and development (R&D) and industrial production of nano-related products.

Iranian Nanotechnology Business Network (INBN) was established in 2008. This network is composed of a group of business advisors and the investors to help and promote the Iranian companies/enterprises active or at least enthusiastic to start their businesses in the field of nanotechnology. The network also facilitates international cooperation and marketing. At the moment, we have more than fifty startup companies and another 15 company's active in production of nanotechnology related materials or products, which are already in the market. Some companies have already been successful in exporting their products to other countries.

Reference

www.en.nano.ir.

Chapter 17

Status of Nanoscience and Nanotechnology in Kenya

Bernard O. Aduda
Department of Physics, University of Nairobi,
P.O. Box 30197–00100, Nairobi, Kenya
E-mail: boaduda@uonbi.ac.ke

ABSTRACT

Nanoscience and nanotechnology deal with the study, manipulation, and the design of materials and devices at the atomic scale. Whereas nanoscience focuses on the new properties, nanotechnology focuses on new devices. The nanoscale science and technology is multidisciplinary and trans-disciplinary. We thus see that nanoscience and nanotechnology bridge the gap between molecular scale and the macroscale. It is now recognized that science founded on the unified concepts on matter at the nanoscale is the new basis for knowledge creation, innovation, and technology integration. Given the current and anticipated impacts of nanotechnology in work efficiency, ways of life and the environment, national (*e.g.* Kenya's) efforts should be focused on how nanotechnology can improve efficiency in manufacturing, energy resources and utilization, reduce environmental impacts of industry and transportation, enhance healthcare, produce better pharmaceuticals, improve agriculture and food production, and expand the capabilities of information technologies. This requires synergistic relationships amongst all the stakeholders right from the beginning of the research and development processes. A study of the Kenya's situation reveals minimal understanding of what nanoscience and nanotechnology, and what their potential benefits are. The teaching of and research in these areas and the accompanying infrastructure are weak and disjointed. Moreover, there are almost nonexistent/ weak specific regulatory tools and policies governing this important but emerging technology and discipline. Gaps presently exist in our scientific knowledge, and coupled with inadvertent outcomes witnessed in other technological advances there are reasons enough for nanotechnology industries and relevant government agencies to invest in understanding possible risks and neutralize them prior to putting the products into the market.

Keywords: *Nanotechnology, Nanoscience, Education, Risks, Benefits, Kenya.*

Introduction

In this article we define nanoscience and nanotechnology and highlight their potential benefits to humanity and also point out pertinent issues that are associated with them–training and education, societal, environmental, safety and health issues, and standardization. We discuss these issues not only globally, but also as they relate to Kenya's situation.

A pertinent question to ask is what is nanotechnology or nanoscience? The widely accepted threshold for a material to be termed as nano is less than 100 nanometers (<100 nm) in at least one dimension. Ordinarily, this is therefore a very small size, and when we talk of nanoscale. Nanoscience and nanotechnology therefore deal with the study, manipulation, and the design of materials and devices at the atomic scale (Balbus *et al.*, 2007). Whereas nanoscience focuses on the new properties, nanotechnology focuses on new devices. We thus see that nanoscience and nanotechnology bridge the gap between molecular scale and the macroscale. It is now recognized that science founded on the unified concepts on matter at the nanoscale is the new basis for knowledge creation, innovation, and technology integration (Roco, 2005).

Nanoscale science and engineering/technology have assumed a critical role in enhancing our understanding of and control over the basic building blocks of matter, giving rise to increased integration in knowledge, technology, and education. Many industries dealing in catalysts, coatings, paints, rubber and tyre products, microprocessor manufacturing, heavy equipment manufacturing, and aerospace are already reaping from the Science and engineering at the nanoscale (Roco and Bainbridge, 2005). Principally the reason for nanoscale science and technology/ engineering research is to advance the larger societal goals such as better understanding of nature at the molecular scale for enhanced productivity, healthcare, and broadening the boundaries of sustainable development and of human capabilities and achievements (Roco, 2003; Roco, 2005).

In the last two decades, research in nanoscale science and engineering/ technology has witnessed exponential growth as evidenced by publications in open literature and patents. Kostoff *et al.*, 2007 show that in 1991, there appeared ~ 10,000 articles as represented in the Science Citation Index/Social Science Citation Index (SCI/SSCI) on nanotechnology and the figures increased to 30,000 in 1998 and to ~ 65,000 in 2005.

Nanomaterials and their Characteristics

We have already alluded to the tremendous research efforts dedicated to nanoscience and nanotechnology during the last two decades. Significant portions of these efforts are focused on understanding the properties of nanomaterials, which are of utmost importance for advanced applications.

Nanostructured materials can essentially be obtained by either "bottom up" or "top down" approaches (Koch, 2007). In the first approach we start with an assembly of atoms or molecules and make them into nanoscale clusters, and followed by consolidation into bulk materials. In the "top down approach, we arrive at the nanostructure by breaking down the original bulk material, *e.g.* by attrition. The specific techniques under these two general approaches can be physical (*e.g.* sputtering techniques) or chemical (*e.g.* techniques based upon the reactions and manipulations of atoms and molecules in a reaction), biological precursors or biomimicry or mechanical (*e.g.* ball milling).

At the nanoscale materials exhibit quite different (exceptional chemical reactivity electronic, mechanical, optical and magnetic) properties from their bulk counterparts, *e.g.*, carbon-based nanoparticles tend to be extremely strong and durable; gold (the most noble of all metals) which is inert in bulk becomes chemically reactive at the nanoscale level. The significantly different physical properties of these novel materials have been attributed to their characteristic structural features in between the isolated atoms and the bulk macroscopic materials. When the dimensions of a system are reduced to the nanoscale level the number of atoms at the surface significantly increases along with the increase in surface area per unit volume. For example, when the dimensions decrease from micron level to nanolevel the specific surface area increases by three orders in magnitude. In such a case, large proportions of the atoms will either be at or near the grain boundaries. The exceptional properties can thus be ascribed to the changes in the total energy and structure of the system. For example, the chemical reactivity, which is a function of the system structure and the occupation of the outermost energy levels, will be appreciably affected at such a length scales, resulting in corresponding change in the physical properties (Satyanarayana *et al.*, 2007).

The properties of nanomaterials can be tailored to numerous engineering applications via atomic-level structural control of materials. However, the real challenge lies in the integration of nanostructured materials into the devices and retaining nanostructures in bulk components.

Consider, for example the use of the sensors in biological, environmental, defense and novel energy applications, fundamental understanding of the working principles of these devices is imperative. Such understanding will include the surface charge dependent and field dependent variation in the conductance of the nanostructures, which is the major basis of the functioning of the metal oxide nanowire, based device configurations.

Applications or Potential Applications

The many publications described earlier cover broad areas, which can be categorized as either medical or non-medical. Specifically, research areas in nanoscience and engineering/technology encompass:

1. Atomic manipulation for design of molecules and supramolecular systems;
2. Computing, sensing, communications, data storage, and display capacities that promise to transform the information technology industries;
3. Controlled interaction between light and matter with relevance to energy conversion among others, that promises ecologically sound production of energy and a reduction in the demand for energy ensuing from increased efficiencies thus ensuring energy independence for many nations;
4. Composite materials with high strength-to-weight ratio, which will guarantee very high performance and energy savings;
5. Exploiting quantum control mechanical–chemical molecular processes for diverse types of chemical and biosensors (catalysis and sensors are the major applications for the nanoscale metal oxide and semiconductor) that find applications in biomedical, chemical, and food industries (Kuzma, 2007).
6. Healthcare and agricultural systems, *e.g.*, chemical storage and transport mechanisms within the organs, miniaturized medical diagnostic and treatment (drug delivery) devices (for medical applications, they are state-of-the-art science vehicles for drug delivery and imaging due to their ability to travel and cross barriers within the body (Soboyejo *et al.*, 2007), biomineralization of nanocrystallites in a protein matrix–highly important for the formation of bone and teeth; animal identification preservation and tracking, etc.
7. Biomimicry–the process of using biological systems as a guide to synthesize the nanoscale materials thereby achieving preciseness and selectivity to the nanostructures synthesized;
8. Nanomaterials and their associated technology may also be the key to better environmental damage control through better products, as well as through specific applications to environmental corrective techniques; examples include adsorbents and catalysts for managing and destruction of contaminants, filters for reducing contaminants in effluents, additives for friction reduction and energy conservation in mechanical applications.

From the foregoing, nanotechnology has assumed a central role for applications to produce probes for biological and chemical separation, sensors and their purification, catalysis and environmental clean-up, energy storage and composites for structural materials, filling materials and coatings (Roco, 2005). There are already areas where nanotechnology is already playing unparalleled role, and many new innovations are still anticipated. It is obvious that nanostructures are set to have profound impact in nanoelectronics, nanodevices and systems, nanocomposite materials (*e.g.* use of the excellent mechanical and elastic properties and low density of carbon nanotubes) combination of nanoscience and technology with biological and therapeutic sciences is anticipated to boost the capacity to provide efficient treatments in otherwise impossible situations (Satyanarayana *et al.*, 2007).

There are however, gaps presently existing in our scientific knowledge, coupled with inadvertent outcomes witnessed in other technological advances are reasons enough for nanotechnology industries and relevant government agencies to invest in understanding possible risks and neutralize them prior to putting the products into the market. Public trust and health will both be boosted when proactive efforts to get nanotechnology right the first time are taken. It is important to note that the public not only safeguards but also serves as the watchdog of every new technology.

Implications of Nanoscience and Nanotechnology and the Kenyan Scenario

A number of issues that have emerged in the era of nanoscale science and technology and these include:

Training and Education

The nanoscale science and technology is multidisciplinary and trans-disciplinary. It has convergence with biotechnology and information technology. This has called for even greater collaboration between and amongst expert from very diverse disciplines. The key challenge for such converging technologies development is the education and training of a new generation of skilled workers versed in the multidisciplinary and trans-disciplinary perspectives requisite for rapid furtherance of the new technologies (Roco, 2005).

The curriculum designs at all levels of education and training in sciences and engineering, and humanities, need to possess the interdisciplinary connectedness that reflect unity in material and information worlds, and in science and humanity curricula. This requires that curricula in sciences, engineering, and humanities should reflect this interconnectedness and be logical and holistic. This would be beneficial not only to the scientists but also to non-technical consumers or funders of the technology.

From the foregoing, a look at the Science, Engineering, and humanities curricula offered at the Kenya's institutions of higher learning, especially at the Universities–both at undergraduate and postgraduate levels–there is hardly any course specifically designed and geared to nanotechnology or nanoscience. Given the current and anticipated impacts of nanotechnology in work efficiency, ways of life and the environment national (*e.g.* Kenya's) efforts should be focused on how nanotechnology can improve efficiency in manufacturing, energy resources and utilization, reduce environmental impacts of industry and transportation, enhance healthcare, produce better pharmaceuticals, improve agriculture and food production, and expand the capabilities of information technologies. This means that synergistic relationships amongst all the stakeholders (the research scientists, productive units in nanotechnology development, venture capitalists, industry, academia, national laboratories, and funding agencies) ought to be developed and enhanced by fostering and forging partnerships right from the beginning of the research and development processes.

Whereas there is no coherently designed curriculum on nanoscience/ technologies in Kenya institutions of higher learning, this does not mean that there is nil training in this emerging technology. There is, for example, an active research group (headed by the author) conducting research on nanoporous dye sensitized solar cells (Waita *et al.*, 2007; Simiyu *et al.*, 2004), nanoporous thin films for photocatalysis (Nyongesa *et al.*, 2007) and also nanoporous water filters. The research activities of this Group are therefore geared to addressing the challenges of energy and provision of clean and safe drinking water–issues of global concern.

It is reported (Pellerin, 2008) that there is Nanotechnology at work in the villages of Obambo and Kadenge in western Kenya, for example, where Sky Hydrants developed by Siemens AG turn brown sludge from the local dam into clear drinking water. The self-cleaning ultrafiltration device consists of three large cartridges that each contains 10,000 minipipes made from thin membranes whose 100-nanometer-diameter openings keep dirt and bacteria from passing through.

Presently Kenya has a weak industrial base but thriving agricultural, livestock and tourism sectors, which though face a number of challenges. Kuzma, 2007, for example has identified a number benefits that nanotechnology/science can impart to these sectors and the accompanying food industry: Agrochemical delivery (*i.e.*, well targeted delivery of pesticides, fertilizers, and other agrichemicals more efficiently); Animal production (*e.g.*, delivery of growth hormone in a controlled fashion, and identity preservation and tracking); Animal or plant health (*e.g.* detecting animal pathogens, such as foot and mouth disease virus, and detecting plant pathogens early); Plant production (*e.g.*, targeted genetic engineering); Sensing (*e.g.* detecting chemicals or foodborne pathogens; biodegradable sensors for temperature, moisture history, etc. monitoring); Environmental control and policing (*e.g.* detecting environmental contaminants early; assessing states of populations or ecosystems; and binding contaminants for easy removal). This incomplete list of potential benefits of nanoscience and technology to Kenya's main economic pillars are immense, especially for improved food production and quality, better animal health, reduced societal/community conflicts such as cattle rustling (as animal tracking will be made easier through use of nanochips implanted in livestock).

In Kenya, The Science and Technology Act, Chapter 250, of the Laws of Kenya, which took effect on 1st July 1977, and is an Act of Parliament the purpose of which is to establish machinery for making available to the Government advice upon all matter relating to the scientific and technological activities and research necessary for the proper development of the Republic; and for the co-ordination of research and experimental development; and for matters incidental thereto and connected therewith. This well-defined intention is contingent on the availability of well-educated and experienced personnel; therefore it is incumbent on the stakeholders to prepare Kenya for effective participation and exploitation of the benefits of nanoscience and nanotechnology.

Societal Implications, Environmental Effects, Health Risks and Effects

History teaches us that the successful introduction of a new technology demands careful attention to the interactions and linkages between the technology and society, which are bi-directional. Specifically, on the one hand, technology alters and confronts social norms, and conversely, the governance structures and values of the society affect advancements in developing the technology. This means that the value systems of the society will have to rapidly change to keep pace with this emerging technology, and also appropriate policies must be put in place to regulate/govern the technologies and emerging issues. There is the risk that regulatory regimes appropriate for one kind of product/industry or nanotechnology development may be inappropriately extended to others. This means that there must be in place a framework for 'good governance' of nanotechnology.

The above complex relationships and effects demand new research efforts on the impact of nanotechnology on sectoral productivity, the use of nanotechnology by industry and geographic area, international competition, the potential effect of different laws regulating nanotechnology across nations, and the potential national return to investment of government support for nanoscience and nanotechnology research. Major projects, in nanotechnology, from inception, ought to involve social scientists, economists and public exponents, in order to guarantee equity in the distribution of contributions and benefits, and to improve potential applications so they can better serve people and thus better achieve success in the market (Roco and Bainbridge, 2005).

Whereas the applications of nanotechnology are becoming widespread and nanotechnology's potential to help achieve cleaner air, water, and soil, more effectively treat disease, and improve energy efficiency and material durability are huge, many of the physico-chemical properties that endow nanomaterials with so much promise are the same which could have adverse effects on human health and the environment (Balbus *et al.*, 2007). As at now our understanding of the environmental effects and health risks associated with nanomaterials use is quite limited (Linkov, *et al.*, 2007).

As already mentioned nanomaterials possess properties and characteristics quite different from there bulk counterparts. Such extraordinary physicochemical properties bring along a concern about the adverse effects of nanomaterials on biological systems, thus there is impetus to understand the toxicity of nanomaterials. As an example, certain nanoparticles are useful in drug delivery systems (targeted therapies) because of their ability penetrate cell membranes. This same characteristic also implies that nanoparticles will also be able to tunnel through physiologic barriers and access body compartments which larger particles and smaller molecules do not easily. Linkov *et al.*, point out that recent studies suggest that inhalation and possible dermal absorption of some nanomaterials may have harmful effects on mammals and a potential for adverse ecological effects. Ultrafine particles, such as nanoparticles can traverse the lungs and enter the systemic circulation leading to a number of side effects. At the cellular level the ultrafine particles have the ability to enter and damage cells' mitochondria. This means that greater delivery of nanoparticles to specific compartments or cellular organelles could result in greater toxicity (Balbus *et al.*, 2007). That the nanoparticles used in drug delivery systems or inhaled may serve, as unintended transporters of toxic molecules must be meticulously studied.

It is known that carbon-based nanoparticles tend to be extremely strong and durable. This raises the obvious question about the biopersistence and accumulation of nanoparticles in the environment and within living organisms. Other worries centre on whether the novel physico-chemical and structural properties of nanomaterials cause unforeseen toxicological behavior at the cellular or organismal levels, and how to effectively manage any potentially harmful nanomaterials characteristics. This calls to question the sufficiency of currently used methods for establishing hazard and exposure and for controlling the ensuing risks. Moreover the existing regulatory tools and policies have limitations yet the rapid developments in nanotechnology outpace the rate of development of regulatory frameworks and even the development of a rigorous, comprehensive scientific understanding of the behavior of nanomaterials in biological systems and the potential for human exposures (Balbus *et al.*, 2007).

Other concerns are that production processes may lead to occupational exposure; consumer products with nanoparticles may also result in exposures with intolerable risks; the disposal of these products, in addition to industrial waste from their manufacture, can result in accumulation of nanomaterials in the environment. Consequently health and safety issues related to Nanomaterials are coming under increased scrutiny.

Despite the concerns above there seem, at present, to be little guidance from regulatory bodies or agencies on nanotechnology. Further, there is a dearth of data or database of toxicity studies on or information about health effects of nanoparticles. These doubts and uncertainties about health effects directly enter into the risk-benefit debates that progressively inform and determine societal responses to, and regulation of, the emerging technologies (Kandlikar *et al.*, 2007).

When we put the above issues in perspective with respect to the Kenyan scenario, we note that the regulatory and governance structures to guide nanotechnology in Kenya are at best non-existent if not very weak. The Kenyan workers (in industrial, agricultural and other sectors) are protected by the Factories Act Cap 514 (1951) of the Laws of Kenya whose preamble state thus "An Act of Parliament to make provision for the health, safety and welfare of persons employed in factories and other places, and for matters incidental thereto and connected therewith". Part vii of this Act on Health, Safety and Welfare–Special Provisions and Rules appears to be the most relevant with regards to safety, health and welfare issues with regards to new technologies such as nanotechnology. Section 55A has provisions, amongst others, that empower the Chief Inspector to "with the approval of the Minister conduct special research, experiments, and demonstrations relating to occupational safety and health as are necessary to explore new problems including those created by new technology in occupational safety and health which may require ameliorative action beyond that which is otherwise provided for in this Act and shall also conduct research into the motivational and behavioral factors relating to the field of occupational safety and health". This section also requires the Chief Inspector to the "prescribe regulations" attendant to the above issues. Implementing such provisions of an Act in an environment characterized by ignorance, lack of preparedness in training, education and infrastructure is a clear recipe for failure in terms of protection of the populace and workers as envisaged under the act. Further, any regulations prescribe will naturally be weak or inadequate or inappropriate. This is exactly where

Kenya is as regards nanotechnology. Worthy of noting is that there is not yet a government concerted effort or programme that addresses the issues of nanotechnology.

Standardization

Some of the major benefits of standardization include fostering interoperability and opening barriers to trade. We have earlier on pointed out that research, discoveries and innovations in nanotechnology are growing at an exponential rate. Maximizing the benefits of these by those involved and the society at large is contingent upon standardization, which has been known to play an significant role in the commercialization of emerging technologies, yet consensus on the necessity of standards has not been reached within the nanotechnology community. Besides, there is need to apply more innovative use of standards. Already there are many nanotechnology products in the market and this need for standards is even more urgent now. Rashba and Gamota state that "The diversity and success of nanotechnology-based applications may well depend as much on the universal ability to control processes at molecular and atomic scales as on how well this burgeoning industry can create common criteria for the fundamental elements upon which new products are based and for architectures that allow for interoperability" (Rashba and Gamota, 2003).

In Kenya, The Kenya Bureau of Standards (KEBS) is a Statutory Organization of Government, established by an act of parliament Chapter 496, in 1974 and is the National Standards Body (NSB) mandated by the Act to perform the following activities amongst others: Standards Development; Product Certification; Quality System Certification; Hazard Analysis and Critical Control Points system certification; etc. To the best our knowledge, KEBS has no standards with regards to nanotechnology products nor has conducted hazard analysis for such products. This together with some of the risks mentioned early has the potential of exposing the citizens to adverse effects due to uncontrolled importation of industrial or medical products.

Conclusion

From the foregoing we note that Nanoscience and Nanotechnology is characterized by: (*i*) Rapid growth; (*ii*) The potential to increase the general economic growth and hence uplift the standards of living of a vast majority of human beings; (*iii*) Capacity for enhancing social connectivity amongst diverse peoples, professions and interest groups; (*iv*) Broad implications in industry, medicine and economy and hence the potential to deal with or solve myriad human problems, ranging from technological and material resources, health to social problems; (*v*) Cuts across several disciplines and therefore requires multidisciplinary and multi-sectoral approaches; (*vi*) It is global in nature and touches virtually the fabrics of human activities; (vii)Just any other new technology, there are issues such toxicity, disposal and standards that must still be addressed.

For the success of nanotechnology measured by its acceptance by society, the following also are key: (*i*) creativity of individual researchers, (*ii*) training of students in nanoscale science and engineering, (*iii*) connections between organizations–local, regional and international, (*iv*) legal frameworks, (*v*) government policies. The success of nanotechnology cannot be determined only by doing good R&D in academic and industry laboratories (Roco, 2003). Addressing nanomaterial risks; educational programmes to ensure safe and responsible development of nanotechnology thus presents a number of challenges. These include ensuring thorough and timely evaluation of nanomaterials prior to commercialization; balancing the benefits as well as the unknown risks of new nanomaterials with the sometimes better-known risks of substances they would be replacing; and applying appropriate safeguards to the production, use, and disposal of engineered nanomaterials in the face of the uncertainties listed above. In the Kenyan context and scenario, nanoscience and nanotechnology is its related issues are still very weak and therefore require urgent attention.

References

Balbus, J.M., Florini, K., Denison, R. A., and Walsh, S. A., 2007. Protecting workers and the environment: An environmental NGO's perspective on nanotechnology. *Journal of Nanoparticle Research, Special issue: Nanoparticles and Occupational Health,* 9: 11–22.

Factories and Other Places of Work Act, Cap 514, of the Laws of Kenya.

Kamat, P.V., 2007. Meeting the clean energy demand: Nanostructure architectures for solar energy conversion. *Journal of Physical Chemistry,* C 111: 2834–2860.

Kandlikar, M., Ramachandran, G., Maynard, A., Murdock, B., and Toscano, W.A., 2007. Health risk assessment for nanoparticles: A Case for Using Expert Judgment, *Journal of Nanoparticle Research, Special issue: Nanoparticles and Occupational Health,* 9: 137–156.

Keller, K.H., 2007. Nanotechnology and society. *Journal of Nanoparticle Research,* 9: 5–10.

Kelly, K.L., Coronado, E., Zhao, L.L. and Schatz, G.C., 2003. The optical properties of metal nanoparticles: The influence of size, shape, and dielectric environment *Journal of Physical Chemistry,* B107: 668–677.

Koch, C.C., 2007. Structural nanocrystalline materials: An overview. *Journal of Materials Science,* 42: 1403–1414.

Kostoff, R.N., Koytcheff, R.G., Clifford G., and Lau, Y., 2007. Global nanotechnology research literature overview. *Current Science,* 92(11): 1492–1498.

Kuzma, J., 2007. Moving forward responsibly: Oversight for the nanotechnology-biology interface. *Journal of Nanoparticle Research,* 9: 165–182.

Linkov, I., Satterstrom, F.K., Steevens, J., Ferguson E., and Pleus, R.C., 2007. Multi-criteria decision analysis and environmental risk assessment for nanomaterials. *Journal of Nanoparticle Research,* 9: 543–554.

Nyongesa, F.W., Nyaga, W.G. and Aduda, B.O., 2007. Electrophoretic deposition (EPD) for photocatalytic applications. In: *Poster presented at the 6th Edward Bouchet-Abdus Salam Institute International Conference*, 23-26 January, iThemba LABS, Cape Town, South Africa

Pellerin, C., 2008. *Nanotechnology could help Billions Gain Access to Clean Water*. http://www.america.gov/st/env-english/2008/September 20080926175127 adkcilerog0.9006006.html (accessed on 30th April 2009).

Roco, M.C., 2003. Broader societal issues of nanotechnology. *Journal of Nanoparticle Research*, 5: 181–189.

Roco, M.C., 2005. The emergence and policy implications of converging new technologies integrated from the nanoscale. *Journal of Nanoparticle Research*, 7: 129–143.

Roco, M.C. and Bainbridge, W.S., 2005. Societal implications of nanoscience and nanotechnology: Maximizing human benefit. *Journal of Nanoparticle Research*, 7: 1–13.

Rashba, E. and Gamota, D., 2003. Anticipatory standards and the commercialization of nanotechnology. *Journal of Nanoparticle Research*, 5: 401–407.

Satyanarayana, V.N.T., Kuchibhatla, A.S., Karakoti, Bera, D. Seal, S., 2007. One-dimensional nanostructured materials. *Progress in Materials Science*, 52: 699–913.

Simiyu, J., Aduda, B.O. and Mwabora, J., 2004. Anthocyanin sensitized nanoporous TiO_2 PEC solar cells prepared by a sol gel process. *Progress in Colloidal Polymer Science*, 125: 34-37.

Soboyejo, W.O., Theriault, C., Mulambo, P., Chi, S., Chanrashekar, R., Oni, Y. and Soboyejo, A.B.O., 2007. *In-vitro* studies of cancer and non-cancer cell death due to hyperthermia and localized drug delivery. In: *The 4th International Conference, The African Materials Research Society (AMRS)*, Dar es Salaam, Tanzania, December 10th–14th.

Waita, S.M., Aduda, B.O., Mwabora, J.M., Granqvist, C.G., Lindquist, S-E, Niklasson, G.A. Hagfeldt, A., and Boschloo, G., 2007. Electron transport and recombination in dye sensitized solar cells fabricated from obliquely sputter deposited and thermally annealed TiO_2 films. *Journal of Electroanalytical Chemistry*, 605: 151–156.

Chapter 18

A Vision for Civic Engagement in Nanotechnology in Mauritius

Hemraj Ramsurrun

Rajiv Gandhi Science Centre, Old Moka Rd, Bell Village, Port Louis, Mauritius

E-mail: h_ramsurrun@yahoo.co.uk

ABSTRACT

Across the world, Nanotechnology is taking a serious option in debates concerning the implications of technological advances on society. Scientists claim that this technology will have a mass effect on all walks of life. Ultimately no single field in science will be spared by the influence of the "next big thing" which is well set to shape the future of mankind and draw the picture for the awaiting industrial revolution. However the situation is different in developing countries especially in the continent of Africa. Mauritius is no exception. It is high time that we take forward an agenda to at least inform the public, who is still lagging behind at latest technologies, about this fast growing technology of the future. This paper presents the vision of the Rajiv Gandhi Science Centre to bring forward a program of informal education in nanotechnology for all stakeholders of the Mauritian population. An insight of Mauritius and nanotechnology is also proposed. Finally a concluding note suggests other possible measures that could be adopted.

Keywords: *Nanoscience, Nanotechnology, Civic engagement.*

Introduction

At the dawn of the 21st century, Science and Technology (S&T) are playing an ever important role in shaping the future of mankind. For countries like Mauritius, modern S&T is even more decisive in its development as a healthy nation as they offer hope for addressing imperative needs of the population in terms of public health, safety, shelter and nutrition. Throughout history, the development of new technologies and their potential impacts is directly linked with the political and social climate and inevitably influencing dramatic societal shifts. For instance, with the advent of new technologies in telecommunications, Internet and others like nuclear power, governments, policymakers, and concerned citizens are bringing important and fruitful adaptations to their growing standards of living. Government entities and policymakers from developing countries willing to bring socio-economic progress are now faced with a mandatory challenge of devising suitable means and ways to promote and popularize new science and technology development and its social applications. Scientific literacy must now be a requirement and a powerful tool for social inclusion in developing countries.

In the last two to three decades, scientists have progressed so far that today matter is being manipulated at an atomic scale and this achievement is marking the birth of a so-called invisible technology: Nanotechnology. Nanotechnologies can be regarded as genuinely interdisciplinary, and have prompted the collaboration between researchers in previously disparate areas to share knowledge, tools and techniques. An understanding of the physics and chemistry of matter and processes at the nanoscale is relevant to all scientific disciplines, from chemistry and physics to biology, and the production of stronger and lighter materials. They are attracting rapidly increasing investments engineering and medicine. Indeed, it could be argued that evolutionary developments in each of these fields towards investigating matter at increasingly small size scales has now come to be known as nanotechnology. Nanoscience and nanotechnologies are widely seen as having huge potential to bring benefits in areas as diverse as drug development, water decontamination, and information and communication technologies from governments and from businesses in many parts of the world.

Governments and industries in the world agree unanimously that Nano promises to have a profound societal impact as it paves way to the next industrial revolution. Undoubtedly, this technology is becoming essential to so many aspects of our economy and our society, which implies that there is a strong, need to promote this new technology among the public. However, scientists and educators are aware that nanotechnology is a complex field incorporating concepts from several disciplines and in such a case it is not so easy to bring people's imagination and understanding to the atomic level. Actions have already been taken in several countries around the world, especially by science centres where a wide range of activities are being conducted to see how the social climate will be influenced with the introduction of Nanotechnology on a large scale in social and commercial applications. It is known that surveys on awareness of nanotechnology have given poor but promising results, especially in countries that have still a long way to go in achieving success in science. If nanotechnology is to transform societies, economies, and personal lives in fundamental ways as many scientists and economists predict, then a proactive engagement among these different entities will be necessary.

Nanotechnology, as a major new area of innovation, is growing much faster as compared to the pace at which public engagement is evolving in science and technology. Thus, new forms of education are needed to prepare members of society for making complex decisions about policy, governance, and values. But the most difficult part is to draw people from different intellectual and social backgrounds into conversations about science and technology and to empower them to take effective and influential decisions. The Republic of Mauritius, as a small island developing state, is quite slow in what concerns keeping abreast of latest technologies and innovations. Although Mauritius is making remarkable socio-economic progress since the last decades, she still lag behind in S&T as compared to other countries. Under these circumstances, it has now become mandatory that one make a significant start in the process of civic engagement in Mauritius in order to contribute towards improving this situation. Nanotechnology, representing the next big thing, can prove itself to a very opportune partner for civic engagement in S&T in Mauritius. Right initiatives needs to be taken by all stakeholders to pay much attention to the societal impacts of this new technology. Indeed, one has to adopt a proactive approach to civic engagement to be able to educate the public about the complexity of this innovation. The Rajiv Gandhi Science Centre (RGSC), under the aegis of the Ministry of Industry, Science and Research is consecrated with the mission of popularizing Science in Mauritius. It rightly expects itself to open its windows and allow fresh knowledge about emerging technologies to invade its corridors and to showcase these developments to its gentle public who is still very young at new technologies such as Nanotechnology. To this end, RGSC has an important role to play in taking forward this agenda and engaging the public in science and the issues that it raises for the society.

Mauritius and Nanotechnology: A Brief Picture

The Republic of Mauritius is a multi-cultural, multi-lingual, and multi-religious country situated in Indian Ocean, about 2500 kilometers from Durban, South Africa. Our island has a native population of about 1.3 million inhabitants, all descendants of immigrants from different parts of the world. According to the Constitution of Mauritius, the population is composed of Hindu, Muslim, Chinese, and the General Population with a variety of languages: English, French, Creole, Hindi, Urdu, Mandarin, Bhojpuri, Tamil, Telegu, Marathi.

The Mauritian economy has undergone remarkable transformations since independence. From a poor country with a high unemployment rate, Mauritius has made its way to a relatively prosperous, diverse and developing state, although not without problems. Initially based on sugar plantation, tea plantation and textile, Mauritius has now projected the ICT and tourism as main pillars for its economy. Recently, the Government of Mauritius has embarked on ambitious projects of transforming the country into a cyber island followed by projects like the Maurice Ile Durable and the Land-based Oceanic Industry. Other sectors like agriculture, health, electronics and construction are well settled and offer opportunities for the future. To cater for these developments, a number of reforms have been carried, especially in the education sector. Moreover, new ministries have been created for Science and Research as well as Renewable Energy.

When it comes to education, the system comprises of 4 components: Pre-primary, Primary, Secondary and Tertiary. A national examination qualifies primary school students for secondary education. At secondary level, students can sit for 'O'-Level and 'A'-Level examinations carried out by the University of Cambridge through the Cambridge International Examinations. Tertiary education in Mauritius is characterized by a wide range of institutions, which provide all levels of tertiary education. Within the public sector, the educational system in general, revolves around the University of Mauritius, the Mauritius Institute of Education, the Mahatma Gandhi Institute, the Mauritius Examination Syndicate, the Mauritius College of Art and the Tertiary Education Commission.In addition to the above publicly funded institutions, private institutions/bodies are presently delivering tertiary-level programmes. The local tertiary education scene also comprises four tertiary education institutions with a regional vocation, namely the University of the Indian Ocean (UIO), the "Institut de la Francophonie Pour L'Entrepreneuriat", the Sir Seewoosagur Ramgoolam Medical College (SSRMC) and the Mauras College of Dentistry (MCD).

Civic education is viewed by educational authorities in Mauritius as an essential and effective tool for preparing young people to be active and responsible citizens, and educating them on the critical issues that govern the future of our country. Civic education is taught across the curriculum in the education system in Mauritius to prepare students to be respected and respectable citizens. This component of the education system has a major role to play in order to increase awareness of the importance of science and technology and issues related to the Mauritian society. In a similar context, the Government of Mauritius has introduced compulsory science on a pilot basis at 'O' level in the form of a subject, namely the "21st Century Science". The Rajiv Gandhi Science Centre, on the other hand, is the only institution in Mauritius that is mandated for the promotion and popularization of Science and Technology. With a view to inculcate creativity and awareness in the minds of our younger generation, RGSC also supplements school education in Science and Technology and related issues.

With the advent of new technologies in the world that is shaping the future of developing countries like Mauritius, science literacy is an inescapable prerequisite for us to progress. If we want to contribute positively towards the development of our country, then every citizen must have a good science foundation while students and young people must acquire competencies and skills in Science and Technology. However, further to studies conducted by the Mauritius Research Council to assess the status of science in Mauritius, it is observed that there is an increasing decline in the number of students embracing science for a career. Much in the same way, as depicted by Prof. S. Jugessur, Pro-Chancellor of the University of Mauritius, our country is lagging behind in terms of research, development and innovations. He further stresses that there should be more willingness to exploit the results of research for economic and social development. This situation, which is in fact very delicate, calls for immediate concern and necessary action.

Nanotechnology and the link with developing countries like Mauritius offer a large room for open discussion. Unlike many powerful nations in the world, it is not so easy to assess how nanotechnology or related issues can really impact on the Mauritian society. In fact, the 'next big thing' has not really taken birth in this island, which aspires to make science and technology a driving force for its socio-economic progress. While many initiatives are being taken in India, Japan, Germany and USA,

Mauritius is not yet embarked on concrete measures to be taken to bring nanotechnology and all its benefits in its society. The situation becomes even urgent when we realize that apart from a group of students and academics, awareness in this area is nearly negligible. However, the future appears to be very bright as Mauritius aims at becoming a nation with a powerful and intelligent human resource.

Going around tertiary institutions in Mauritius, one can find few encouraging signs. For instance, research on nanocarrier and drug delivery systems is being carried by academics of the University of Mauritius (UoM). A forthcoming conference to be organized jointly with the USA on biomaterials and nanomedicine, is also awaiting the go-ahead for it to take place in November 2009. Nanotechnology is also being considered to be included in some programmes that are offered at UoM. At secondary education level, Nanotechnology has already been included in the syllabus of Chemistry, which indicates that our science students, although not all, are now studying the basics of this topic. Academics at the Mauritius Institute of Education, an institution that offers teacher training, are working on programmes to explain nanotechnology to teachers. From a different approach, the RGSC notes a growing interest in this field among secondary school students. This is justified by the number of projects received relating to Nanotechnology in one of its national competitions, the Science Communication Contest (SCC). The SCC provides lower secondary school students with an opportunity to make a presentation on any topic related to S&T. On the other hand, the private sector seems to lack momentum in investing on Nanotehnology products and services. Today one can note only one or two companies in Mauritius as authorized suppliers of products and services that are directly or indirectly related to Nanotechnology.

In a country where there is a strong need to boost up science, we cannot expect much from its population to know or understand about Nanotechnology. A situation of this sort is arising in Mauritius most probably because of a considerable lack of popularization and awareness on this topic. Being new, this field seems to be quite unclear to us especially about its potential impact on our economy and livelihood. Education on this topic, whether formal or informal, is not yet emphasized fully by science educators. May be the necessity to bring down the imagination of the learners to the atomic level is hindering the progress of efforts placed by educators. In order to inform or educate properly, a wide range of resources in terms of school items, equipment for demos, documentations etc, must also be acquired although this can be highly expensive. Most importantly, Mauritius, like many other developing countries, is experiencing difficulties to keep pace with technological advances. While other nations have started to dedicate part of their budget to S&T, Mauritius will probably take some more time to follow these steps. To invest in Nanotechnology on a large scale and to achieve fast results in research and development requires exorbitant funding which our country may not afford at this stage.

However, the future is very bright for a country like Mauritius that contains a large pool of intelligent and gifted minds. We are a peaceful nation with lots of potential to emerge on a regional basis given our social and political stability and also efforts being placed by our government to educate its population in general. Although, scientists in the world believe that a paradigm shift is necessary at the social level in order to benefit from advances in S&T, Mauritius, with its high level of education, will easily adapt itself to changing conditions in what concerns science and society. Our economy provides room for investment in many areas of S&T, which are either directly or indirectly related to Nanoscience and Nanotechnology. Nanotechnology has major applications in areas like medicine, electronics, food, energy, air quality, clean water supply, fabric and others. All these disciplines have a place in the economy of Mauritius and have enough prospects for Nanotechnology. Tourism is the major driving force of our economy. Nanoscience and nanotechnology can be used to boost up this sector by providing high-tech tourist services in terms of health and body-care, food processing, cosmetics, clean water supply and garments. Similarly, the Mauritian public is very fond of electronic devices like computers, mobile phones and household appliances. This is good sign for Nanotechnology to improve the electronics industry of Mauritius. The government has recently created a ministry for renewable energy, which implies that important applications of nanotechnology can affect this sector. Generally, our industries can well accommodate this new technology. However, the main questions to answer are how and when. As a first initiative, one must embark on popularization activities to give the Mauritian public an insight of this technology

A Vision for Civic Engagement in Nanotechnology in Mauritius

In today's knowledge-based society, to achieve our economic growth targets, we need to continue investing in the renewal and growth of our science and technology human resources. This was precisely highlighted by Hon D. Gokhool, Minister of Industry, Science and Research in the Government of Mauritius. Developing countries like Mauritius need to innovate and inspire from neighbouring developed nations in areas like Science and Technology. To this end, important decisions and actions have to be mined with a view to achieving success. An essential component of this mission is the role that Science Centres can play in bridging the gap between Science in the laboratory and the public. Activities in line with the promotion and popularization of Science are therefore critical and unavoidable catalysts in the process of bringing scientific knowledge to the doorsteps of the public. In fact, Science centres around the world are fantastic places for the public to engage itself in the process of learning and discovering science and technology. Having hands-on experiences with exhibits already available is an excellent means to inculcate creativity in the minds of the people, from the young to the elderly.

The Rajiv Gandhi Science Centre is the only organization in Mauritius, mandated for the promotion and popularization of Science and Technology. In line with its mission and vision, the RGSC is well embarked on a wide gamut of activities to cater for all layers of the Mauritius population. With the advent of nanoscience and nanotechnology and in respect of the little or negligible awareness in this field, the centre already foresees a series of action lumped together in the form of a strategic plan to kick-start a long process of engaging the nation in nanotechnology. With the full-fledged support of its parent ministry, RGSC aspires to become an umbrella institution in the social and commercial prosperity of nanotechnology in Mauritius.

International Conference on NANO

The development of new technologies inevitably has societal implications. Nanotechnology, on its own, will have a massive effect on all walks of life, not just science. The RGSC, at the very outset, has a central goal of engaging the government, industry partners and the general public in dialogue on nanotechnology. As a first step in this direction, the centre intends to invite international scientists and experts in the field of nanoscale science and nanotechnology to discuss and raise awareness on the societal and ethical implications of this technology. As an expected result, this initiative will enable effective encouragement on strategic planning aimed at building capacity for the development, dissemination and education of nanoscale science and nanotechnology in Mauritius. Government, policy-makers, industry representatives, educators and interested professionals/citizens will have the opportunity to engage in discussions on issues pertinent to this field. Programmes can be in the form of workshops, seminars and conferences. Such an initiative willstart a new avenue for the country and its population in Science and the masses would benefit.

Collaboration with Leading Organizations in NANO

While in the world, specialized centers on Nanoscale Science and Engineering are being launched, the RGSC, on its turn, emphasizes to build on institutional partnerships and integrate in networks of like-minded organizations. This will extremely facilitate towards reaching our objectives of sharing knowledge and expertise in nanotechnology and bringing those innovations to the Mauritian public. Through collaborative programmes with international institutions, important forums can be organized for our educators (both formal and informal) to discuss and devise means for educating the society. This will in turn equip our promoters of nanotechnology with necessary skills and knowledge to face the growing challenge of demystifying nano for the Mauritian public.

The Nanogallery

Science centers provide opportunities for everyone to experience science and understand its concepts. Visual impression helps in facilitating the process of grasping various facets of scientific knowledge. The RGSC has, since its inauguration in 2004, set its way in a very promising direction in its endeavor of communicating science and technology on a local basis. As its core attraction, the 4000 sq. m building holds five permanent galleries and one Science Park that contain a full variety of interactive exhibits, informative panels, educational software and multimedia projections. These exhibits, while illuminating important scientific knowledge and concepts encourage visitors to enjoy hands-on experiences. In this context, a gallery on nanotechnology will be a key resource at the center to demystify nano for the public by simultaneously engaging them in fun and self-learning experiences. This will be a major step ahead to boost up nano in the island, as essential basic information will be readily available to the public. As the need states, the center is seriously considering the development of its NANO gallery in its programme of revamping its galleries. Basically, this exhibition area will explore the potential applications and recent development in the field of nanoscience as well as its implication to the society. To this end, RGSC looks forward for fruitful collaborations with leading institutions in nano popularization and exhibit development.

NANO for Students

The student community at all levels represents a major percentage of our target audience. Enlightening their growing minds and to give them maximum opportunities to exhibit their innate talents in S&T, is one of our leading priorities. It is of utmost importance that our future professionals be exposed to science, technology and innovations right at the early age. Among its current activities destined to the student population, RGSC regularly organizes science project-based competitions to help them develop creativity, teamwork, inquiry, research and problem-solving skills. Eminent competitions in this line are: National Science Challenge, Science Communication Contest, Junior Science Adventure and the new Science Project Competition. These competitions are either open-ended or thematic and cater to students at all levels. The RGSC envisages organizing its competitions under the theme of Nanoscience and Nanotechnology. This will bring important awareness in the field at an early age in the life of those individuals who aspire to lead the country. This will also support evaluation of the progress of nano in Mauritius. Other concurrent activities for students include workshops, science shows, demonstrations, small lectures/talks, film shows and science for kids. The RGSC has in the pipeline to develop a package that contains a wide spectrum of these small activities geared towards connecting our students with nano. The objective is none other than stimulating creativity and expanding the imaginative capacities of our children the better understand the technology of the future.

NANO and the Community

It was a first time event in Mauritius when the RGSC organized a Science Festival in 2007. This initiative of the centre amazingly brought more than 15,000 people in its premises to enjoy a full series of attractions and hands-on activities that was proposed. Over 4 days of intensive popularization campaign, the Science Festival 2007 showcased the history in the field of Science, Technology and Innovations. Various stakeholders of S&T were called to exhibit how science is contributing towards the Mauritian economy.

Following the success of the Science Festival 2007, the RGSC can now envisage the organization of a Nano Festival for the first time in Mauritius. Although this may not be for the immediate future as a national event, the RGSC can plan the NANO Festival with the main objective of engaging the public much in the same way as the Science Festival 2007. A wide gamut of activities will be proposed including demonstrations, shows and hands-on exhibitions. All under one roof, the objectives are clear-cut: to make the public accept nanotechnology as the technology that will shape our destiny. Other reasons of holding this event are increasing rate of change in nanotechnology, introduction of new nanotechnology–enabled products, and diversity of such products. Undoubtedly this will have a profound influence of the development of nano in Mauritius as important clues to its social and commercial implications will be obtained. Industry partners will seize this opportunity to evaluate prospects for nanotechnology in Mauritius.

On the other hand, RGSC, in its vision to close the gap between science and the public, has often organized public outreach activities at the level of communities. An eminent example is night sky observations in small villages. Connecting nano with the community is a concept that appears difficult to realize and success in this direction will not be for tomorrow. But the hopes are here as Mauritius is gaining in terms of literacy level. This will facilitate the process even in villages. Popularization talks, demos, shows and hands-on practices on nano that will be carried at community level, will increase the number of proponents of nano, especially in villages.

Housewives and elderly people have a special role to play in connection with education in the family. These people are the closest to our kids and adolescents who need constant support and guidance during their education process. In science popularization, this sector of our population has been neglected although these people can become fruitful advocates of S&T in Mauritius. These are matured people who have undergone experiences in the troughs and crests of life and can have big influence on engaging their busy family members in S&T. The only requirement is giving incentives to these people to get engaged in the process of Science popularization and promotion. Nanotechnology is an ideal arena where these people can be engaged to propagate knowledge in the family and community. By sufficiently informing and training housewives and elderly or retired professionals, these people can become helpful promoters of nanotechnology in the family and consequently lead the show in social organizations, civic groups and women councils.

NANO and the Media

Mauritius is a democratic country where media has a large influence on the opinion of its audience. After television, print media are the most important source of information for consumers. With the help of our media partners, bringing awareness on nanotechnology can be facilitated and time saving. Mauritius, through the Ministry of Education Culture and Human Resources, has recently launched the "Knowledge Channel" which is offered by the Mauritius Broadcasting Corporation. Several stakeholders are involved in production of small video programmes aimed at supplementing education at all levels. Seizing this opportunity whereby the media is being solicited to give a helping hand towards improving quality of education, the RGSC aims at developing TV and Radio programmes in Nanotechnology. This will largely increase our audience and catalyse the diffusion of basic information on Nanotechnology. Similarly, the press and other print media will be involved in devising and implementing suitable means for promoting and popularizing Nanoscience and Nanotechnology.

NANO and the Industrial Sector of Mauritius

The Government of Mauritius has recently created a Ministry of Industry, Science and Research which has the vision of making our industrial sector globally competitive, environmentally sustainable and socially responsible, propelled by Science, Technology and Innovation. As part of its mission, the ministry aims at enhancing the global competitiveness of the industrial sector and at the same time promoting science and research for wealth generation, job creation and sustainable development. With this laudable initiative, the Nano sector represents a highly promising direction in which our industrial sector can achieve global competitiveness. By promoting research and development in nano on a local basis, our industrial sector will undeniably benefit from the applications of nano in terms of job creation, sustainable development and wealth generation. At a later stage, our country can envisage on implanting a nano industry much like the tourist or textile industries with a view to support its economy. The RGSC can have a leading role towards taking forward this agenda although it may require a decade to see important developments in this direction.

Conclusion

Nanotechnology is considered to be the key to opening the future and the world is witnessing inestimable progress in this field. With the help of the "next big thing", a new era of inventions is being predicted that will have crucial impacts on the life and lifestyle of human beings with plenty of opportunities to solve problems that we are faced with. In light of this new concern, Mauritius has no other choice than to sensitize it and start on an agenda to allow this new technology to take birth. To achieve tangible results, civic engagement is a major requirement. The success of engaging the nation in Nanotechnology will rely on several factors. One of the key requirements being the extent to which science popularizing agents commit themselves in devising means and ways to inculcate knowledge. Society at large needs to be encouraged and enabled to play an active role in the process of learning and understanding science. The RGSC will have to work together with its stakeholders to see how it can face the growing challenges of engaging the public in nanotechnology and start achieving results through implementation. The vision of the centre is already generating a positive impulse towards the government, industry partners and policy makers and influencing them to go on board to give a huge support to a project of this dimension. With the applications of nanotechnology revolving around major sectors of our economy, we believe that our country has a very bright future. As much emphasis is being placed on science, technology and innovations, our little population is ready adapt itself to changing environmental conditions as may be justified by the "Maurice Ile Durable" project which aims at engaging the government, the public and industry partners in making Mauritius a more sustainable island. The population is already adapting itself with a paradigm shift towards sustainability. We believe that Nanotechnology will be an important partner, although not for tomorrow, in achieving the goals set under this nationwide program of sustainability. At such point in time where Mauritius lies at the crossroad of major technological developments, Civic engagement in nanotechnology will also urge the population to become more innovative, visionary and develop its potential with a view to improving quality of life and making Mauritius the "Maurice Ile Durable" for the decades to come.

Acknowledgement

I would like to thank Dr. Maulloo, Director of the Rajiv Gandhi Science Centre for his insightful advices, consultations, and discussions.I am grateful to the Ministry of Industry, Science and Research and the Rajiv Gandhi Science Centre Trust Fund Board for their laudable support.

References

Dimension, Bimonthly News Journal of the Association of Science-Technology Centres January/February 2008, September/October 2008, January/February 2009.

Donald C. Maclurcan, September 2005, "Nanotechnology and Developing Countries, Part 1: What Possibilities?"

K. W. Lim, Singapore, "Nanotechnology in Developing Countries".

Mauritius Research Council, May 2004, "Teaching and learning of Science in Schools" (Republic of Mauritius)

Nanotechnology Articles. http://www.understandingnano.com/nanotechnology-articles.html

"Positive image of nanotechnology in the media" http://www.chemie.de/news/e/89487/

Chapter 19

Present Status of Nanotechnology Research in Myanmar

Moe Zin Win[1] and S. Theingi[2]

[1]*Materials Science and Materials Engineering Research Department, Kyaukse, Myanmar*

E-mail: moe.2006@gmail.com

[2]*Deputy Director General, Department of Technical and Vocational Education, Yangon, Myanmar*

E-mail: theingi.sa @gmail.com

ABSTRACT

Myanmar, one of the developing countries from South-East Asia, introduced nanotechnology 2005. The Union of Myanmar is the second largest country in South-East Asia, with the total land area of 261,228 square miles. Myanmar enjoys favorable climatic conditions and a rich diversity of topography. The population of the country is estimated at 54.3 million with the population growth rate of 2.02 per cent. The energy crisis that has affected the world in the recent years has not spared Myanmar either. One of the best ways to produce clean-energy is wind-mill construction to meet the growing energy demand, and permanent magnet production is essential for the construction of wind-mill. When we tried to produce the permanent magnet, nanotechnology was found most essential for metal forming. We discuss the present situation and future prospects of nanotechnology in Myanmar as a developing country.

***Keywords**: Myanmar, Energy, Wind mill, Permanent magnet, Nanotechnology.*

Introduction

Nanotechnology is the creation and utilization of materials devices, and systems through the control of matter on the nanometer-length scale. The aim of nanotechnology is to learn the exploit these properties and efficiently manufacture and employ the structures. Control of matter on the nanoscale already plays an important role in scientific disciplines as diverse as physics, chemistry, materials science, biology, medicine, engineering and computer simulation. Current research in nanotechnology requires multi-disciplinary knowledge in science and engineering (Roco, *et al.*, 2000).

General Overview of Nanotechnology

The World Nanotechnology Market (2006) reported the detailed and updated overview of the nanotechnology market worldwide. The key findings are that more than $9.5 billion was spent worldwide on nanotechnology research and development in 2005, which showed a considerable rise as compared to 2004. Out of the total R&D expenditure, major contributions were from various governments and corporations which together contributed nearly $9 billion. The rest of the R&D spending was from venture capital firms.

By 2008, the total global demand for nanoscale materials, devices and tools, it was projected, would cross $28 billion mark. The U.S. Nanotechnology market was predicted to reach $ 13 billion by 2008 and cross $19.8 billion by 2013 showing a strong growth.

The market for textiles using nanotechnologies, it was projected would cross the US $13 billion mark by 2007. By 2012, the market is expected to reach US $115.2 billion. Governments of nearly 30 countries are actively participating in R&D activities. Because of the rapid growth, the market is expected to cross $1 trillion mark by 2015.

United States has the largest share of global investment in Nanotechnology. The US market had a share of 27 per cent in 2005, followed by Japanese market with more than 24 per cent of share (The World Technological Market, 2006). The western European market also had a quarter of the market share with major investment in countries like Germany, UK and France. Other countries like China, Russia, South Korea, Canada and Australia hold the rest of the share.

Nanotechnology is a disruptive technology and has influenced a large number of industrial segments such as materials, electronics, and medicine. It is expected to penetrate and influence various commercial markets. Seeing the prospects of the technology, massive investments are being made in nanotechnology. Developed as well as developing nations around the world are promoting nanotechnologies its use will bring down costs of products considerably and will make them more effective. It can also find answers to problems like pollution and water purification hence it will eventually become an indispensable part of any country's progress in almost all aspects of economic and social life.

Nanotechnology is in its formative years, and currently only basic nanostructures can be created with some control. Most of the nanotechnology-based solutions are still in the early stages of R&D. In order to move to practical applications in various industries, nanotechnology will have to perform at high precision levels, achieving superior levels of efficiency compared to current standard micro- or macro-scale, automated instruments. Nanotechnology face challenges like toxicity and safety concerns, technical issues, lack of standardization and public awareness. Also there are communication and cultural barriers because of lack of a common language between the research communities (The World Technological Market, 2006).

However, by year 2007, nanotechnology was expected to show a significant impact on industries including healthcare products, chemicals, computers, and the telecom industry. Right now, the nanotechnology is used in making lighter and stronger plastics, which can be used in cars, carbon nanotube reinforcements that are used in tennis rackets and many other consumer market products. In near future, there will be more changes in making of the existing and new products. By 2010, we can have 3D nanosystems like robotics, 3D networking. However, for molecular nanosystems we have to wait till about 2020.

Nanotechnology will bring new functions and properties to improve and develop new products and application possibilities. There is a possibility of creating more powerful and advanced computers, improved medical treatments, and practically any type of extraordinary item with nanotechnology in future- (The World Technological Market, 2006).

Role of Nanotechnology for Developing Countries

At the nanoscale, it is possible to combine material science with biological sciences and even bionics, thus offering very attractive possibilities in electronics, medicine and across many other industries such as food, energy, environment and even space technology that could have innovative applications for human development.

For developing countries, nanotechnology can open up new avenues for research for rapid progress in priority areas. As one of the developing country, Myanmar has tried to advance research and development in nanotechnology just as many developing countries such as China, India, Malaysia and Thailand. Myanmar's national strategic framework in nanotechnology is under preparation to be submitted to the government by the Technology Promotion Committee, under the Ministry of Science and Technology whose organization chart is shown in Figure 19.1.

All universities under the Department of Technical and Vocational Education will continue to play a key role in the development of nanoscience and technology. For Metallurgical Engineering and Materials Science Department, the undergraduate program is a five-year course. During the first two year, students will have to learn the basic science subjects and some engineering subjects. The third and fourth years of the five year course are designed such that the students develop a deeper understanding of the subjects that are being taught. In the final year of the five-year course, the curriculum and syllabus is more advanced and in accordance with the current trends. The students learn some subjects concerned with nanotechnology such as Characterization of Materials, Nanomaterials and Materials Technology and Materials Aspects in Design.

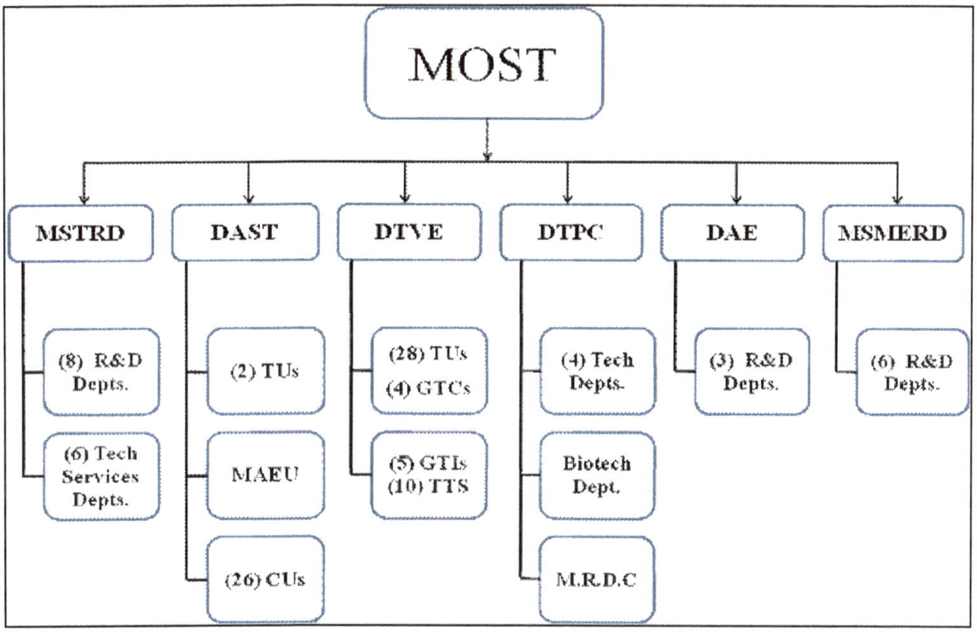

Figure 19.1: Organization Chart of Science and Technology in Myanmar

MOST: Ministry of Science and Technology; MSTRD: Myanmar Scientific and Technological Research Department; DAST: Department of Advanced Science and Technology; DTVE: Department of Technical and Vocational Education; DTPC: Department of Technology Promotion and Coordination; DAE: Department of Atomic Energy; MSMERD: Material Science and Materials Engineering Research Department; R&D Depts.: Research and Development Departments; TUs: Technological Universities; MAEU: Myanmar Aerospace Engineering University; CUs: Computer Universities; GTCs: Government Technological College; GTIs: Government Technical Institute; TTS: Teacher's Training School; MRDC: Myanmar Research and Development Centre.

Current Trends of Materials Science and Technology in Myanmar

Under the supervision of the Ministry of Science and Technology, there are four different areas of the field of Materials Science and Technology as follows:

1. Materials Science
   (*i*) Magnetic Materials
   (*ii*) Dielectric Materials
   (*iii*) Structural Materials
   (*iv*) Refractory Materials
2. Ferrous Metallurgy
3. Rare-Earth Metal Extraction
4. Foundry Engineering

The future development focuses on Materials Science and Technology will be on Magnetic Technology to get a balance of the research between past, present and future Materials Technology.

General Overview on Magnetic Behaviour of Materials

Most materials interact with a magnetic field, just as dielectrics interact with an electric field. Magnetic materials are used to operate electrical motors, generators, transformers, to store and retrieve information on magnetic tape or in computers, to serve as actuators and sensors, to focus electron beams, to assist in medical diagnostic devices, and for a host of other applications. The most widely used magnetic materials are based on ferromagnetic metals and alloys such as iron, nickel, and cobalt or ferromagnetic ceramics, including various ferrites and garnets- (Askeland, 2003).

Magnetic behavior is determined primarily by the electronic structure of a material, which provides magnetic dipoles. Interactions between these dipoles can be modified by composition, microstructure, and processing of these basic materials (Askeland, 2003).

Magnetic materials are vital components of most electrical machinery. Modern-day devices are dependent on the understanding magnetism and magnetic materials. Often these magnetic materials are concealed in subassemblies and are not directly exposed to the end user. For over more than one hundred years there have been usages of permanent magnets. Since early 1940's to the present there have been developments in materials and manufacturing methods that have improved the properties of permanent magnets. These developments have resulted in a common use of magnetic devices. For example permanent magnets are used in signs, telephones, cameras and devices for medical applications (Chan and Olli, 2008).

Finally, magnetic materials are used to make strong permanent magnets. Strong permanent magnets, often called hard magnets, require the following:

1. High remanence (Stable domains)
2. High permeability
3. High coercive field
4. Large hysteresis loop
5. High power (or BH product)

For hard, or permanent, magnetic materials, the domain remain almost completely aligned when the field is removed, large coercive field are required to randomize the domains, and a large hysteresis loop is observed. This condition provides the material with a high power (Askeland, 2003).

The ancient Greeks and Chinese were the first to find and use naturally occurring iron ore called magnetite. One of the earliest applications of permanent magnetic material was the lodestone that led the mariners to safe harbour. Progress in understanding magnetism was limited until William Gilbert published "de Magnete" in 1600 (Chan and Olli, 2008).

Gilbert was the first to apply scientific methods into the studies of magnetism and he is also credited as the first to discover that the earth is a giant magnet. Pierre Weiss postulated in 1907 the first modern theory of permanent magnets (Chan and Olli, 2008).

The conclusion was that these are comprised tiny individual magnetic domains. Between 1820 and through the early 1920's advances were of theoretical nature in understanding magnetism. In 1921 development beyond magnet steels began and this led to introduction of cobalt chrome steel. In 1935 subsequent work has led to an introduction of hardened alnico (aluminium nickel cobalt), Cunife (copper nickel iron) and platinum cobalt in 1936.

The alnico magnets opened new doors compared to the steel magnets. Stronger magnets were introduced in the 1970's and 1980's. They became known as the rare earth family of magnets. They made it possible to make some devices smaller and at the same time more powerful. The elements found in these, samarium or neodymium, are in fact both readily available (Chan and Olli, 2008).

Rare–earth magnets such as neodymium-iron-boron magnets (Nd-Fe-B) offer new design possibilities and opportunities to improve the performance of electrical machines. However, the characteristics of Nd-Fe-B magnets alter significantly with increasing temperature, causing the second quadrant characteristic to become non-linear, resulting in irreversible demagnetisation, recoverable only by remagnetisation (Rizk and Nagrial, 1997).

Because of their outstanding energy product, the minimum magnet volume required to establish a specified flux density in the machine's air gap could be acceptable both from economical and size points of view. The high remanence means that a higher gap flux density, resulting in higher efficiency of the machine (Rizk and Nagrial, 1997).

More than twenty years after the first rare-earth magnets were developed; the discovery of Nd-Fe-B system opened up the possibility of lower cost, high-energy product materials. In their anisotropic, fully dense form, they can now have energy products up to 40 MGOe, with high remanence (>1Tesla) and linear second quadrant B-H characteristics. Two separate groups made the initial discovery, Sumitomo in Japan, with a powder metallurgical method of manufacture of fully dense anisotropic products, and General Motors in the US who set up the Magnequench Company to produce powders for bonded magnets. The sintered products have taken a significant proportion of the SmCo market, especially where high energy and reduced magnet volume are crucial. The reduced materials cost together with improved magnetic performance have enabled the sintered Nd-Fe-B products to capture significant proportions of the SmCo market, the Alnico market and in new applications. The problems of corrosion resistance and temperature stability have caused some concern, and the difficulty (or high cost) in forming accurate shapes in sintered products has restricted the utilisation and hence, limited the scale of production. Motors, generators and actuators have always featured strongly in the application of magnets, and forecasts indicate that they will continue to have the most potential for the exploitation of Nd-Fe-B. Indeed, numerous companies already state that Nd-Fe-B is the material of

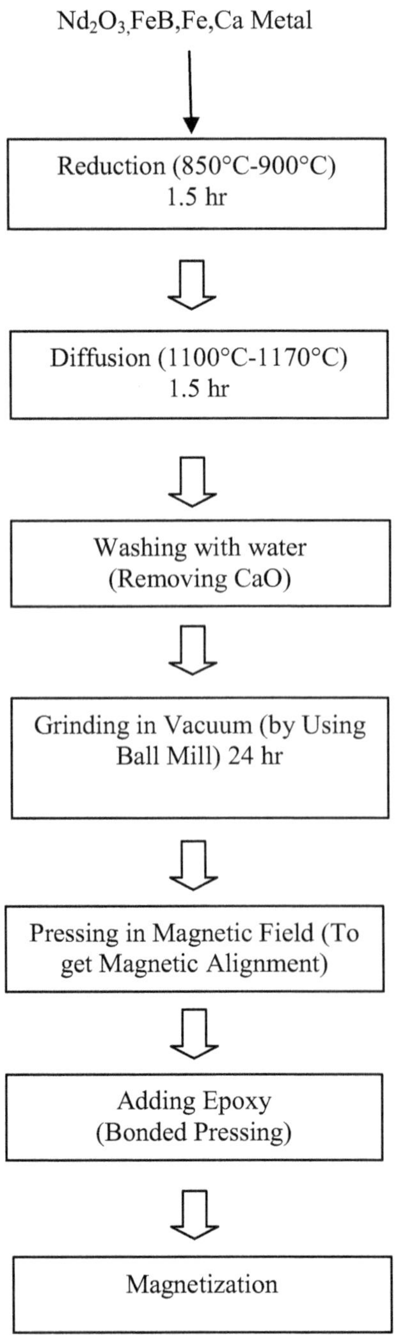

Figure 19.2: Schematic Flow Diagram of $Nd_2Fe_{14}B$ Permanent Magnet Production

choice for new designs of machine. A trend, which will undoubtedly continue, both for cost sensitive and technologically demanding applications, as the cost of Nd- Fe-B continues to decrease and anisotropic bonded versions emerge (Rizk and Nagrial, 1997).

Current Status of Permanent Magnet Production in Myanmar

For Myanmar, $Nd_2 Fe_{14} B$ permanent magnet is successfully produced at the Materials Science and Materials Engineering Research Centre under the Ministry of Science and Technology. The permanent magnet is used in Mini Hydro Mills and Wind generator which are essential for renewable energy production. It is also used in industry, medical apparatus, computer parts, electronic instruments etc, After 2004, it became popular to produce $Nd_2 Fe_{14} B$ permanent magnets in Myanmar.

By producing permanent magnet locally, we can use in Mini Hydro Turbine and Wind Turbine at low cost. In such way, we can enhance the living standard of the people who live in rural areas. Because of growth in the use of renewable energy there is reduction in emission of carbon dioxide leading to reduced global warming.

There are three methods for the production of $Nd_2 Fe_{14} B$ permanent magnet. These are;

1. Direct Reduction Diffusion Method
2. Precipitation Method
3. Powder Metallurgical Process.

The method used for our production is Direct Reduction Diffusion Method. This process is shown with the schematic flow in Figure 19.2.

It is very difficult to make $Nd_2 Fe_{14} B$ permanent magnet which has the strongest magnetic force. $Nd_2 Fe_{14} B$ is very fine size as 0.2 μm single domain. High Energetic Ball Mill/ Jet Mill/ Planetary Ball Mill are used to pound particle size (2-3 μm). In our study, there are no such kinds of machines to get the finest size particles. It takes several hours to get 10 μm particle size using Vacuum Ball Mill in the Research Centre. Therefore, our production is in half-way of nano scale.

Conclusions

The current status of nanotechnology in Myanmar is introduced. In this paper, nanotechnology background is studied and an overview of nanotechnology is presented. Our research work dealing with nanotechnology, permanent magnet production for generators has been introduced. Then, the lack of technology transfer and the weakness of nano scale apparatus are highlighted. Myanmar, one of the developing countries from South-East Asia is trying to develop and improve current situation of Nanotechnology.

References

Askeland, D.R. and Phule, P.P., 2003. *The Science and Engineering of Materials*, 3rd edn. Thomson Learning. McGraw Hill Inc., US.

Chan, A. and Olli, T., 2008. *Permanent Magnetic Materials*. http://www.jyu.fi/static/fysiikka/vuosikertomusket/annual2008.pdf/, Uppsala University, October.

Roco, M.C., Williams, R.S. and Alivisatos, P., 2000. *IWGN Workshop Report Vision for Nanotechnology in the Next Decade*, Springer, pp. iii.

Rizk, J. and Nagrial, M., 1997. http://www.itee.uq.edu.au/ School of Mechantronic, Computer and Electrical Engineering, University of Western Sydney, Nepean, Neodymium-Iron-Boron Magnet in Electromechanical Applications, P.O. Box 10, Kingswood, NSW 2747, AUSTRALIA.

The World Technological Market, 2006. pp. 1-4.

Chapter 20

Nanotechnology: Present Status and Future Prospects in Nigeria

Hammanga. Zainab
Deputy Director, Advanced Materials Division,
Raw Materials Research and Development Council,
(RMRDC) No. 17 Aguiyi Ironsi Street, Maitama, Abuja, Nigeria
E-mail: Zainabu1408@yahoo.com

ABSTRACT

The paper presents the efforts of the Raw Materials Research and Development Council in the Development of A Blue Print for the Development and Utilization of New and Advanced Materials for the country. It discusses the present status, challenges and future prospects of Nanotechnology in Nigeria. Nanotechnology is a new field in Nigeria, and systematic efforts are being made by both the academia, research institutes and government to create awareness and interest in nanotechnology development. Currently, emphasis is on capacity building and research in some of the R&D centres. The Nigeria Nanotechnology Initiative, an integrated effort involving inter disciplinary collaboration between scientists, technologists and entrepreneurs has identified priority areas in medicine, agriculture, energy and water purification that can be met by developing strong capabilities in Nanotechnology within the shortest possible time. The potential impact of Nanotechnology could be enormous, adding value to Nigeria's vast local raw materials, which are presently being exported in raw forms only to be transformed into finished products that are imported at exorbitant prices. In line with the global trends in which economic prosperity is knowledge driven, Nigeria while making giant strides in developing its natural resources, is focusing on New and Advanced Materials, and Nanotechnology development as a strategy in the application of science, technology and research otherwise the raw materials of today may be irrelevant to the needs of tomorrow.

Keywords: *Nanotechnology, Raw Materials Research and Development Council, Nigeria Nanotechnology Initiative.*

Introduction

The potential of Nigeria to develop nanotechnology and nanomaterials exists since the country is richly endowed with abundant raw materials such as agricultural, metallic and non–metallic resources (Okey, 2004). The country is also richly endowed with a population of over 140 million, about 100 universities and many other tertiary institutions (Inegbenebor and Babalola, 2009), as well as petroleum and natural gas. It is on this premise that the government has increased emphasis on the development of traditional raw materials to serve as feedstock for the production of new and advanced materials.

However most of these resources are in their primary forms and Nigeria, like most of the other developing countries is still exporting her raw materials at very low prices, invariably dictated by the end users, only to import back same in the form of finished or semi finished goods at much higher prices.

With the current country's efforts at formulating polices and programmes to promote effective and efficient utilization of her raw materials, the National Science and Technology Policy identified such cutting edge technologies as Advanced Engineering Materials, Biotechnology and Nanotechnology as its focus for industrial development. Several agencies under the Federal Ministry of Science and Technology are now engaged in Research and Development efforts in new and advanced materials. These agencies include the Raw Materials Research and Development Council (RMRDC), National Agency for Science and Engineering Infrastructure (NASENI), National Biotechnology Development Agency (NABDA), Sheda Science and Technology Complex (SHESTCO) and Engineering Materials Development Institute (EMDI).

Current Status of Nanotechnology Development

Nanotechnology is a new field in Nigeria and since the Country is building this technology from scratch, the emphasise is to build it right by emphasizing on some nanotechnology research and development capacity building. There are some research at academic level in some of the R&D centres.

The Nigerian Nanotechnology Initiative (NNI)

The NNI was launched in 2006 to coordinate the research and development of nanoscale science, engineering and technology in the country and NNI team involves interdisciplinary collaboration between scientists, engineers and technologists, working at multiple centres, both within and outside the country. The objective of the NNI is to develop an integrated Nigerian effort that will enable Nigeria to become a major player in nanotechnology within the shortest possible time, develop and disseminate nanotechnology through coordination and the interactions. The program also includes linkages between universities, research institutes and industry. The initiative is being coordinated by NASENI within the Federal Ministry of Science and Technology.

The NNI has resulted in the identification of priority development areas and the scope cover areas that are relevant to Nigerian and global needs in the following areas:

- ☆ Medicine (disease detection and treatment)
- ☆ Energy (solar cells and light emitting devices)
- ☆ Agriculture (drought resistant seeds and food preservation and genomics)
- ☆ Water purification and testing (nano-porous filters and bio-oxidants)

Research and Development Activities

At the moment there are limited Nanotechnology R and D activities going on in tertiary and research institutes, and few private laboratories. Effort is being made to encourage the provision of special resources for the development of infrastructure that will also provide linkages between R&D institutions and funding organizations especially for Nanotechnology applications.

Some R&D activities that have been identified by the Nigerian Nanotechnology Initiative are in the following areas:

- ☆ Nanomedicine (disease detection and treatment- cancer, malaria, HIV, tuberculosis, sickle cell anemia).
- ☆ Nano porous and nano structured materials for water filtration and agriculture (Nanofilters and bio-oxidants).
- ☆ Nano electronics–solar cells and light emitting devices.

Presented below are some aspects of in-Country efforts.

Currently a National Laboratory at the Engineering Materials Development Institute (EMDI) is working on the development of the following:

1. Advanced Organic Materials in Electronic Device fabrication for energy production (organic electronic such as polymer light emitting devices, LED/ photo voltaic PV cells).
2. Key areas in nanotechnology such as nanomedicine for disease detection and treatment, nano-engineered materials for improved agriculture, nanotechnology for energy and nano porous materials for water filtration.
3. Advanced bio-materials used in medicine for detecting and curing diseases such as cancer, malaria and tuberculosis, viral targets and sickle cell anemia, using nano particles and nano fibers.

In addition to promoting research in nanotechnology, regular workshops are organised to sensitize and educate Nigerians. These workshops include special sessions on nanoelectronics, nanobiotechnology, nanogenomics and nanofilteration.

A lot has been achieved through international collaborations in niche areas of research with the US/Africa Materials Institute (USAMI) in Princeton University. The idea is to build on existing strengths in people and resources by coordinating the interactions between Nigerians at home and abroad, identify those already working in the area of nanotechnology and involve them in capacity building.

Manufacturing Activities/Investment Projects

This is a new technology area in which manufacturing activities are not presently prominent.

Information Management

There is a dearth of information on activities in the various areas of Nanotechnology development. To address the information gap, RMRDC set up a team of experts drawn from the various fields of New and Advanced Materials both in the public and private sectors to produce a "Blueprint on the Development and Utilization of Advanced Materials including Nanotechnology and Nanomaterials" in Nigeria. A field study on the status of Nanotechnology is part of the recommendations of the blueprint. This information could then be disseminated for the benefit of all stakeholders. As recommended by the Experts Meeting, an information portal on Advanced Materials is being developed by the Advanced Materials Group of RMRDC to drive its developmental efforts.

Efforts of RMRDC at Nanotechnology Development

The Raw Materials Research and Development Council (RMRDC) is an agency of the Federal Government of Nigeria vested with the mandate to promote the development and utilization of Nigeria's industrial raw materials. It originated from the recommendations of a workshop on Industrial Materials organized by the Manufacturers Association of Nigeria (MAN) and the Nigerian Institute of Social and Economic Research (NISER) in July 1983. It commenced operation on February 10, 1988 and it is today, Nigeria's focal point for the development and utilization of the nation's vast industrial raw materials.

RMRDC operates a network of co-coordinating offices in all the 36 states of the Federation through which information, consultancy and technical support needs of investors, researchers and industrialists are met. In addition, the Council (RMRDC) maintains collaboration with the Organized Private Sector (OPS), international bodies and research institutions to strengthen the Public-Private-Partnership (PPP) strategy of the current economic reforms programme of the Federal Government of Nigeria.

The Council does not carry out by itself, any research at the Laboratory level, but assists and supports institutions (research and industry) in undertaking these assignments.

In fulfillment of its role as the national focal point on new and advanced materials development as conferred on it at the Sub-Regional Workshop on New Materials for Economic Community of West African States (ECOWAS) Countries in 1993, the Council collaborates with all relevant stakeholders in both the public and private sector to create awareness and provide support for the development and utilization of New and Advanced Materials in the Country.

Other important initiations undertaken by the Council include Council's participation at the Workshop on "Implication of New and Advanced Materials Technology For the Economies of ESCWA Countries" (Economic and Social Commission for Western Asia) in Damascus, Syrian Arab Republic in 1992, where the Council presented the paper "Advanced Materials–Nigeria's position".

In line with the recommendation among member countries–the Council co-hosted the ECOWAS regional conference on new and advanced materials in partnership with UNIDO from 16-19 November 1993, in Lagos, themed "New Materials Emerging

Vector for Sustainable Industrialization and Development in the ECOWAS Sub-Region." The sub-regional conference conferred on the Council to be the focal point on Advanced Materials development.

In recognition of the role of Advanced Materials in industrial development and the manufacturing sector, the Advanced Materials Unit under Research Evaluation and Monitoring Division of the Council was elevated to a full-fledge division in February 2007. The new division took off with a sensitization seminar with the theme, "Advanced Materials: A Challenging new World of Products and Services" aimed at stimulating national awareness on the development and trends on new materials and their impact on Nigeria's economy and social development which took place at Abuja on the 10th of June 2008.

The newly created Advance Materials Division also set up a team of experts drawn from the various fields of Advanced Materials drawn from both the public and private sectors to produce a Blueprint on the Development And Utilization of Advanced Materials (including Nanotechnology and Nanomaterials) in Nigeria. The blueprint which is expected to serve as a spring board for all national programmes, projects and collaborations is to culminate into a National Plan for implementation by the Science, Technology and Cognate Institutions in the Country.

The document, which has spelt out possible intervention strategies in the short, medium and long term, also assesses the status, potentials and scope of activities in the development of nanotechnology and nonmaterial. It examines their possible impacts and future trends on industrialization and the economy, advices on ways to increase awareness on the latest development in other countries, examines opportunities available and how best to exploit them and recommends priority projects and policy initiatives for their development.

Other areas addressed by the Blueprint include:

- ✰ *Manpower Development*: Identifies available manpower and manpower needs in the development of Nanotechnology. Data of experts also being compiled.
- ✰ *Facility Acquisition/Development*: Assesses infrastructural facilities and needs for Nanotechnology.
- ✰ *Research and Development Activities*: Identifies research groups, collaborations, on-going R&D activities, and project areas of focus for future interventions in research institutes, universities and private laboratories.
- ✰ *Manufacturing activities/Investment Projects*: Identifies current and potential manufacturing activities and investment projects.
- ✰ *Information Management*: Identifies the information gaps and offers solutions.
- ✰ *Policy Issues*: Assesses current policy issues and identifies relevant policies that are critical to the development and utilization of Nanotechnology.
- ✰ *Recommendations*: Makes appropriate recommendations accompanied by implementation strategies. Spells out future prospects for Nanotechnology development in Nigeria.

- ☆ Identifies a program of work and action plan for Nanotechnology development for the country.

The blueprint is to be subjected to a stakeholders forum which will include researchers (from research institutes, universities and polytechnics), students working in the area of New and Advanced Materials, representatives of the Organized Private Sector, representatives of government agencies, and industrialists for their inputs before implementation.

Future Prospects and Challenges for Nanotechnology Development

Prospects

The impact of Nanotechnology in the Country could be enormous. Within the context of Nigeria, Nanotechnology has the potential to stimulate a renaissance in Nigerian science, while providing innovative solutions to problems in a number of areas. These include: new and affordable approaches to Nanomedicine, Nanobio-engineered seeds, plants and animals for improved agriculture, Nanoscale thin films for solar cells and light emitting devices for alternative energy and Nanostructures porous materials for water purification. Nanoscience and Nanotechnology will also provide an existing new frontier that could promote interdisciplinary collaboration between researchers in universities, research institutions, technologists and entrepreneurs across different parts of the Country.

Challenges

1. Insufficient funding, the yearly allocation to Science and Technology Development has been lower than what had been recommended and considered adequate under the National Science and Technology Policy. This is 2.5 per cent of GDP or 5 per cent annual budget (Olaniyan, 2004).
2. Inadequate number of high quality trained human resource (scientists and technologists). Need for re-engineering of schools curriculum to include more science content covering all areas of New and Advanced Materials, and Nanotechnology to actively promote manufacturing technology?
3. Lack of basic infrastructure leading to a huge gap between technology development and its application for the benefit of the society, the strengthening S and T infrastructure to deliver know- how of utmost importance.
4. Poor interactions among academia–industry and other stakeholders which led to poor linkages between researchers, technologists and industry.

Recommendations

The following are some of the recommendations identified by the Blueprint on the Development and Utilization of Advanced Materials (including Nanotechnology and Nanomaterials) in Nigeria:

1. To develop focused programmes on Nanoscience and Nanotechnology in line with national priorities that would promote the development of specific projects in Nanomedicine, Nanoelectronics, Nanostructured materials, and Nanotechnology for agriculture.
2. R&D activities should be increased to focus on practical applications of nanotechnology products and nanomaterials, especially in agriculture, energy and health care development.
3. Basic infrastructures required by recognized research centers for the promotion of nanotechnology should be provided through:

   Targeted manpower development of scientists/technologists and development of well-equipped laboratories.

   - ✰ Institution of incentives such as bursary and scholarship awards in universities and polytechnics to encourage students to specialize in potential high technology areas.
   - ✰ Initiating, supporting and funding specific R&D in Nanotechnology.
   - ✰ Strengthening existing Nanotechnology programmes in research institutions and universities.
   - ✰ Establishing institution based Centers of Excellence in Nanotechnology that would promote commercialization of a research results.
4. Manpower development drive to be vigorously pursued by conducting workshops, short courses, seminars and on-the-job training. Other measures will involve staff exchanges, study tours and fellowships.
5. Greater and regular stakeholder interactions to enlighten on trends in this field and also rub minds with experts on their R&D activities. This will also serve to link researchers with technologists from both academia and industry.
6. There is a need to develop local and international networks to facilitate technology transfer.
7. There is the need to have a national reference database, which RMRDC is spearheading.
8. Need for the mobilisation of substantial capital by establishing National Science and Technology Funds, increasing budgetary allocations, and raising intervention funds through tax deductions. Increased level of participation could be also being obtained through the provision of incentives to industry to fund projects in Nanotechnology.

Conclusion

As rapid economic advances and wealth creation have become knowledge-based, private sector-led, S&T-driven, and government-facilitated, the need for interactions amongst researchers, technologists, industry and policy makers remain a critical success factor, to which Nigeria, and indeed, other developing countries will meet the key elements of the Millennium Development Goals (MDGs) which basically seek to eradicate poverty and hunger.

The desire of the developing nations (Nigeria inclusive) to be technologically and economically self-reliant could be achieved by using nanotechnology to provide low cost energy, improved and cheaper agricultural products, efficient drug delivery and a lot that would improve standard of living.

Acknowledgements

I wish to express my profound appreciation to the Centre for Science and Technology of the Non-Aligned and Other Developing Countries (NAM and ST Centre) for accepting and sponsoring me to participate at the International Workshop on Nanotechnology–Present Status and Future Prospects in Developing Countries, 18-20 May, 2009, Kashan, Iran.

I would also like to acknowledge and appreciate the efforts of Prof. Arun P. Kulshreshtha, Mr. M. Bandyopadyay, Dr. Saeed Sarkar, Mr. Ali Reza Roodsaz, Dr. M. Molanejad and all the others who have worked tirelessly towards the hosting of a successful workshop. May the Almighty God reward you abundantly.

References

Ayo, D.B., 2008. General concepts, prospects and challenges of industrial application of advanced materials: A challenging new world of products and services in Abuja. In: *A Sensitization Seminar on New and Advanced Materials,* Organised by the Raw Materials Research and Development Council (RMRDC).

Babalola, P.O. and Inegbenebor, A.O., 2009. Nigerian capacity for the development of advanced engineering materials. In: *International Seminar Raw Materials Research and Development Council Technology Exposition Souvenir.*

Blueprint for the Establishment of New Materials Division in Raw Materials Research and Development Council 1992. *Study Review Committee RMRDC.*

Draft Blueprint for the Development and Utilization of New and Advanced Materials. Study Review Committee (2009) RMRDC.

http://tech.groups.yahoo.com/group/nigerian–nanotechnology–initiative. Accessed 4/29/2009

http://groups.yahoo.com/group/nigerian–nanotechnology–initiative/summary. Accessed 4/29/2009.

Odigure, J.O., 2007. *Developing a Framework for Advanced Materials Programme in Nigeria: A Study.* Review Committee RMRDC.

Okey, O., 2004. *Sourcing and Utilisation of Nigeria Raw Materials for Industrial Development Proceedings of the Nigeria Material Congress.*

Olaniyan, S.B., 2004. Funding and commercialization of science and technology research findings. In: *1st Conference of Ministers on Science and Technology, Abuja-Nigeria, organised by Economic Community of West African States (ECOWAS)* in collaboration with the Federal Ministry of Science and Technology, Nigeria.

Report and Proceedings of The Sub-Regional Workshop on New Materials for ECOWAS Countries organised by UNIDO/ RMRDC 16th–19th November, 1993.

Chapter 21

Sri Lankan Nanotechnology Initiative

Ajith P. de Alwis
Science Team Leader, Sri Lanka Institute of Nanotechnology (SLINTEC)
Professor of Chemical and Process Engineering, University of Moratuwa, Sri Lanka
E-mail: ajithd@susnanotec.lk

ABSTRACT

The paper presents the current Sri Lankan Nanotechnology Initiative (NNI) in Sri Lanka from its inception to current status. The future direction the developments are to take place is also presented. All aspects from Nanotechnology Policy, Institutional framework in place to meeting critical human resource development needs are discussed and presented.

Keywords: *Sri Lanka, Nanotechnology, NANCO, SLINTEC, Policy.*

Introduction

As a country with abundant resources and human talent, Sri Lanka has failed to take advantage with respect to emerging technology platforms. With examples such as Singapore which learnt lessons from Sri Lanka, now positioned quite well in the economic scene. Sri Lanka struggles to make ends meet in many ways. Industrial sector has ignored to its cost the important investments via R&D. R&D in Sri Lanka has been largely uncoordinated especially when considering targeted value-added goals. It was felt with nanotechnology being considered the next industrial revolution, that Sri Lanka again cannot wait and see the ship sail by! Hence the NNI!

Nanotechnology Initiative in Sri Lanka is an interesting case of a group of enlightened Sri Lankans abroad who saw an opportunity for Sri Lanka to forge ahead in economic development. The idea after the influencing of an equally enlightened group of individuals in Sri Lanka to consider the emerging field of nanotechnology concept when proposed with sound data, was well taken by the Ministry of Science and Technology (MS &T) along with the National Science Foundation (NSF), which spearheaded the drive with the blessings of the HE President of Sri Lanka. A key event was the presentation to HE President of Sri Lanka by Prof Ravi Silva (Univ. of Surrey) in November 2005. The President was briefed in detail on the potential of this emerging technology in uplifting the nation, and the presentation received his fullest attention. NSF then took the lead role in moving the initiative forward.

As a developing country one can question the serious immersion into a technology drive which is yet to deliver real results at an affordable cost elsewhere. The reasons that apply in developing economies when deciding on a nano initiative as stated in Maithripala (2007) has its relevance in Sri Lanka as well. The reasons appear to be two fold. First, it is viewed it as an opportunity to leap frog their technologies and economies to be on par with that of developed countries. Second the visualisation of the great potential of nanotechnology for solving the pressing social needs.

National Nanotechnology Initiative

NSF initially setup a task force with the plan to harness nanotechnology in Sri Lanka. NSF comes under The Ministry of Science and Technology and is the lead agency for supporting research and development. NSF in this regard was assisted by a number of professionals who had first hand understanding of the subject. It was interesting to note the extent to which the human resource availability was present in the country to support the preliminary drive.

The Cabinet memorandum titled National Nanotechnology Initiative was presented to the Cabinet by the President and the Minister of Science and Technology and the approval given on 23rd August 2006. The objectives were to generate a critical mass of personnel supported by necessary facilities to innovate at the national level. To promote nanotechnology-based research, develop industries and to attract best Sri Lankan expertise, both here and abroad, was an objective from the inception.

Currently a five-year government funded project costing Rs 5.5 billion (US $ 0.055b) supporting the establishment of a research institute and a nanoscience park is underway. The project is a collaborative venture between government and private sector with five leading private sector enterprises sharing 50 per cent equity along with government's contribution. The industries joined with the understanding of the potential benefit for their competitive position to be derived with developments. In parallel the NNI recognizes the need for an effective intellectual property mechanism. The research institute established is the Sri Lanka Institute of Nanotechnology (SLINTEC) which has a panel of international advisors and is currently moving ahead with a three-year plan to realize some key objectives of the NNI.

A key strategic component of the current phase of developments is as follows. Establishing two public private companies with BOI (Board of Investment) status, with shared ownership between public and private. NSF holds the government share. NANCO, which is the second company, is to address the long term needs by establishing the nanoscience center and the industrial estate (Nanoscience Park).

SLINTEC is to address the immediate needs of industries and to enable positioning of the nanotechnology drive through application oriented research. At present SLINTEC is coming into operations and NANCO is established with land allocated for the park. Both organizations have their respective Boards of Management with equal representation of public and private sectors. Self-sustenance within 5 years is also a key objective for SLINTEC and NANCO. This phase of developments received cabinet approval on 10th October 2007.

Sri Lankan Nanotechnology Policy

In parallel NSF has moved forward to establish a Nanotechnology Policy, which is also nearing completion having moved through the first phase of public consultation. This policy will be a sectoral policy element of the National Science and Technology Policy. This process is driven by the National Committee on Nanotechnology established by the NSF.

The policy is currently at the final draft stage. The drafting was done by a committee setup for this process by the National Science Foundation, which spearheaded the National nanotechnology initiative. As policy is formulated after the initial steps of the NNI, policy writers had the luxury of seeing some the results of the NNI. Hence a sense of urgency had been built into the Policy, which is well embodied in the Vision statement. The Vision statement is: "Rapidly enhance the competitiveness of Sri Lanka's industry and economy by capturing the opportunities afforded by the developments and innovations in nanotechnology".

The policy is a sector specific policy and the implementation follows the recommendation of the National Science and Technology Policy.

The draft policy has the following set of objectives;

1. Research and Development (R and D) and rapid commercialization are the key factors in building the momentum required to take nanotechnology into the economic realm. Excellence in R&D is essential to providing funding to bring together world-class researchers and provide them with state-of-the-art research facilities, monetary incentives and a conducive environment to carry out research.
2. To provide research grants to investigators and interdisciplinary teams of investigators with emphasis on targeted industry oriented research.
3. To generate a pool of Sri Lankan IP and safeguard intellectual property rights that would add to the nation's competitiveness.
4. To provide an institutional framework for national decision making, networking and establishing regional and international co-operation in nanotechnology.
5. To transfer knowledge generated into innovative products and services which can provide the competitive edge to local industry.
6. Promote and adapt the use of nanotechnology based materials and services to suit the needs of our society and industry within an ethical framework.

7. Establish a regulatory framework to manage the impacts of nanotechnology on public safety and the environment.
8. Disseminate information on nanotechnology and products to the general public and coordinate a public information system.
9. To create a world-class environment for nanotech research, development and commercialization including a human resources.

It is expected that the policy would give significant emphasis to moving research rapidly to deployment in the industry. This is another lesson learnt from lack of industrial deployment of local research and it is felt that no longer should this be tolerated nor could be afforded in the long term interest of the nation.

A Unique Public-Private-Partnership (PPP)

Sri Lanka has for a long time cried out for public private partnerships in research and development. As shown in Table 21.1 private sector participation in research and development efforts had been low.

Table 21.1: Growth of R&D Expenditure (GERD) in Sri Lanka

Year	*GDP Current Prices (Rs.m)*	*GERD Rs. (US$)*	*GERD as Percent of GDP*	*Total Population (million)*	*GERD per Million Population Rs. million*
1966	7,529	19.8 (4.1)	0.30	11.5	1.7
1975	11,100	45.1 (6.4)	0.40	13.5	3.3
1984	142,700	257.0 (9.7)	0.18	15.6	16.5
1993	499,800	649.0 (13.1)	0.13*	17.6	36.8
1996	769,900	1,410.0 (23)	0.18	18.3	77.0
2000	1,258,000	1,810.0 (22.9)	0.14*	18.4	98.4
2004	1,800,750	3,807.5 (40.9)	0.21	19.4	196.2

Source: National R&D Surveys Sri Lanka 1996 (NARESA), 2000 and 2004 (NSF).

The NNI has changed that situation dramatically. The giants from industry have committed 50 per cent of the funding required for SLINTEC and pledged infrastructure support as well as human resource commitments. As a result of this commitment SLINTEC became operational from September 1, 2008.

Major industries that supported the PPP were Brandix and MAS (two industries that dominate the textile sector), Hayleys (Pioneers in Activated Carbon and a Sri Lankan multinational company), Dialog (A telekom operator with the largest market capitalization in the country) and Loadstar (A solid tyre manufacturer which enjoys 30 per cent global market share in the sector). It is important to note that all these companies had been noting the international developments with respect to their individual sectors and understood well the importance of innovation driven developments for their future growth. These companies together contribute to about 7-8 per cent of Sri Lanka's GDP.

Table 21.2: R&D Expenditure by Source of Funding 1984-2004
Rs. million

Source of Funding	*Year*		
	1984	*1996*	*2004*
Government	239.4 (93.2 per cent)	981.0 (69.6 per cent)	2,571.3 (67.5 per cent)
Private	17.4 (6.8 per cent)	21.5 (1.5 per cent)	21.9 (0.6 per cent)
Foreign	-	324.5 (23.0 per cent)	861.8 (22.6 per cent)
Other*		82.6 (5.9 per cent)	352.5 (9.3 per cent)
Total	256.8 (100.0 per cent)	2,647.7 (100.0 per cent)	3,807.5 (100.0 per cent)

Sources: National R&D Surveys Sri Lanka 1996 (NARESA) and 2004 (NSF)

*Other: Funds generated by the institution itself by providing services etc.

Initially the expectation was that the government was proposed to be the sole sponsor which the government treasury rejected. Treasury was insisting that a 50:50 partnership with the private sector be established. It was then up to the NSF backed up by the Ministry of Science and Technology to convince the private sector to this cause, which was successfully realized. Subsequently the role played by the private sector was in rapid mobilization. Private sector was to ensure the startup of Sri Lanka Institute of nanotechnology and 50 per cent of its investment was to support human resource recruits and other operational expenditure.

Human Resources

Another malaise that afflicted and continues to affect the economy is the Brain Drain which is quite significant. Sri Lanka is perhaps the only country in the region, which provides free education up to first-degree completion at a state university. However, over the years many of the graduates have sought greener pastures abroad. It is really important that this is halted and reversed and a conducive climate for Brain Gain is implemented. SLINTEC and the NNI hope to contribute in this regard as well. Policy envisages setting up of a state-of-the-art Research Center and a Nanotech Park and a conducive framework for operations. This double development would be a strong message to aspiring individuals. It is heartening to note the questions coming to SLINTEC even at this early stage inquiring about prospective opportunities and with interest in returning back to the island. Currently Sri Lanka is not meeting the general numbers suggested by UNESCO for S &T personnel per million of population and for a specific sector such as Nanotechnology the gap is significant. For the initial stage the presence of about 70 scientists and engineers with some exposure in Nanotechnology has been a tremendous benefit. A significant number of this 70 had ideal experience and expertise with direct PG qualifications of relevance to Nanoscience and Nanotechnology.

Intellectual Property Rights

NNI from its inception was quite aware of the importance of IP. With the formation of SLINTEC the interest had been confirmed with the institutional cadre expanded to cover IP issues. Currently the National Intellectual Property Act is in operation. However, the nation observes only sporadic patent activity and the number commercialized from among those that gets patent rights is even less. This is a serious shortcoming in this era of knowledge economy. It is seen that Sri Lanka has really not concentrated on deriving benefits from using its intellectual capital. The past recorded data on patent, design and trademarks submissions to the Sri Lankan Office indicates the present situation which is far from satisfactory (Tables 21.3a–c).

Table 21.3a: Industrial Property Statistics–Marks

Year	*Applications*			*Registrations*		
	Resident	*Non Resident*	*Total*	*Resident*	*Non Resident*	*Total*
1997	2891	2035	4926	656	1109	1765
1998	2982	1397	4379	700	974	1674
1999	3114	1906	5020	264	708	972
2000	3317	2321	5638	294	761	1056
2001	2818	1705	4523	186	585	771
2002	3745	1613	5358	155	549	704
2003	3623	2139	5762	359	601	960
2004	2343	3600	5943	853	975	1828
2005	4587	1900	6487	544	992	1536
2006	4821	2392	7213	741	901	1642
2007	3382	2835	6217	212	499	711
2008	2895	3012	5907	907	1001	1908

These values indicate the current situation in the country. It is apparent that these need to be reversed if the country is to move ahead in any significant manner. NNI addresses this situation as well in trying to foster an innovation culture linking universities and industry. In fact IP generated revenue is a factor under consideration for the sustainability of the initiative.

Sri Lanka Institute of Nanotechnology (SLINTEC)

SLINTEC was set up to expedite the nanotechnology movement. It had been felt that science centre cum estate would take a minimum of three years even under the best of conditions as this was to be a 100 per cent state owned enterprise. Hence as an interim measure and to develop the research and development base SLINTEC was set up. SLINTEC is to address the immediate needs of the industries. SLINTEC was provided with 17000 sq feet building space by one of the industrial partners (MAS holdings) which enabled speedy implementation. SLINTEC is now functional and becoming active on a range of developments for the targeted industries. It has also taken on board a dissemination function so as to stimulate interest in Nanotechnology.

Table 21.3b: Industrial Property Statistics–Designs

Year	*Applications*			*Registrations*		
	Resident	*Non Resident*	*Total*	*Resident*	*Non Resident*	*Total*
1997	175	24	199	172	20	192
1998	297	30	327	268	15	283
1999	154	25	179	150	19	169
2000	187	10	197	179	06	185
2001	520	26	546	482	11	493
2002	345	40	385	253	13	266
2003	386	42	428	365	23	388
2004	254	50	304	224	40	264
2005	257	47	304	269	88	357
2006	477	39	516	422	45	467
2007	328	24	352	8	6	14
2008	333	56	389	85	4	89

Table 21.3c: Industrial Property Statistics–Patent

Year	*Applications*			*Registrations*		
	Resident	*Non Resident*	*Total*	*Resident*	*Non Resident*	*Total*
1995	75	114	189	64	95	159
1996	50	114	164	98	107	205
1997	81	193	274	65	96	161
1998	54	158	212	44	97	141
1999	119	248	367	78	101	179
2000	71	250	321	59	169	228
2001	120	236	356	71	104	175
2002	123	202	325	59	54	113
2003	95	189	284	63	52	115
2004	120	195	315	103	85	188
2005	149	211	360	64	116	180
2006	153	270	423	68	69	137
2007	151	279	430	54	37	91
2008	209	241	450	89	70	159

Initial investment on equipment for the research centre was carried out through 12 special area specific groups organized under the NSF consultative committee (eg. Group on Activated Carbon, Electronics, Ceremics and Metals etc.). Specific research prioritizations were also carried out at the end of the exercise to short-list from the equipment identified. This provided the backdrop to the equipment finally selected to SLINTEC. Figure 21.1 shows the current location of SLINTEC. SLINTEC occupies the bottom floor of this apparel facility perhaps an indication "there is certainly room at the bottom!"

SLINTEC has an international advisory panel that has excellent credentials in the nanoscience and nanotechnology, all of who are Sri Lankans. Of these 04 scientists 02 is from UK, 01 from Ireland and 01 from USA. This is considered to be a great advantage. SLINTEC would also co-opt sector specific international advisors. This is part of the country's attempt to reverse the brain drain and develop a brain gain mechanism. In addition SLINTEC has 02 sector specific advisors. SLINTEC also has developed its own internal vision statement–To be the leading research and innovation platform for sustainable nanotechnology in Asia.

Selection of Projects

SLINTEC is to concentrate on few selected projects to deliver results, generate capability and trust with the industry by demonstrating the power of the new technology. SLINTEC has identified that projects would be screened for sustainability.

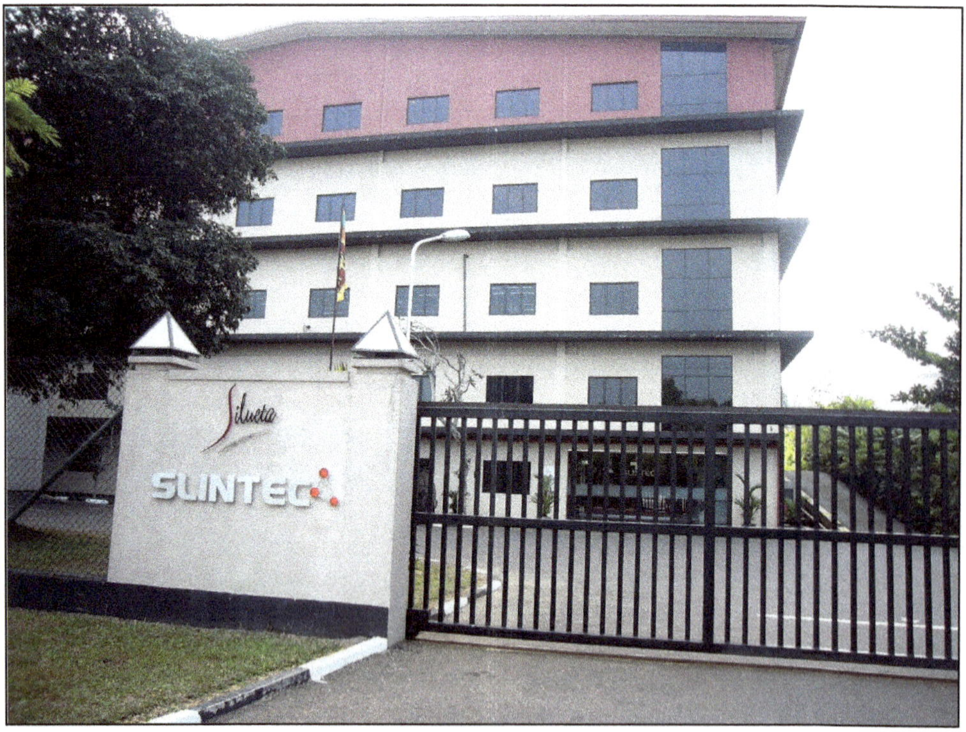

Figure 21.1

All projects submitted are subjected to this process with the proponents having to justify the proposal on environmental, social and economic acceptability. It is hoped that even with a newer technology and with a mandate to generate new concepts and developments the focus for sustainability and the societal needs of millennium development goals would not be forgotten. As a further measure to cement this within the DNA of the organization SLINTEC has signed on for UN Global Compact (perhaps the first organization with a responsibility to nanotechnology to do so) as well as stating sustainable nanotechnology as the operating mantra. From its inception there are 11 projects lined up from industry and all had been wetted under this framework.

NANCO

This is the Nanotechnology Company established with the state and private sector coming together though the asset base is 100 per cent state owned. NANCO is to address the long-term needs by establishing the Nanoscience Centre and the Industrial Estate (Nanoscience Park). The state has already allocated a 60-acre land closer to Colombo to set up an industrial estate dedicated to Nanotechnology. The state has extended its full support usually extended to FDI by registering NANCO as a BOI (Board of Investment) enterprise and has placed its management under the private sector. All investors would receive the typical benefits extended by the GOSL to investors who would meet BOI requirements.

Conclusion

"May you live in interesting times" is a Chinese greeting. This is quite apt considering the Sri Lankan Nanotechnology Initiative. From an idea with some focused lobbying the initiative has grown into a reality and has taken a national dimension with a mission to serve. What awaits are the initial results that would go a long way in establishing the growth and deployment of advanced technology in an economy that was relying more on labour and luck for its existence.

It is certainly becoming a fact that in Sri Lanka the S &T personnel are likely to live through 'interesting times' due to the National Initiative in Nanotechnology. However, country's economy stands to gain only through what is achieved by the initiative!

References

Maithripala, D.H.S., 2007. Why Nanotechnology, and How? *Proc Abstracts, TECHNO Engineering and Technology Conference 2007, Nanotechnology for Sustainable Development*, Colombo.

NSF, 2007. *Nanotechnology Consultative Committee Report.*

Vitharana, T., 2008. Sri Lanka nanotechnology initiative. *Proc Abstracts, TECHNO Engineering and Technology Conference 2007, Nanotechnology for Sustainable Development*, Colombo.

www.nipo.gov.lk (accessed on 30-04-09).

www.nsf.ac.lk

www.ingramcontent.com/pod-product-compliance
Ingram Content Group UK Ltd.
Pitfield, Milton Keynes, MK11 3LW, UK
UKHW022212051026
15624UKWH00001B/132

9 789351 241225